I0028465

NEW MILK.

"Meeleck, Come! Meeleck, Come!"
Here's New Milk from the Cow,
Which is so nice and so fine,
That the doctors do say,
It is much better than wine.

This wholesome beverage is carried all around the city by men in carts, wagons, and very large tin kettles, as we see in the cut. The cows are pastured on the Island of New-York, some along the New-Jersey shore, and large droves on Long Island. Milk sells from 4 to 6 cents per quart, delivered at our doors every morning in the winter season, and twice a day in summer.

In warm weather, one may see large churns, mounted on a wheel-barrow, pushed along by colored men, mostly from Bergen, on the Jersey shore, crying BUTTER MEE-LECK! WHITE WINE! WHITE WINE!

SELLING MILK IN EARLY DAYS
(From *The New-York Cries;* published by Mahlon Day, New York, 1836)

MODERN DAIRY PRODUCTS

Composition • Food Value • Processing
Chemistry • Bacteriology • Testing
Imitation Dairy Products.

BY

LINCOLN M. LAMPERT
Formerly, Supervisor, Dairy Laboratory
California Department of Agriculture

THIRD EDITION
Completely Revised and Enlarged

CHEMICAL PUBLISHING COMPANY, INC.
NEW YORK

New York

ISBN 0-820-60360-0
2nd Printing, 1984

Printed in the United States of America

Preface

The contents of this book reflect information gathered through work and long association with persons in the academic, regulatory, and commercial areas of the dairy industry. The objectives of the two earlier editions have been retained, namely, to present reliable information, in a non-technical manner, on the composition, nutritive value, manufacture, chemistry, and bacteriology of milk and dairy products. The general aspects of the subjects are treated in the first portion of the book, while more detailed information pertaining to specific products are discussed later, without stressing highly technical or theoretical procedures.

Many educational institutions today place less importance upon the processing of dairy products than upon food technology in general. Workers in the dairy industry often acquire their knowledge by experience and on-the-job training. It is hoped that this book will prove of value to these employees by introducing them to the broad aspects of the dairy industry and the possibilities of bringing in new techniques.

The progressive executive, milk sanitarian, regulatory official, home economist, dietitian, and instructor in agricultural and vocational schools will find items of interest in the book. Some subjects treated may be difficult to locate elsewhere even though the reader may have access to a technical library. The references given at the end of each chapter have been selected to guide the reader, not only to original data, but to recent publications pertaining to the specific subject of interest. Economic factors, especially the impact of imitation and substitute products, have been felt throughout the dairy industry. Although they are not dairy products, the importance of imitation products warrants the inclusion of a chapter dealing with them.

I am grateful to those who have reviewed sections of the manuscript. Thanks are also due to the companies and organizations that generously supplied trade information and illustrations. The encouragement and help given by my sister, Josephine Lampert, has done much to make this book possible.

April, 1975 Lincoln M. Lampert

Contents

chapter 1

Milk as a Food

Milk is one of the few foodstuffs consumed in its natural state. It is the only article in the diet, with the exception of honey, whose sole function in nature is to serve as a food. Practically everything else we eat fulfills some other function in the animal and vegetable world. Thus most vegetables are roots or leaves of plants, fruits are parts of plants containing seeds, and meat comes from the bodies of animals.

In Biblical times, the ideal home was in a "land flowing in milk and honey." Today a common expression used in reference to nutrition is that "milk is the most nearly perfect food." It is the one foodstuff upon which all nutritionists agree concerning its value for the growth and development of children and young animals. Not only is it the most important food during early childhood, but in one form or other it continues in our normal diet throughout life. In sickness especially, it is the one food most often called upon to sustain and nourish the body. No other single substance can serve as a complete substitute for milk in the diet. Its value was established when the first mammal nursed its young, but only in recent years have the reasons for this superiority been recognized. In these discoveries the various branches of chemistry, biology, and bacteriology have had an important part. From the very beginning milk has served as a foundation for the study of nutrition.

Milk is defined to be the lacteal secretion, practically free from colostrum, obtained by the complete milking of one or more healthy cows. In general, milk is the secretion of the mammary glands of animals that suckle their young.

National Attitudes on Milk

The use of milk in the diet depends not only upon its availability, but also upon the economic and cultural attitudes of the people. Some unusual traits are found among certain African tribes. In Kenya, the Hamite people keep the milk of a special cow for the sole use of an expectant mother. In Somalia, the mother does not consider her milk suitable for her infant and so uses goat's, camel's or cow's milk; in contrast to this custom, in Zaire (Congo) only the mother's milk is given to the child.

TABLE 1:1

Milk and dairy products: Per capita production and consumption in 17 countries, 1971 and 1972
U.S. Department of Agriculture, October 1973

(Pounds)

Country	Production of fluid milk		Consumption: Milk equivalent 1/		Fluid milk and cream		Butter		Cheese		Canned Milk 2/		Dried Milk 3/	
	1971	1972 4/	1971	1972 4/	1971	1972 4/	1971	1972 4/	1971	1972 4/	1971	1972 4/	1971	1972 4/
Canada	824	812	768	760	280	278	15.0	14.7	11.2	11.5	13.0	12.3	5.0	4.7
United States	580	583	558	561	259	258	5.1	4.9	12.1	13.2	11.8	11.1	5.5	4.7
Austria	973	976	741	766	344	342	12.1	12.9	9.0	9.6	3.8	4.0	1.6	1.8
Belgium	791	822	770	713	157	145	16.5	14.8	18.4	18.2	6.9	6.0	2.5	1.9
Denmark	2,022	2,107	955	961	405	401	19.2	18.9	22.8	24.5	5/	5/	4.0	4.4
Finland	1,530	1,545	1,230	1,246	504	510	31.3	30.9	10.2	12.2	5/	5/	4.3	4.7
France	1,189	1,238	957	988	201	195	18.0	19.0	30.8	32.1	3.8	3.7	1.4	1.3
Germany, West	761	768	712	666	166	157	17.5	15.7	12.1	12.8	17.1	17.0	2.6	2.7
Ireland	2,747	2,904	1,216	1,200	468	464	27.6	26.9	5.5	6.1	5/	5/	5/	5/
Italy	376	384	427	428	157	163	3.9	3.5	19.8	20.4	7.3	7.1	1.4	1.6
Netherlands	1,400	1,498	693	677	331	323	5.1	4.2	19.1	20.0	22.9	22.7	3.2	3.8
Norway	1,026	1,016	993	985	541	542	11.5	10.9	20.2	20.6	5/	5/	5/	5/
Sweden	783	801	759	788	368	368	11.2	12.4	19.2	19.6	1.5	1.2	13.3	16.6
Switzerland	1,111	1,153	954	970	358	357	15.9	15.6	21.2	22.7	5/	5/	3.4	4.1
United Kingdom	502	535	850	804	342	342	18.0	15.9	12.4	11.9	8.7	7.9	4.5	4.8
Australia 6/	1,297	1,248	877	863	313	310	20.3	19.2	9.0	9.2	11.6	12.2	2.3	3.6
New Zealand 7/	4,568	4,740	1,210	1,209	396	397	39.4	39.1	9.5	9.6	8/ 4.6	8/ 5.8	5.7	4.8

1/ Fluid milk and dairy products in terms of whole milk equivalent, fat-solids basis. 2/ Mainly condensed and evaporated milk but includes canned skim milk. 3/ Includes both dry whole milk and nonfat dry milk. 4/ Preliminary. 5/ Not available or quantity negligible. 6/ Year ending June 30. 7/ Year ending May 31. 8/ Includes dry whole milk.

The total milk production in 36 of the world's major dairy countries in 1973 was estimated to be 759 billion pounds.

The use of dairy products in a particular country tends to follow a traditional pattern from year to year, but may vary appreciably from one country to another. As shown in Table 1:1, Finland, Ireland and New Zealand rank high in the use of butter, but low in the consumption of cheese; France, Netherland and Sweden rank high in consumption of cheese, but lower in the use of butter. In most European countries the consumption of canned milk is relatively low, but high in the Netherlands. In India, where much of the populace is underfed, there are many cattle, but other than buffalo milk, little milk is available. Hindu culture deems the cow a sacred animal, not to be killed. The cows are undernourished and would yield little milk; many are so old they have lost their ability to produce milk. The feed these animals consume deprives the younger cows of the food they could use to produce milk, which the people would use were it available.

Regional and climatic influence also determines the kind of animal whose milk is used. The cow is adapted to the temperate zones. In southern Europe the milk of goats and sheep is used; the Lapps of northern Europe use the milk of the reindeer. In southeast Asia the milk of the water buffalo is used. The peoples of Europe and in those regions where they have migrated, such as North America, Australia and New Zealand, are the great users of cow's milk and its products. (Table 1:1). The Soviet Union is the world's top producer of milk (83.2 million metric tons); the United States is second in production with 53.2 million tons.[11] Finland is the leading user of milk and milk products, with a whole milk equivalent consumption of 1,246 pounds per person in 1972. Ireland and New Zealand are close seconds, with 1,220 pounds each per person. New Zealand was the leading consumer of butter; France, of cheese, the Netherlands of canned milk, and Sweden of, of dried milk.

Approximately one-third of the milk produced in the countries listed in Table 1:1 was consumed as fluid milk, 36 percent was utilized for butter production, 15 percent for cheese production, ten percent was utilized for canned milk, dry whole milk and ice-cream production. About 7 percent of the milk produced was fed to livestock.

The high incidence of milk intolerance among Negroes, American Indians, and the Asiatic peoples is discussed in the section on lactase in Chapter 4.

Dairy Industry in the United States

In the United States the dairy industry ranks high in importance, according to the United States Department of the Interior, only the products of the iron and steel, meat, and automobile industries are sources of greater wealth or income than the dairy industry.

The economic value of the dairy industry reaches beyond the wealth produced from milk and milk products. The conservation of soil fertility is of prime importance to agriculture. Experience has shown that the manure obtained by keeping a dairy herd is a practical method to maintain soil fertility. If the farmer grows his own feed, the cow efficiently converts this to milk and returns to the soil much of the material needed to maintain fertility. This is not always obvious, because pasture growth is grazed off day by day and there is no measurement of the return until the forage is converted into milk and meat products. About one-half of the beef and nearly all of the veal is obtained from dairy herds. Much of this meat supply goes into the manufacture of processed products, such as sausages and canned meats. In 1973 dairy farmers received about $8.1 billion from the sale of milk and cream and about $2 billion more from the sale of cows, calves, and heifers from their herds.

The number of cows has decreased from 14.1 million in 1969 to 11.3 million in 1974, but total milk production has remained about constant (120 billion pounds in 1972). During these same years, the average milk production per cow increased to 10,271 pounds per year, varying from 13,406 pounds per cow in California to 6,383 in Mississippi.

About one-third of the milk produced in the United States in 1973 originated in the northeastern quarter of the country. Wisconsin, with about 18.4 billion pounds and New York, with about 9.7 billion pounds, furnish a little over one-sixth of the nation's milk supply. The 1973 return to the Wisconsin dairy farmers was $11.06 billion, the New York producers received $704 million. This return is in contrast to that of Minnesota, for example, where most of the milk is used for the manufacture of other dairy products, rather than for consumption as fresh milk and cream. Minnesota produced 9.6 million pounds of milk, about the same as New York, but the return was $568 million, some $136 million less. Further discussion of the classification, grading and pricing of milk is given in Chapter 10.

In 1972, 43.5% of the total United States milk production was used as fluid milk and cream. According to the Milk Industry Foundation, 19.0% was used in the manufacture of butter, 19.2% for cheese, frozen milk products took 9.3% of the total production and 2.5% went into the manufacture of evaporated and condensed milk, 3% was used on the dairy farm and the rest, 3.5% went into dry whole milk, creamed cottage cheese and miscellaneous uses.

A significant decrease in the consumption of butter in the United States has occurred since 1964, when it was at a record level for the postwar period. The decline in milk equivalent consumption in the United States is related to the decrease in butter consumption, a major component of the milk equivalent figure. A definition and listing of milk equivalents is given in Chapter 20.

More than four-fifths of all milk sales in the United States are made through stores, especially chain stores which process their own dairy products.

A considerable quantity of milk is consumed by those that drink coffee. In the United States, some 355 million cups of coffee are consumed daily. According to the Wall Street Journal, less than 32% of young people, under 19, drink coffee. It is least popular among college students, who favor milk as a beverage. The use of coffee cream has decreased, about two-thirds of coffee drinkers use a

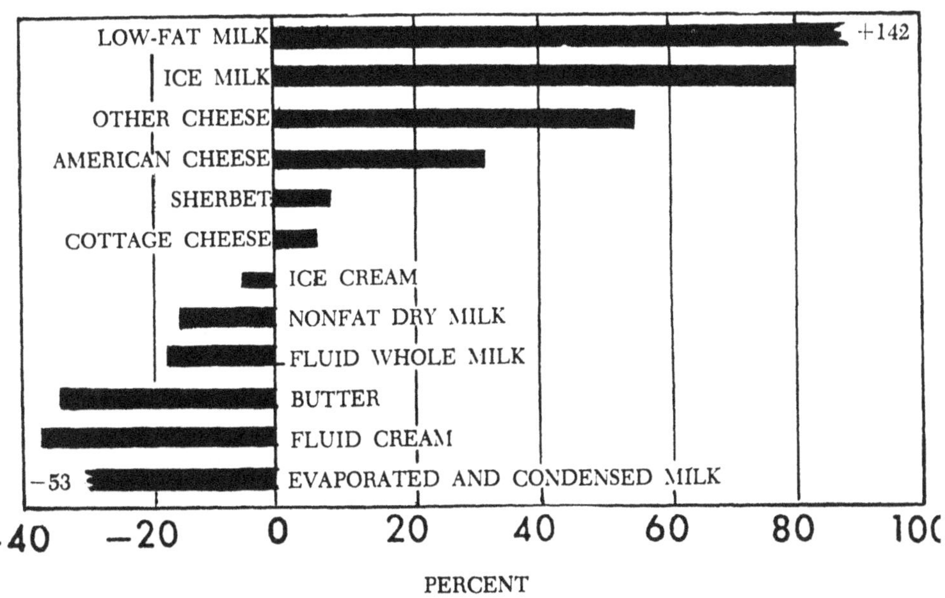

Change in dairy products sales per capita 1960 to 1970

Fig. 1:1 (1971 Handbook of Agrigultural Charts, Agricultural Handbook 423, USDA, Nov. 1971.)

"coffee whitener", which has made a deep inroad in the use of milk products. These creaming mixtures are described in Chapter 21.

A nation-wide survey reported in 1973 by the United States Department of Agriculture indicated considerable misinformation by consumers concerning the fat content of dairy products. Regular milk (about 3.5% fat) was thought to contain 20% or more of fat. The fat content of ice cream was also grossly over-estimated. Low-fat milk which often is labeled to indicate fat content was the only product correctly stated.

DAIRY CATTLE

In the United States, several breeds of cattle are recognized as dairy breeds; that is, they are especially well selected for production of milk and milk fat. There are associations for each of the principal breeds and these establish the requirements for the registration of pure-bred dairy cattle. Animals whose parents are not registered or that do not meet registration requirements are known as "grade" animals. In addition to the breeds described here there are several dual-purpose breeds, used for production of both milk and meat. The Short-Horn is most numerous of this type. There are about 75,000 registered Milking Short-Horn cows in the United States. The milk yield is about 11,000 pounds per year, containing about 410 pounds of fat.

The composition of the milk of the various breeds is given in Table 1:3. Variations in the composition of milk among animals of the same breed can exceed the differences in the average composition of the milk of the different breeds.

Ayrshire

The Ayrshire breed originated in the County of Ayr, in the southwestern part of Scotland, about 180 years ago. They have since spread over northern Europe and are an important dairy breed in Canada and New Zealand. They were first imported into the United States around 1822. There are now about 400,000 Ayrshire cattle in the United States, with the number increasing, especially in the far western and southern states. New York and surrounding States have the largest populations of registered Ayshires. About 14,000 were registered in the United States in 1972. The Ayrshire grazes easily and will thrive on pasture where other breeds, such as the Holstein, would have difficulty in finding suitable forage.

Ayrshire cattle vary in color from mostly white to nearly all brown or cherry-red, with any combination of these colors. Black markings are undesirable. Usually the tail is white. The horns are large and tapering, first turning outward, then forward and back, giving the head a distinctive appearance. The mature cow weighs about 1,200 pounds. The average yearly yield of milk from the cows in a well-managed herd is about 13,000 pounds, which contain about 500 pounds of milk fat. The milk of the Ayshire cow is white and the fat globules are small.

Brown Swiss

The Brown Swiss breed, introduced in this country in 1869, comes from Switzerland and it is probably one of the oldest known breeds. They are brown in color, the shade varying from a silver gray to dark brown. Streaks of white on the belly or legs are considered objectionable. The nose, tongue, tips of the horns and switch are always black. The mature cow weighs about 1400 pounds. The average yield of milk of cows on official test is about 14,000 pounds a year, and contains about 570 pounds of fat. Brown Swiss Cattle are reported to be able to withstand sudden changes in weather without significant change in milk production.

They have a long, productive milking lifetime, producing well at most altitudes and in all climates. In 1973, 14,000 cows were registered in the United States where the total registered Brown Swiss population was about 200,000.

Guernsey

The Guernsey is one of the Channel Island breeds, where it was developed around the year 1650 from the Fromont du Leon of Brittany and the Normandy Brindle cows. Since 1824 no cattle have been imported into Guernsey and those brought in during the German occupation in World War II have been slaughtered. The Guernsey probably ranks second to the Holstein as the most numerous dairy breed in the United States.

Channel Island cows were imported into the United States over 150 years ago but they became mixed with common herds. The first record of pure bred Guernsey cattle in the United States dates from an importation to a farm in Massachusetts about 1931. Mention of a cow imported in 1815 from the Island of Alderney is made in the records of the American Jersey Cattle Club.[7]

In all, only about 13,000 head of Guernsey have been imported into the United States, but from these we now have about 265,000 living, registered Guernsey cattle registered in the United States.

The Guernsey is fawn colored, with clearly defined white markings.

A cream-colored nose is desired by most breeders, rather than a black muzzle. The mature cow weighs about 1,100 pounds. The average yield of milk is about 11,311 pounds, containing about 523 pounds of milk fat. The milk is rich and golden in color. Some consumers prefer milk from the Channel Island breeds and will pay a premium for it.

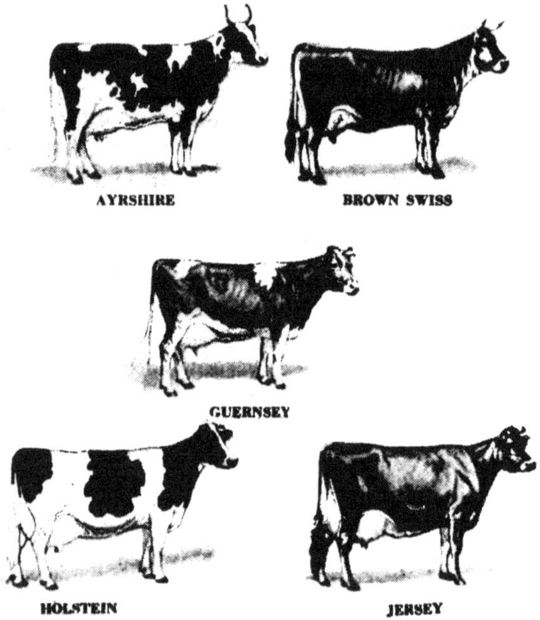

Fig. 1:2. Various Breeds of Dairy Cows.

Holstein-Friesian

Holstein-Friesian cattle, the largest of the dairy breeds, were developed in the Netherlands, probably over 1000 years ago. They are called *Holsteins* in the United States; in other countries they usually are called *Friesian*. They are the predominant dairy breed in the United States and Canada. In 1973, there were about 10 million registered Holstein-Friesian cows, comprising over 70% of the pure-bred dairy cattle in the United States, and producing about 85% of the total milk production.

The mature cow weighs from 1,100 to 1,750 pounds. The color is black and white, with the color boundaries sharply defined, rather than blended. No solid-color animal can be registered. The Holstein cow prefers to graze on level, rich pasture land. The cows do not withstand hot weather so well as do some other breeds, such as the Jersey.

The volume of milk produced is large, with a lower fat content than the other dairy breeds. The average yearly yield is about 12,000 pounds, with about 410 pounds of milk fat. The milk is not so highly colored as that of the Guernsey or Jersey cows. Milk from heavily producing herds, especially if cows are first coming into milk production, may have less than 3.5% of fat and a correspondingly lower solids-not-fat content.

Jersey

Jersey cattle originated on the Island of Jersey, one of the Channel Islands. Cattle from any of the other Channel Islands are known as Guernsey. No cattle have been imported to the Island of Jersey, except for immediate slaughter, since 1789. The Jersey breed is the smallest in size of the popular dairy breeds. In the United States, the mature cow weighs about 1000 pounds, but those on the Island of Jersey generally weigh between 800 and 900 pounds.

Jersey cattle were first imported into the United States in 1815. There are now about 2¼ million living Jersey cattle, registered and grade.

The usual color is fawn or cream, but shades of gray, brown, and almost black are found. Often they resemble the Brown Swiss in color. In 1972 the average yearly yield of milk for cows on test was about 10,166 pounds, containing about 515 pounds of fat. Like that of the Guernsey cow, the milk is rich in color and high in fat content. There is little difference in the yield and fat content from these two breeds. Jersey milk contains a little more fat, but the Guernsey cow has somewhat more total milk production.

When compared to the other breeds, under similar conditions, the Jersey cow has the reputation of being the most economical producer of milk fat. They fare better than the Holstein on rough or scanty pasture, and they can endure warm weather better than can some other domestic breeds. There is evidence to show that the Jersey cow has a longer and more productive life than the other dairy breeds.

Secretion of Milk

In its form and structure, the udder of a cow is a cutaneous gland. In the primitive Australian mammal, the egg-laying echidna, the mammary glands actually look like sweat glands. This animal has no teats and the milk oozes out of the glands, much like sweat. The young lick the milk from the hairs upon which it collects.

Over the centuries, cows have been selected for high production of milk, so that today the dairy cow has the best developed mammary glands of any animal. A good cow will produce in one year about ten times its own weight in milk, and exceptional cows may produce twenty-five times their weight.

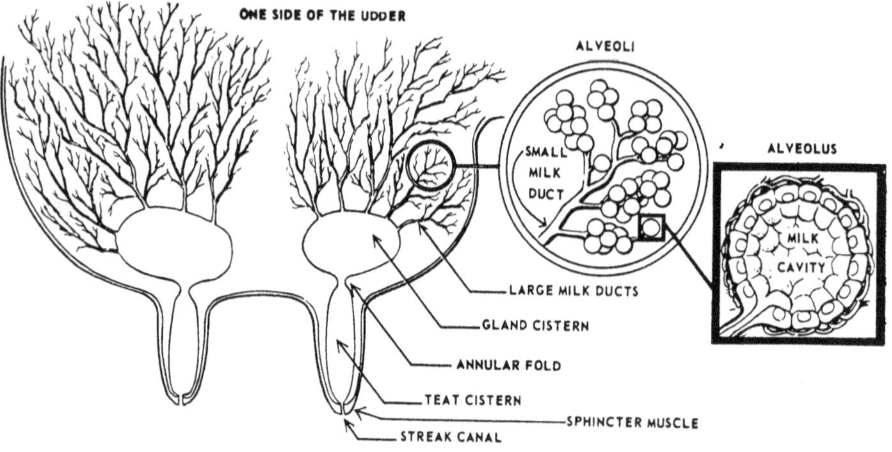

Fig. 1:3. Courtesy of University California, Davis.

The cow's udder consists of four distinctly separate glands, known as the quarters. Each quarter is provided with a teat, the hollow, interior portion of which is known as the *teat cistern*. The teat cistern extends upward into the body of the gland and is connected with the gland cistern, which may vary in capacity from less than one pint to about one quart. Radiating from the walls of the cistern are numerous tubes or ducts which branch out or divide innumerable times (Fig. 1:3). The ducts are very small in that portion of the quarter removed from the cistern, but become larger as they approach and enter the gland cistern.

Some secretion of milk occurs in the lining of the ducts, but the principal secretion takes place in the alveoli, the enlargements at the very end of each of the smallest branches of the duct system. Milk is secreted more or less continuously between milking periods. It may be slowed or stopped completely by the pressure of the milk accummulated in the alveoli. Frequent milking decreases this pressure and favors milk production. At the milking time, the cow ejects or *lets down* her milk. This is a reflex or involuntary action, controlled by sensory nerves which carry the message to the pituitary body in the brain. This message may be originated by the sucking calf or by the washing and manipulating of the udder by the milker. The pituitary gland then furnishes to the blood stream a hormone known as *prolactin* which is an important factor in the growth of the udder and milk production[8]. Another hormone, *oxytocin*, causes the alveoli to contract and thereby squeeze the milk from them into the ducts through which it travels to the cisterns and teats.

Conditions that frighten or anger the cow at milking time interfere with the function of the pituitary body so that its hormones are not released into the blood stream and the result is that she does not yield her milk. Oxytocin gradually is destroyed in the blood stream, and therefore the milking operation must be completed before its influence is lost, which usually is within ten minutes.

Most of the milk obtained at a single milking is present in the udder at the time of milking. About half of this milk is stored between milkings in the milk cistern and ducts, while the balance is stored by stretching the udder. As the udder is stretched, the pressure within it increases, thus restricting its blood supply and so reducing the rate of secretion.

Milk is a secretion and in its composition if differs greatly from that of the blood from which it is derived. For example, milk fat, casein, and lactose, which are synthesized in the udder, are not found elsewhere in the body. Very large amounts of blood must be passed through the udder in the production of milk; it is estimated that a cow producing six gallons of milk will pass the equivalent of over ten tons of blood through the udder.

A number of substances are not normal constituents of milk may be eliminated from the body by the milk. In this category are many drugs and certain flavor-producing substances, such as may be present in onions, turnips, and silage that the cow may consume. Of recent interest is the elimination in the milk of radioactive fall-out materials, such as strontium-90 and of pesticidal chemicals, such as DDT, which might be present in the cow's feed. This subject is discussed in more detail in Chapter 8.

An unusual occurrence, possibly associated with the transfer through milk of an injurious substance, was reported from Tasmania. School children there showed an increased incidence of goiter, even though their diet was supplemented with iodine. This was attributed to the use of milk from cows on pasture that contained marrow-stem kale. This plant, with others such as cabbage and turnips, contains an antithyroid or goiter-producing substance, capable of being transmitted through the milk.[3]

Research that tends to refute the hypothesis that milk from cows consuming feed of the mustard family may cause goiter in persons who drank it, at least in the cases of adults, was done in Finland under the guidance of Dr. A. I. Virtanen. They found that the cow could not consume enough of the feed to cause her milk to hinder the uptake of iodine in human beings[1].

COMPOSITION OF MILK

Milk has a very complex composition. This differs, not only among the various species but it also may vary much within any species or individuals within that species. Some of its constituents, such as milk fat, milk sugar, and casein are not found elsewhere, either in the body or in nature. Milk is practically the only foodstuff that contains all of the different substances known to be essential for human nutrition. Nevertheless, milk alone is not a complete food for many animals after they have grown beyond the suckling stage. Some of the essential nutrients such as iron, copper, and manganese, and some of

TABLE 1:2

Average Composition of the Milk of Certain Mammals[a]

	Water %	*Protein %*	*Fat %*	*Lactose %*	*Ash %*
Cow	See Table 1:3				
Human	87.60	1.20	3.80	7.00	0.21
Ass	89.08	1.98	2.45	6.04	0.45
Buffalo	82.44	4.74	7.40	4.64	0.78
Camel	87.67	3.45	3.02	5.15	0.71
Cat	83.05	7.00	4.50	4.85	0.60
Dog	74.55	3.15	10.20	11.30	0.80
Elephant	85.63	3.20	3.12	7.42	0.63
Ewe	80.87	5.40	8.05	4.78	0.90
Goat	87.33	3.50	4.15	4.20	0.82
Llama	86.55	3.90	3.15	5.60	0.80
Mare	89.18	2.60	1.59	6.14	0.49
Porpoise	41.28	11.20	45.80	1.15(?)	0.57
Rabbit	68.50	12.95	13.60	2.40	2.55
Reindeer	63.38	10.25	22.42	2.50	1.45
Seal	34.00	12.00	54.00	none	0.53
Sow	80.63	6.15	7.60	4.70	0.92
Vixen	81.86	6.35	6.25	4.23	1.31
Whale	69.80	9.43	19.40	?	0.99
Zebu	86.20	3.00	4.80	5.30	0.70

[a] Compiled from various published analyses.

TABLE 1:3

Range [a] and Average Percentage Composition [b] of Milk of Different Breeds of Cows

Breed	*Water*			*Fat*			*Protein*			*Lactose*			*Ash*			*Total Solids*	*Solids-not-Fat*
	Min	Max	Aver	Min	Max	Aver	Min	Max	Aver	Min	Max	Aver	Min	Max	Aver	Aver	Aver
Ayrshire	84.24	89.44	87.42	2.92	5.66	3.85	2.92	4.58	3.32	2.41	6.11	4.70	0.58	0.58	0.71	12.58	8.73
Brown Swiss	—	—	86.93	—	—	3.95	—	—	3.43	—	—	5.00	—	—	0.73	13.07	9.16
Guernsey	82.12	87.93	85.89	3.65	7.66	4.70	2.65	5.45	3.75	3.57	5.78	4.86	0.60	0.85	0.75	14.16	9.36
Holstein	82.38	89.28	88.12	2.72	6.00	3.52	2.44	6.48	3.30	3.96	5.71	4.70	0.56	0.86	0.69	12.21	8.69
Jersey	82.32	89.04	85.60	3.28	8.37	5.05	2.98	5.83	3.75	2.73	5.66	4.84	0.57	0.82	0.76	14.40	9.35
Short Horn	—	—	87.42	—	—	3.70	—	—	3.32	—	—	4.82	—	—	0.74	12.58	8.88

[a] Overman, Sanmann and Wright, *Ill. Agr. Expt. Sta. Bull.* 325:1929.

[b] Various authorities, especially Armstrong, *J. Dairy Sci.* 42:1 (1050); Macy, Icie G. *et al.*, The Composition of Milks, *Nat. Acad. Sci., Pub.* 254 Nat. Res. Council, Washington, D.C. (1953).

the vitamins, are not present in sufficient amount or in the proper proportion to supply the requirements of complete nutrition. At birth, normal infants have a sufficient reserve of the mentioned minerals to last them until their diet can include foods which contain them, but certain vitamins, such as ascorbic acid and vitamin D often must be added to the infants' diet.

Comparison of Milk of Various Mammals

The average composition of cow's milk, as well as the composition of that of other mammals, based upon a study of over 300,000 chemical analyses published by various authorities, is given in Tables 1:2 and 1:3.

The variation in the composition of milk of different kinds of animals seems to be related to differences in the stage of development of their young at the time of birth. Some animals, such as cows, goats, horses, and sheep, are comparatively well developed and practically able to take care of themselves within a short time after birth. Others, such a rabbits, pigs, dogs, and cats, are comparatively helpless at birth and much bodily development is needed before they are capable of caring of themselves. The milk must furnish to the second group more body-building nutriments, such as protein and minerals, whereas substances that supply more energy, such as fat and carbohydrate (lactose), are needed by animals of the first group. Human milk cannot be classified in this manner, but its composition is nearest to that of the equine animals. The human child grows more slowly than the young of animals used for milking.

Colostrum

Colostrum is the secretion of the mammary glands during the first few days of lactation after giving birth. It differs from normal milk in composition, flavor, and odor. The odor is strong and the flavor bitter. A temporary increase in fat occurs shortly after calving. For about the first three days the colostral secretion will coagulate upon the application of heat. Normal composition of the milk occurs about five days after parturition. This gradual change apparently accustoms the suckling animals to the use of normal milk. Colostrum contains more solids than the subsequent milk.

Colostrum is very rich in globulins which serve as the carrier of antibodies which protect the suckling animal against disease-producing organisms. The composition of colostrum is given in Table 1:4.

TABLE 1:4

Transition from Colostrum to Normal Milk
(Compiled from available data)

Time after calving	*Casein* %	*Globulins* %	*Fat* %	*Lactose* %	*Ash* %	*Chloride* %	*Total Solids* %
At once	5.00	11.07	6.55	2.90	1.22	0.152	26.74
6 hours	3.50	6.60	7.82	3.29	0.97	0.150	22.18
12 "	3.12	2.86	4.10	3.88	0.88	0.148	14.84
18 "	3.00	2.14	4.00	3.75	0.85	0.122	13.74
24 "	2.61	1.91	3.64	3.82	0.85	0.120	12.83
36 "	2.86	1.32	3.58	3.68	0.84	0.118	12.10
72 "	2.77	1.10	3.52	4.41	0.84	0.120	12.64
5 days	2.74	1.00	3.55	4.79	0.83	0.130	12.91
10 days	2.62	0.68	3.57	4.92	0.82	0.128	12.61

Some Factors that Affect the Composition of Milk

A number of factors may influence the composition of milk. Among these are the breed and individuality of the cow, time of milking, season of the year, and the cow's feed and living conditions.

Each quarter of the cow's udder may produce milk of somewhat different composition. It has been noted that the milk from the quarter first milked has the highest fat content, and that drawn from the last quarter has the lowest. As the milking proceeds, the fat content of the milk increases slightly, so that the *strippings*, the last portions drawn, are highest in fat. The first portion obtained may contain about 1% of fat while the strippings may have over 7%. It has been suggested that the higher fat content of the strippings probably is due to the presence of large fat globules which pass with difficulty through the milk ducts. Another theory holds that only certain cells produce fat and that the decrease of pressure in the udder during milking causes an increased activity in these cells.

In addition to a variation in fat content during milking, it also has been observed that there is an increase in the solids-not-fat during the milking period. For example, the fat and solids-not-fat in the first-drawn milk was 2.4 and 8.3%, the middle portions had 3.8 and 8.6%, and the strippings contained 8.9 and 8.2%, respectively.[6]

Breed:

The breed of the cow has a great influence upon the composition of her milk. The greatest difference is found in the fat content. Guernsey and Jersey cows yield milk of high fat content whereas that of the Holstein and Ayrshire breeds is relatively low in fat. The lactose and ash content show less variation. Table 1:3 shows the range and average composition of milk from different breeds of cows. The breed has also an influence upon the color of the milk.

Time of Milking:

The time of milking has some effect upon the composition. If an equal length of time elapses between the morning and evening milkings there is no consistent difference, but if the cow is milked in the morning and then late in the evening, the morning milk may contain from 0.5 to 2.0% more fat. If the cow is milked three times a day, the noon milking usually shows the highest fat content. The amounts of protein and lactose in the milk of an individual cow vary but little from one regular milking to the next.

Seasonal Variation:

The dairy farmer often notes that the fat content of his herd's milk begins to decline in the late spring and generally rises again in the fall. This usually has been attributed to the change in feed from grain in winter to green pasture in the spring. Although this change may have some influence, the principal reason appears to be the temperature of the environment. The fat content of the milk decreases as the weather becomes warmer and increases again with the approach of winter.

The fat content also tends to increase towards the end of the lactation period. If this comes in the fall or winter, the rise in fat content of the milk may be very noticeable. The solids-not-fat content of the milk generally follows the variation for the fat content.

Age:

The age of the cow has a slight but definite effect on the composition of her milk. Studies made in England show that over a period of about ten years the fat content decreases about 0.2% and the solids-not-fat content about 0.5%.[2]

Disease:

The milk from a diseased cow may vary greatly from normal milk. Generally there is an increase in the content of fat and salt and the lactose content is diminished. Milk from a diseased udder tends to approach the composition of cow's blood-serum, especially in that its albumin content is increased.

Feed:

The yield and composition of milk are affected by the amount and kind of feed consumed by the cow. This is to be expected since the constituents of the milk ultimately are derived from the feed. Milk fat is the most variable component of milk. If the cow is fed a high-energy, low-fiber diet, the milk fat content may drop from its normal amount to as little as 1%.

When the cow's diet contains much concentrate and is low in roughage or if the diet is high in ground, pelleted roughage, the feed is fermented more rapidly in the rumen than is a conventional diet. During fermentation in the rumen, a conventional diet yields about three times as much acetic acid as propionic acid. A diet, such as first mentioned, favors the production of propionic acid rather than acetic acid. Acetic acid is important in the synthesis of milk fat while propionic acid is used in the formation of blood sugar. The change in fermentation end-products therefore affects the metabolism of the cow, resulting in a reduced fat content of the milk and an increase in the body-weight of the cow.

Overfeeding does not increase the normal flow of milk, but underfeeding has a pronounced effect. A great decrease from the normal ration will cause a rise in the fat content but the quantity of milk produced is materially decreased. Considerable variation may occur in the protein and carbohydrate content of the cow's feed without having much effect upon the composition of the milk, but the total output will be reduced if these components of the feed are reduced materially and thereby effect the cow's well-being.

Fat-rich feed or the inclusion of various fats or oils in the cow's diet has a definite effect upon yield, composition, and properties of milk fat. Some fatty acids that are not normal constituents of milk fat may appear in it if they are present in the feed. The introduction of polyunsaturated fatty acids is discussed in Chapter 2. The presence of certain fish oils, such as menhaden and codliver oil, in the feed will cause the fat content of the milk to decrease markedly. In general, the feeding of fat-rich foods is not a practical way to increase the fat content of the milk.

The inclusion in the feed of compounds such as thyroprotein or iodated casein may result in a small increase in fat production, but this is a temporary condition. Prolonged use of these preparations is dangerous, especially during warm weather, or for cows in advanced gestation because they cause changes in the animal's metabolism.

COLOR OF MILK

The natural color of milk varies from a bluish white to a brownish yellow, depending upon the amount of fat and solids-not-fat present. The white or milky appearance is due to the colloidal dispersion of the fat globules, calcium caseinate, and calcium phosphate in the milk. The size of the fat globules also somewhat influences the color, as does the breed and the feed of the cow.

The principal substances that actually impart a yellowish color to milk are the pigments carotene and riboflavin (vitamin B2). The greenish-yellow color of whey is due to the presence of riboflavin; in milk this color is masked by the other constituents present. Green feed increases the carotene content of the milk and its color. It is not practical from the commercial standpoint to influence the color of milk by altering the cow's feed. For example, feeding up to forty pounds of carrots per day to a cow was found to give but a slight increase in the color of her milk.

Guernsey and Jersey breeds can transfer more carotene from their feed to the milk-fat than can Holstein, Aryshire, and other breeds. The increased content of carotene and fat, and larger fat globules in milk from Guernsey and Jersey cows, are responsible for the deeper color of their milk compared with that of other breeds. Goat milk is described in Chapter 12. Like goat milk, buffalo milk has little color; its carotinoid content varies from 0.25 to 0.48 microgram per gram, compared to up to 30 micrograms for cow's milk.[10]

High heat-treatment, such as may occur in the manufacture of evaporated milk, imparts a brownish coloration to the product (see Maillard reaction).

The discoloration of milk by bacterial action is discussed in Chapter 7.

ENERGY VALUE OF MILK

The utilization of food by the body is essentially a chemical process which may be compared to the consumption of fuel by an engine. The body uses a complicated process of combustion, because the fuel is not only needed to run the engine of the body but also to become part of the body itself. Fats are the most efficient sources of energy; they furnish nearly twice the amount of energy per unit of weight as do the carbohydrates or proteins.

Whether it is produced by an engine or by the body, energy may be measured in terms of heat units and usually is so expressed in studies of food values. The unit commonly used is the *large calorie*, which is equivalent to the amount of heat needed to raise the temperature of a kilogram (2.2 pounds) or water from 15° to 16°C. This is practically the same amount of heat needed to raise the temperature of four pounds of water one degree Fahrenheit. Until the discovery of vitamins, much stress was placed upon the number of calories contained in the diet, and complete and adequate nutrition was thought to depend upon whether or not the diet furnished sufficient calories. Today it is known that other factors must be considered too, but the calories are just as important as ever.

Certain factors control the exact energy requirements of an individual. For example, a man of large stature needs more food than does a person of small stature. Children need food energy for growth as well as for the maintenance of bodily health and for supplying the energy for their activities. The energy requirement of an individual tend to decline with advancing age. At absolute rest, a person needs only enough energy for the work of the necessary life processes. This usually is called *basal energy* and averages about seventy calories per hour for the average adult. It corresponds roughly to the energy needed to keep two fortywatt electric lamps burning. During sleep the energy requirement is somewhat less. As women average about four-fifths the size of men, their basal energy requirement is about fifty-five calories per hour. A normal, well-nourished infant consumes about fifty calories in its daily diet for each pound of body weight. Somewhat more than this amount is used in the first few months of life but the caloric intake decreases as growth slows down.

The amount of energy that various foods yield to the body has been studied in great detail. The values found by the American scientists, Atwater and Bryant, about sixty years ago are often used. These values are four calories per gram of carbohydrate, four calories per gram of protein and nine calories per gram of fat. It has been found that these factors do not apply accurately to individual food stuffs, and modified values have been published.* Some of the recommended values, especially pertaining to dairy products, are

TABLE 1:5

Energy Value of Food Components
(Calories per Gram)

Foodstuff	*Protein*	*Fat*	*Carbohydrate*
Milk	4.27	8.79	3.87
Milk Fat	—	8.79	—
Vegetable fat	—	8.84	—
Animal fat	—	9.02	—
Eggs	4.36	9.02	—
Sugar	—	—	3.87
Vegetables	3.11	8.37	3.99
Fruits	3.36	8.37	3.60

*See *Energy Yielding Components of Food and Computation of Calorie Values*, Nutrition Division, Food and Agriculture Organization of the United Nations, May, 1947.

shown in Table 1:5. Recommended daily caloric allowances are given in Chapter 6.

By the use of these factors, the amount of nutrient energy of any foodstuff of known composition may be calculated. As an example, the calorific value of milk of average composition can be calculated from the data given. The mineral matter and water have no calorific value. As the composition is given in percentages, there are present 3.66 grams of fat, 3.42 of protein and 4.92 of lactose per hundred grams of milk. The calorific values are:

3.66×8.79 calories=32.17 calories from the fat
3.42×4.27 calories=14.60 calories from the protein
4.92×3.87 calories=19.04 calories from the carbohydrate
65.81 calories from 100 grams milk.

For most practical purposes, the more easily calculated values obtained with the Atwater and Bryant factors may be used; in the preceding example, these factors give the calorific value of the milk as 66.30 calories.

One pint of milk of average composition weighing 487 grams will furnish about 323 calories. If the fat content is higher than 3.66%, the calorific value will be materially greater.

If only the fat content is known, the calorific value of milk may be estimated by means of the following formula:

Calories per pint=52.65x(percentage of fat+2.42)

Compared to some other foods, milk is low in energy value owing to the large amount of water present. As shown, a pint of milk, which contains about 14 ounces of water, supplies about 320 calories, whereas a pound of bread, which contains about 5.7 ounces of water, supplies about 1180 calories. If all the water were evaporated from the milk and if the bread also were water-free, the figures would be 2446 calories per pound of dry milk, and 1829 calories for the dry bread.

The fat supplies nearly one-half of the total calories in milk, the lactose about 25%, and the protein about 21%. Fat ordinarily constitutes about 4% of the weight of milk; in this 4% is concentrated nearly one-half of the total calorific value of milk.

HANDLING AND TRANSPORTING MILK

During recent years there has been a significant change in the method of handling milk at the producer's level. Holding and hauling milk in cans is rapidly giving way to the use of refrigerated tanks on farms and for hauling. This change has considerably increased the cost of equipment used by the milk producer and by the hauler who transports the milk to the dairy plant. The procedure has proved to be practical and for those that are able to install bulk tanks, their economy and advantages offset the additional cost. The overall cost of bulk handling often may mean a saving of one-half the cost of when cans are used.

As will be shown, the use of bulk tanks has improved the quality of milk because it is held and shipped under conditions that minimize contamination and bacterial growth.

The first farm tank was introduced in California, but now it is in general use throughout the United States, especially in the areas where much milk is produced. By 1969, about 233,500 tanks were in use in the United States, ranging from 37,660 in Wisconsin to 35 in Alaska. In Canada, over 20,800 were in use, about two-thirds of them in Ontario. The farm tank has been an influence in the trends toward fewer and larger dairy farms. The equipment is economical only where production of milk is large. Usually a herd of at least thirty cows is needed to warrant the use of a farm tank.

The change to bulk handling of the milk has also been accompanied by changes in the milking barn, such as automated feed handling and pipe line milking. Another requirement is that roads and bridges must be strong enough to carry the heavy trucks as they travel to and from the dairy farms. At first, dairy plants received milk either in cans or by tank. Now the tendency is to avoid producers that use milk cans, and many plants will deal only with dairy farmers that have tanks. Often a premium is paid to producers that use bulk tanks. Under these conditions the small-scale dairy farmer cannot compete with the large producer and must either enlarge his herd or go out of the dairy business.

Another change caused by the farm tank is that the farm becomes the place of purchase for the milk, rather than the dairy plant, as when cans are used. Now the milk hauler measures the milk in the farm tank, judges its quality, and takes the samples for the fat test upon which payment will be made. This has required the employment of a specially well-trained and reliable person for this work.

About one-half of the bulk tanks in use hold from 250 to 500 gallons, but smaller ones, as well as tanks that hold up to 7000 gallons, are in use. Farm tank operation makes it unnecessary to gather milk more often than every other day, because the milk is cooled quickly to about 35° to 40°F, and the opportunity for subsequent contamination is minimized. The tank is made of stainless steel and is equipped with agitators to keep the milk in gentle motion, so that it is cooled quickly and uniformly. It is cooled by direct expansion refrigeration or by an "ice bank" type of cooler in which water cooled to about 33°F is circulated around the side walls and bottom of the tank. Bulk tanks are equipped with a recording thermometer to insure proper temperature control of the milk. Often the tank is equipped with an automatic cleaning device, by which rinse water and sanitizing solutions are sprayed over all of the inner surfaces of the tank.

The quantity of milk in the tank usually is measured by means of a dip stick. This is a metal rod, calibrated specifically for the tank with which it is used. The milk is measured to within one gallon on tanks that hold up to 500 gallons, and to two gallons on larger tanks. Some tanks have glass sight-tubes, which indicate the level of milk in the tank and from which the contents may be measured.

After examining, measuring, and sampling the milk, the hauler

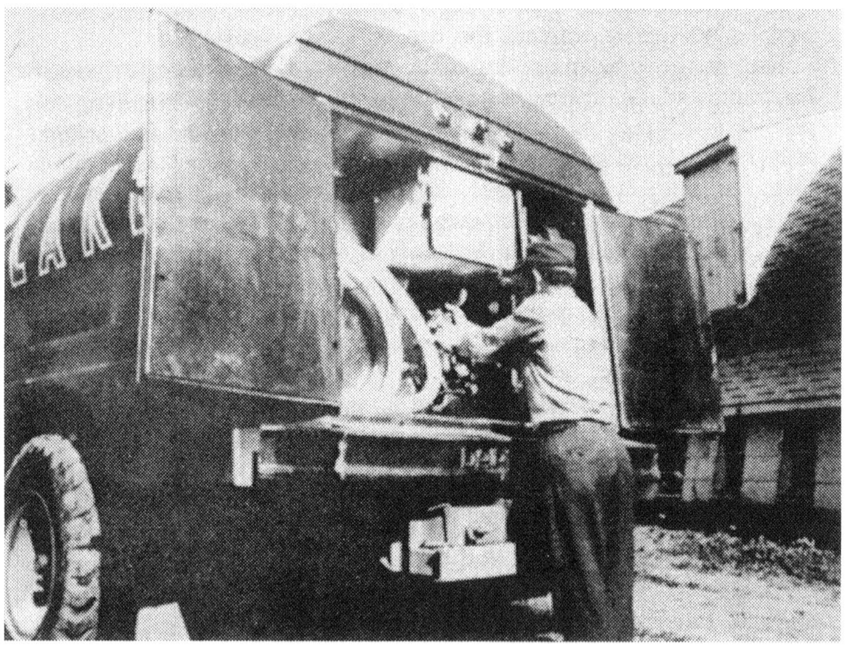

Fig. 1:4. Preparing to load bulk milk at a farm. (Courtesy, Univ. Minnesota).

TABLE 1:6

Storage Life of Dairy Products

Product (commercial pack)	*Approximate storage life at specific temperatures*	*Critical or dangerous storage conditions*
Butter (in bulk)	1 month @ 40°F. 12 months @ —10°F.	Above 50°F., or damp or wet storage.
Butteroil (sealed, full tins; maximum moisture 0.3%)	3 months @ 70°F. 6 months @ 50°F. 9 months @ 32°F.	Above 75°F.
Anhydrous milk fat (sealed, full tins; maximum moisture 0.2%)	3 months @ 70°F. 6 months @ 50°F. 9 months @ 32°F.	Above 75°F.
Ghee (sealed, full tins)	6 months @ 90°F. 9 months @ 70°F. 18 months @ 40°F.	Above 90°F.
Cheddar cheese	6 months @ 40°F. 18 months @ 34°F.	Above 60°F. or below 30°F.
Processed cheese	3 months @ 70°F. 12 months @ 40°F.	Above 90°F. or below 30°F.
Grated cheese (in moisture-proof pack)	3 months @ 70°F. 12 months @ 40°F.	Above 70°F., or above 17% moisture in the product
Sterilized whole milk	4 months @ 70°F. 12 months @ 40°F.	Above 90°F. or below 30°F.
Nonfat dry milk, Extra Grade (in moisture-proof pack)	6 months @ 90°F. 16 months @ 70°F. 24 months @ 40°F.	Above 110°F.
Dry whole milk, Extra Grade (gas pack; maximum oxygen 2%)	3 months @ 90°F. 9 months @ 70°F. 18 months @ 40°F	Above 100°F.
Sweetened condensed milk	3 months @ 90°F. 9 months @ 70°F. 15 months @ 40°F.	Above 100°F. or below 20°F., or dampness sufficient to cause can rusting.
Evaporated milk	1 month @ 90°F. 12 months @ 70°F. (cases to be inverted every 2 months) 24 months @ 40°F.	Above 90°F. or below 30°F., or dampness sufficient to cause can rusting.

Compiled by Dairy and Poultry Division, Foreign Agricultural Service from material supplied by Standardization Branch, Consumer and Marketing Service and Eastern Utilization Research and Development Division of Agricultural Research Service, USDA, FAS M-172, 1966.

pumps the milk into the truck tank. Here it may lose its identity since it may now be mixed with the milk from other producers. The truck tank is made of stainless steel, is insulated, and may carry from 1500 to 6000 gallons of milk. A lighter-weight truck with reinforced plastic tank, up to 4000 gallons in capacity, is also in use. Compartments on the truck hold the milk pump, motor, electric cord for making connections, plastic hose for transferring the milk, equipment for taking samples and a refrigerated or insulated chest for holding the samples.

An adaptation of the bulk-handling procedure is the milk dispenser. This is a refrigerated cabinet in which three- or five-gallon milk cans are placed. From the bottom of the can a short rubber or plastic tube passes through a clamping device. This serves as a spigot from which the milk is drawn into glasses for serving. A modification of this equipment makes use of a disposable five-gallon polyethylene bag, equipped with a dispensing tube. The bag is contained within a strong corrugated paper box. The plastic bag, which is sealed and sanitary, is filled at the dairy and delivered under refrigeration to the consumer. The bag eliminates the use of expensive cans and the necessity of cleaning and sterilizing them.

The milk dispenser is widely employed in restaurants and other eating places. A similar dispenser is available for homes where much milk is used. An industry report claims that more than 150,000 milk dispensers are now in use.

The sanitary aspects of farm tanks and other bulk milk handling equipment is discussed in Chapter 7.

REFERENCES

1. Agricultural Research, U.S. Department of Agriculture, June, 1964.
2. Bartlett, A., Variation in solids-not-fat content of milk, *J. Dairy Research*, 5:113 (1934).
3. Clements, F. W., A thyroid blocking agent as a cause of endemic goitre in Tasmanisa, *Med. J. Australia*, 2:369 (1955).
4. Dairy Products and Sugar in Coffee in the United States; Pan-American Coffee Bureau, New York (1964).
5. Engel, H. and Schlag, H., *Milchwirtschaft, Forsch.*, 2, 1 (1924).
6. Gardner, R. R., Thomas, W. R. Willard, H. S., Proceedings, 37th Meeting, Western Division, Amer. Dairy Sci. Assoc., (1955).
7. Hill, Charles L. The Guernsey Breed, F. L. Kimball Co., Waterloo, Iowa (1917).
8. Kon, S. K., and Cowie, A. T., Milk, The Mammary Gland and its Secretion; Academic Press, New York (1961).
9. Parrish, D. B., et al, Composition of Colostrum and Early Milk of Holstein Cows. *J. Dairy Science*, 33, 457 (1950).
10. Singh, B. S., Yadow, P. C., and Pathak, R. C., Indian, *J. Dairy Sci.*, 16:121 (1963).
11. United Nations, Statistical Yearbook, New York, 1972.

chapter 2

Milk Fat

The popular definition for fat is *a greasy or oily substance*. To the chemist, the terms *fat* and *oil* are nearly synonymous, but ordinarily the fats are solid and the oils liquid at room temperature. They are wide-spread in Nature and are found in practically all vegetable and animal matter. The more technical term *lipid* may be used for the fats as well as for the associated fat-like substances such as lecithin and cholesterol. The term *butterfat* often is used incorrectly for milk fat, but true butter fat, i.e., the fat obtained from butter, differs in composition from the fat as it exists in milk. Fat is the most variable constituent of milk.

Milk Fat Globules

Milk fat and the associated lipid material are present in milk in the form of myriads of small, individual globules predominantly in a liquid state at 37°C (98.6°F). About 1500 billion or more may be found in a pint of milk. It is evident that the globules of fat are very small; smaller, in fact, than some bacteria. The globules vary from about one-tenth to twenty microns in diameter, and average about three microns in diameter. Bacteria often are from two to five microns long. (One micron is approximately 1/25,000 of an inch).

Milk of high fat content usually contains globules that are larger than average in size. The milk from Jersey and Guernsey cows contains more of the larger-sized globules than does milk from the other dairy breeds. Toward the end of the lactation period, there is a tendency for the fat globules to decrease in size.

Attached to the surface of each fat globule is a layer of protein and phospholipid material, the so-called *milk fat membrane*. This adsorbed layer protects each globule, so that it maintains its identity and does not combine or coalesce with others to form a larger globule or mass of fat.[18,24] The membrane appears to have two layers; the outer one contains enzymes originating in the mammary gland, the inner layer consists of a phospholipid-protein complex.[27] The approximate amount of phospholipid material in the membrane is about 35% lecithin, 30% cephalin and 30% sphingomyelin. These are discussed later in this chapter. There is evidence to show that vitamin A and carotenoid pigments as well as certain enzymes, such as phosphatase, are not dissolved in the fat itself, but probably are associated with the surface membrane of the fat globule.[27] The rising of the fat globules to form the layer of cream on milk is described in Chapter 13.

When milk or cream is churned, the mechanical agitation to which the fat globules are subjected, causes the enclosing film to break. This enables the fat in the individual globules to combine to form a mass of butter which separates from the buttermilk. The membrane material of the fat globule is retained in most part in the buttermilk, but a portion of it, especially that part rich in phospholipide material, rather than protein, passes into the butter. The disruption of the milk fat membrane is associated with certain flavor defects which may occur in milk and other dairy products, (See Chapter 10).

Fatty Acids of Milk Fat

Milk fat is made of about 12.5% of glycerol and about 85.5% of fatty acids, by weight. In most of the fatty acids, the carbon atoms are linked to one another in the form of a chain, at the end of which is the acidic or carboxyl group (—COOH) which characterizes the organic acids. Each of the principal fatty acids contains an even number of carbon atoms, and this is typical of the natural fats and oils. Very small amounts of fatty acids with an odd number of carbon atoms have been found in milk fat. A number of investigators have shown that about one-half of the fat in milk arises from the fat in the cow's diet, the balance is synthesized in the mammary gland. Acetates and derivatives of butyric and other acids in the blood stream are the source of the fatty acids and glycerol synthesized in the mammary gland. Bacterial fermentation of the cow's feed that occurs in the rumen produces these fatty acids. It is from these fatty acids that milk fat contains the lower fatty acids which are not found in human milk.

Milk fat probably contains a greater number of different fatty acids than any other food fat; at least 52 have been identified. These are combined with glycerol in units of three fatty acid molecules to each one of glycerol, forming tri-glycerides.[10,14,31] There are thousands of possible ways in which the fatty acids may be combined, but in most instances, these combinations appear to be definitely directed rather than randomly distributed. The tri-glycerides form about 98-99 percent of the milk fat.[12,24,34]

Di- and monoglycerides which contain one or two molecules of the same fatty acid linked to the glycerol are present in milk fat. Up to about 0.5% of the diglyceride and only about 0.04% of the monoglyceride may be present.

In the saturated fatty acids, each carbon atom is attached to the neighboring carbon atom by a single bond. Thus, butyric acid is a saturated, 4-carbon compound.

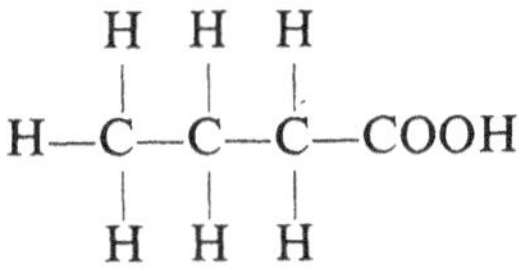

$$\begin{array}{ccccccc} & & H & & H & & H \\ & & | & & | & & | \\ H & - & C & - & C & - & C - COOH \\ & & | & & | & & | \\ & & H & & H & & H \end{array}$$

Stearic acid has a similar structure with a chain of 18 carbon atoms. Oleic, linoleic, and linolenic acids also have 18 carbon atoms each, but they have one, two, and three double bonds, respectively.

When two adjacent carbon atoms in the chain each lacks a hydrogen atom they are said to be unsaturated. The chemical bonding is then made by a double link between the carbon atoms, forming an unsaturated fatty acid.

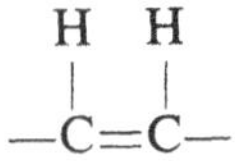

$$\begin{array}{ccc} H & & H \\ | & & | \\ -C & = & C- \end{array}$$

These usually are liquid at room temperature.

A simplified designation usually is used to show the number of carbon atoms and the degree of saturation of a fatty acid. Thus 18:2 indicates linoleic acid, with 18 carbon atoms and two unsaturated bonds.

Many of the fatty acids have common names, such as shown in Table 2:1, but all have a chemical name which expresses, in Greek, the number of carbon atoms, as well as the degree of unsaturation. Thus, attached to the word giving the number of carbons, is the suffix-*anoic* for saturated fatty acids; *enoic* for the unsaturated ones. If more than one double bond is present, the suffix is preceded by the designation for the number of double bonds; thus, for two double bonds we have *dienoic*, for three, *trienoic*. Fatty acids with more than one double bond are called *poly-unsaturated* or *polyethenoid*.

Table 2:1 gives the fatty acid content of some animal fats; Table 2:2 for vegetable fats.

When fats that contain polyunsaturated fatty acids, are heated, as may occur when foods are heated or deep-fried in the fat, the polyunsaturated fatty acids may become saturated.

Butyric acid is found in no natural food fat other than milk fat. This fatty acid, as well as the 6-, 8-, and 10-carbon fatty acids have a strong, characteristic odor and their release during the decomposition of milk fat is the cause of rancid flavor and odor in milk products. The characteristic flavor developed in cheese made from goat's milk, as well as that made from ewe's milk, is related to their high content of these fatty acids.

Of the fatty acids in milk fat, from 60 to 70% are saturated, 25 to 35% are unsaturated, and about 4% are polyunsaturated. The saturated fatty acids present in the largest amount are myristic, palmitic, and stearic; The chief unsaturated fatty acids are oleic, linoleic and linolenic. Arachidonic acid, which has four double bonds, is present up to 0.14%.

TABLE 2:1
Fatty Acid Content of Milk Fats

		Weight Percentage				
Fatty Acid	*Carbon Atoms*	Cow (11,14)	Goat (12)	Buffalo (13)	Mare (15)	Human (23)
Butyric* (Tetranoic)	4	3.57	3.0	5.8	0.4	0.4
Caproic* (Hexanoic)	6	2.22	2.5	0.6	0.9	0.1
Caprylic* (Octanoic)	8	1.06	2.8	0.9	2.6	0.3
Capric* (Decanoic)	10	1.88	10.0	1.0	5.5	1.7
Lauric (Dodecanoic)	12	2.96	6.0	1.6	5.6	5.8
Myristic (Tetradecanoic)	14	11.20	12.3	9.0	7.0	8.6
Palmitic (Hexadecanoic)	16	25.24	27.9	35.2	16.1	22.6
Stearic (Octadecanoic)	18	11.90	6.0	15.3	2.9	7.7
Arachidic (Eicosonoic)	20	0.22	0.6	0.1	0.3	1.0
Oleic* (Octadecenoic)	18	30.0	21.1	20.5	42.4	36.4
Linoleic* (Octadecadienoic)	18	2.8	3.6	1.5	—	8.3
Linolenic* (Octadecatrienoic)	18	0.5	—	—	—	—

*These fatty acids are liquid at room temperature.

TABLE 2:2
Fatty Acid Content of Some Vegetable Oils[30,42]
(Percent by weight)

Fatty Acid	*Coconut*	*Corn*	*Soy Bean*	*Cottonseed*	*Safflower*	*Peanut*
Caproic	0.4	—	—	—	—	—
Caprylic	8.0	—	—	—	—	—
Capric	7.0	—	—	—	—	—
Lauric	53.0	—	—	trace	—	trace
Myristic	16.9	trace	0.5	1.1	0.3	1.4
Palmitic	8.5	11.8	12.2	21.1	4.0	8.3
Stearic	2.0	1.9	1.2	2.3	2.0	3.1
Oleic	4.2	30.4	22.3	22.8	17.0	52.0
Linoleic	1-3	40-50	60.5	45-48	70-74	20-28
Linolenic	trace	1.2	3.1	3.4	2.7	2.8

The fats of plant origin in the cow's diet contain large amounts of polyunsaturated fatty acids. These fats are hydrolyzed in the cow's rumen. The liberated fatty acids are then converted into saturated fatty acids by the action of the organisms in the rumen. [28]

Experiments have been made to increase the degree of unsaturation in milkfat. In Australia, investigators homogenized safflower seed oil with sodium caseinate solution, followed by a treatment with a 37% solution of formaldehyde and then spray drying. The resulting coated oil drops pass through the rumen unchanged. The caseinate coating is digested in the acidic environment of the abomasum or fourth stomach. Here the unchanged polyunsaturated fatty acids are absorbed. An increase of up to ten times the amount of unsaturated fatty acid normally present occurs in the milk fat. Similar results have been obtained by workers in the United States Department of Agriculture.

The milk from cows given this type of feed develops an objectionable oxidized flavor. The cream produces a light-colored butter, which tends to be soft and sticky. [32]

Traces of free fatty acids, in widely varied amounts, are present in milk. Generally, in a complete milking, the first milk drawn contains the larger amount, and this decreases, so that the last drawn from the udder has the lowest.[36] As stated previously, the fat content of the milk as it is drawn from the udder, changes in the reverse order.

The analytical chemistry for the detection of foreign fats in butter makes use of the fact that the proportion of fatty acids of low molecular weight in milk fat is fairly constant.[13,16,25] Other fats and margarine contain little or none of these fatty acids, so a deficiency of them is an indication that the butter is adulterated with a foreign fat.

A number of compounds, known as *lactones*, derived from the fatty acids are found in trace amounts in the fat of processed milk products, especially in heated milk fat. These compounds contribute to the flavor of dairy products. Depending upon the particular lactone present and its concentration, the flavor may be desirable or objectionable.

Defects of flavor associated with the decomposition of milk fat are discussed in Chapter 10.

Minor Components of Milk Fat

Other than the free fatty acid that may be associated with the milk fat, there are present, in small amount, a number of other compounds. Among these are the phospholipids, sterols, tocopherol (vitamin E), carotenoids, and vitamins A and D.

Phospholipids

Most of the phospholipids are glycerides of long-chain fatty acids, combined with phosphoric acid and the nitrogen-containing compound, choline. Choline also is part of the vitamin B complex and it is essential for the metabolism of fats, cholesterol, and for growth.[4] Sphingomyelin, found in small amount in milk, contains *sphingosine*, a nitrogenous alcohol, instead of glycerol.

Lecithin:

Lecithin is the best known phospholipid and the principal one found in milk. It is also found in egg yolk, the tissues of the nervous system of animals, and in nearly all vegetables, especially soy beans. Milk contains about 0.03% of phospholipids, chiefly lecithin, sphingomyelin and cephelin. Milk fat contains about 0.19% of phospholipids.[13,27]

Upon separation, about 70% of the phospholipids in milk go into the cream; when the cream is churned about one-half of the phospholipid as present are retained in the buttermilk. As the buttermilk represents but a small part of the original volume of milk—perhaps 2-3%—a considerable concentration of phospholipid, about 0.18%, is present. It is believed that this contributes to the desirable flavor of buttermilk.

When freshly prepared, the isolated phospholipids of milk are colorless and odorless, but quickly oxidize in the air, turning brown and acquiring an unpleasant odor. This transformation appears to be associated with the development of the "oxidized" and "cardboard" or "cappy" flavor in milk.

Lecithin, usually derived from soy-beans, helps to emulsify fat with the aqueous portion of a product, and is used in the manfacture of margarine, chocolate, and some emulsifiers for ice cream. As milk contains but little lipoid phosphorus (the phosphorus portion of the fat) compared to egg yolk, the amount of egg yolk added to ice cream or egg nog may be measured by an analysis for the lipoid phosporus content.[9]

Unsaponifiable Portion of Milk Fat

When a fat is saponified and the resulting soap is extracted with ether, the material dissolved in the ether is called the unsaponifiable portion, because the soap, itself, is not soluble in ether. In milk fats the unsaponifiable matter consists mostly of sterols. Sterols are crystalline alcohols, characterized by a complex structure related to the bile acids, hormones of the sex and adrenal glands, and to vitamins D_2 and D_3. In vegetable fats, the sterols, known as *phytosterols*, differ from those of animal origin, especially in that they are not absorbed from the intestinal tract.[35] The phytosterols in the cow's feed do not pass into the milk fat.

Cholesterol:

The principal sterol found in milk is *cholesterol*. This sterol, either free or esterified—that is, combined with fatty acids—is found in all the body tissues, especially the brain and nervous system. Even though none may be contained in the diet, the body synthesizes about four grams of cholesterol a day. The ordinary diet furnishes about one-half gram a day. It is necessary for the formation of bile, which is needed for the digestion of fats. It is evident that persons on a vegetable diet must synthesize all the cholesterol that their body needs.

Milk contains about 0.015% of cholesterol; pure dry milk, about 0.36%.[13] Egg yolk contain much more cholesterol than does milk fat; about 1.5%. Next to egg yolk, cream cheese and cheddar cheese contain relatively high cholesterol content (See Chapter 19). In the author's method, the amount of cholesterol in the product is taken as a measure of the egg yolk in ice cream or egg nog.[21,3]

The concentration of cholesterol is higher in the milk fat globule membrane than in the fat portion. It has been proposed that cholestrol is an agent in the synthesis of milk fat in the udder from lipids in the blood stream.[8]

MILK FAT IN NUTRITION

No specific amount for fat in the daily diet has been recommended by nutritionists, and little information is available on the subject. The body can make use of the carbohydrate, protein, and fat in the diet interchangeably and so use one to compensate for the lack of another. The total intake of each should be such that the protein in the diet is not used to supply caloric energy that should be furnished by the carbohydrate and fat in the diet. A diet in which the fat supplies 20% of the total calories appears to furnish an adequate supply for human requirements. Certain unsaturated fatty acids, such as linoleic and arachidonic, are essential to the normal, healthy growth of experimental animals, but the human need has not been demonstrated.

Fat is the only form in which the body stores most of its energy requirement. Between 35 and 40% of the fat in the average American diet is derived from meat and milk products. These contain mostly saturated fatty acids; the polyunsaturated fatty acid content rarely exceeds 4%.

In the United States, the consumption of vegetable fats has increased at the expense of fats of animal origin. For example, in 1940, soybean oil furnished about 8% of dietary fat, by 1964 it reached about 50%. (Fig. 2:1)

Milk fat is more easily digested and larger amounts of it can be absorbed without producing a digestive disturbance than any other common, edible fat. Milk is coagulated in the stomach and as the fat is retained within the particles of curd, it cannot form large masses of fat which would be difficult to digest. Most common edible fats of low melting point have about the same degree of digestibility.[7] About 97% of the milk fat ingested is utilized by the body; for lard, beef fat, and mutton fat the figures are 97, 93, and 88%, respectively. Mutton fat has a relatively high melting point. Fats that are composed of short-chain fatty acids are readily absorbed by the digestive tract.

During recent years, numerous articles in the popular press have stressed the belief that a high level of cholesterol in the blood leads to heart disease, and that fat in the diet—especially the over-use of saturated fats—increases the amount of cholesterol of the body. Neither supposition has been proven, but much investigation leads to the belief that there may be some connection between lipids in the diet and the development of *atherosclerosis*, especially in overweight persons. Atherosclerosis is a type of arteriosclerosis (hardening of the arteries) involving abnormal cholesterol metabolism. A thickening of the inner walls of the blood vessels occurs, accompanied by a fatty degeneration of the tissues, associated with an increase in their cholesterol content. Although it is not a new disease, atherosclerosis

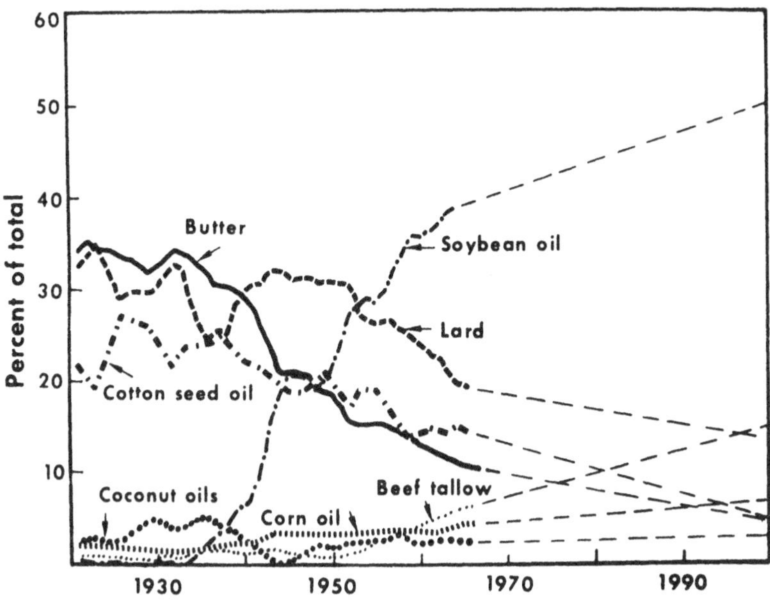

Fig. 2:1. Relative importance of major sources of food fats and oils.[33]

probably is today responsible for more disability and death than any other disease.

There is no agreement whether or not persons afflicted with atherosclerosis have a higher-than-normal cholesterol content in their blood plasma (normal for the average adult male is 150 to less than 200 mg per 100 ml. of plasma, but increases with advancing age). Present opinion considers that the commonly existing average blood lipid levels in developed societies are too high. In one investigation it was found that the average serum cholesterol level was 146 mg/100 ml for males age 10; 174 mg/100 for males age 30; and 185 mg/100 ml for males age 50.[43]

The Framingham, Massachusetts Diet-Heart Study, covering some ten years, suggests a cautionary note with respect to hypotheses relating diet to serum cholesterol. No relationship was found between the incidence of coronary heart disease and diet. "There is no discernible association between reported diet intake and serum cholesterol level", according to this report.[17]

Experiments with laboratory animals show that any procedure that increases cholesterol in the plasma will in time produce atherosclerosis. In man, a sustained rise in blood cholesterol and of certain large fat-bearing constituents in the blood does predispose to the development of atherosclerosis. A detailed study, extending over a period of 20 years, indicates among other factors, there is a relationship between a high level of serum cholesterol and the risk of developing coronary disease.[40] In Norway, a group of men were studied over a period of five years.[41] Some used a diet materially reduced in cholesterol and saturated fats; soy bean oil was used to replace other fats in the diet. The average reduction of serum cholesterol in this group was 17.6%(52 mg.) whereas the group on a normal diet had a reduction of 3.7% (11 mg.).

An investigation based upon three years of observations indicates that "substitution of polyunsaturated for saturated fat in the diet results in an absolute increase in excretion of cholesterol and bile acids in essentially all subjects studied". [38] It is believed that the polyunsaturated fat diminshes reabsorption of cholesterol from the digestive tract and increases mobilization of cholesterol deposits. The replacement of most of the calorie requirement by carbohydrates instead of dietary fat may increase the blood serum triglycerides and lead to coronary heart disease.

A detailed investigation made in Germany led to the conclusion that no significant increase in cholesterol occurs in persons with arteriosclerosis when all the fat in the diet is supplied by butter.[39] The diet contained 2500 calories, 25% from butter, 60% from carbohydrates and 15% from milk protein.

Until factual knowledge is available, which will reconcile various contradictory findings, it appears fruitless to speculate concerning the connection of cholesterol with atherosclerosis.[2] Qualified investigators state that individuals that are not overweight, and that have no history of cardio-vascular disease, need no special dietary precaution in this regard. Where loss of weight is indicated, it would appear advisable for the diet to furnish not over 30% of its total calories from a fatty source. The saturated fats in the diet (meat fat, egg fat, butter) could then be substituted by a source of unsaturated fatty acids, such as are supplied by vegetable oils (safflower, cotton seed, soy bean, or corn oil). [19,20]

A similar recommendation states that the diet should supply no more than 35% of the total calories from fat.[44] Of this amount, less than 10% of the total calories should come from saturated fatty acids and up to 10% from polyunsaturated fatty acids. The average daily intake of cholesterol is recommended to be about 300 mg, but care must be taken to assure adequate protein intake if severe cholesterol restriction is undertaken.

The only other dietary factory known to decrease the cholesterol content of the blood is niacin, which may cause side reactions, such as flushing and pruritus.[22]

The increased and prolonged ingestion of large amounts of vegetable oils rich in unsaturated fatty acids, may increase the need of the body for alpha-tocopherol (vitamin E). It is estimated that whereas an intake of as little as 5 mg may suffice for the adult having a low store of linoleic acid, this requirement may increase to 30 mg if the intake of oil is large.[37]

If a food-stuff label contains a statement of cholesterol content, it must be given in terms of milligrams per serving and milligrams per one hundred grams of food, according to the United States Food and Drug Administration.

REFERENCES

1. Albrink, M. J., Meigs, J. W., and Granoff, M. A., "Weight Gain and Serum Triglycerides in Normal Men," *New Engl. J. Med.*, 266, 484-489 (1962).
2. Committee to Review Research Developments in the Field of Diet as Related to Heart Disease. *Med. J. Australia*, Vol. 1 (No. 7): 309, (1967).
3. Cook, J. H. and Mehlenbacher, A., Determination of egg yolk in egg white, *Anal. Chem.*, 18:785 (1946).
4. Du Vigneaud, V., A Trail of Research: Cornell University Press, Ithaca, N.Y. (1952).
5. Edmonson, R. F., U.S. Agricultural Research Service, E.R.R.L. Pub. 3786, Oct. 1972. Also see *J. Dairy Sci.*, 55, 677, 1972.
6. Hansen, H. P. and Shorland, F. B., The branched-chain fatty acids of butterfat, *Biochem. J.*, 50:102 (1951); *ibid.*, 55:662 (1953).
7. Hoagland, R. and Snider, G. G., *Nutritive properities of animal and vegetable fats; U.S. Dept. of Agr. Technical Bull.* 725, (1940).
8. Homer, D. R. and Virtanin, A. I., Cholestrol in the Milk of Cows on Normal and Protein-Free Feeds, *Acta Chem. Scand.* 20, 2321 (1966).
9. Horrall, B. E., A study of the lecithin content of milk and its products, *Purdue Agr. Expt. Sta. Res. Bull.* 401 (1935).
10. Jack, E. L., The fatty acids and glycerides of cow's milk fat, *Agr. and Food Chem.*, 8:377 (1960).
11. Jack, E. L. and Henderson, J. L., Fatty acid composition of glyceride fractions separated from milk fat, *J. Dairy Sci.*, 28:65 (1945).
12. Jack, E. L., Henderson, J. L., Hinshaw, E. B., Distribution pattern of fatty acids in glycerides of milk fat, *J. Biol. Chem.*, 162:119 (1946).
13. Jack, E. L. and Smith, L. M. Chemistry of milk fat: A review, *J. Dairy Sci.*, 39:1 (1956).
14. Jensen, R. G., Quinn, J. G., Carpenter, D. L., and Sampugna, J., Liquid Gas Chromatography in the Analysis of Milk Fatty Acids, *J. Dairy Science* 50, 119-126 (1967).
15. Kartha, A. R. S., The glyceride structure of natural fats, *J. Am. Oil Chemists Soc.*, 31:85 (1954).
16. Kenney, M .A., A survey of United States butterfat constants: 11: Butyric Acid. *J. Assoc. Offic. Agr. Chemists*, 39:212 (1956).

17. Diet and the Regulation of Serum Cholesterol, Framingham Diet Study, Section 27, 1970.
18. King. N., *The Fat Glouble Membrane*; Commonwealth Agricultural Bureau, Farnham Royal, Bucks., England (1955).
19. Kinsell, L. W., Nutrition in vascular diseases, Borden's *Review of Nutritional Research*, 17, No. 1, 1956.
20. Kinsell, L. W., *et. al.*, Fat metabolism, *J. Clin. Endocrinol. and Metab.*, 12:909 (1952).
21. Lampert, L. M., Cholesterol as a measure of the egg yolk in ice cream, *Ind Eng. Chem., Anal. Ed.*, 2:159 (1930).
22. Lever, W. F. and Waddell, W. R., *J. Invest. Dermatol.*, 25:233 (1955).
23. Longenecker, H. E., Composition and structural characteristics of glycerides in relation to classification and environment, *Chem. Rev.*, 29:201 (1941).
24. McCarthy, T., Patton, S., Evans, Laura, Structure and synthesis of milk fat, II: Fatty acid distribution in the triglycerides of milk and other animal fats, *J. Dairy Sci.*, 43:1196 (1960).
25. *Methods of Analysis* (9th Ed.); Association of Agricultural Chemists, Washington, D.C., 1970.
26. Morrison, L. M., Diet and atherosclerosis, *Ann. Int. Med.*, 37:1172 (1952).
27. Morton, R. K., The lipoprotein particles in cow's milk, *Biochem. J.*, 57, 231 (1954).
28. Patton, S. and Kester, E. M., Saturation in Milk and Meat Fats, *Science*. 156, 1365-66, 1967).
29. Prentice, J. H., through *Dairy Science Abstracts*, 31, 353, 1969.
30. Rice, F. E., Nutritional evaluation of the replacement of the fat in whole cow's milk by coconut oil, *J. Agr. Food Chem.*, 8:488 (1960).
31. Ryhage, R. Identification of fatty acids from butterfat using a combined gas chromatograph-mass spectrometer, *J. Dairy Res.* 34:115 (1967).
32. Scott, T. W., et al, Production of polyunsaturated milkfat in domestic ruminants, *Austral. J. Sci.*, 32:291 (1970).
33. Rasmussen, C. L., The Changing Scene of Fats and Oils, U.S. Department of Agriculture, Agri. Res. Service, 74-43, Sept. 1967.
34. Sommer, H. H., The fat emulsion of milk from a chemical standpoint, *Milk Dealer*, Oct. 1951 (also in *Cherry Burrell Circle*, 1951).
35. Schoenheimer, R., *Science*, 74:579 (1931).
36. Thomas, W. R., Harper, W. J., Gould, I. A., Free fatty acid content of fresh milk as related to portions drawn, *J. Dairy Sci.*, 37:717 (1954).
37. *What's New in Food and Drug Research*, Food and Drug Research Laboratories, New York, December 1960.
38. Wood, P. D. S., Shioda, R. and Kinsell, L. W., Dietary regulation of cholesterol metabolism. *Lancet*, 2:604, (1966).
39. Lembke, A. Frahm, H. and Greggersen, H., Butter as the preponderant fat in the diet of arteriosclerotic persons, XVII International Dairy Congress, E/F 587-594 (1966).
40. Dawber, T. R., Kannel, W. B. and Lyell, L. P., An approach to longitudinal studies in a community: The Framingham Study. *Ann. New York Acad. Sci.*, 107:539 (1963).
41. Leren, P. The effect of plasma cholesterol lowering diet in male survivors of myocardial infarction. *Acta Med. Scand.*, Suppo. 466 (1966).
42. Kinsella, J. E., Butterfact complexity results from variety of fatty acids, *American Dairy Review*, 82, March 1969.
43. Wynder, E., Hill, P.: Blood lipids; how normal is normal? *Prev. Med.* 1:161-166 (1972).
44. American Heart Association, *Diet and Coronary Heart Disease*, New York (1973).

chapter 3

The Milk Proteins

Proteins, which are supplied in our diet by foods such as lean meat and cheese, are complex organic compounds of carbon, hydrogen, oxygen and nitrogen, sometimes containing sulfur and phosphorus. The word *protein* is derived from the Greek *proteios,* meaning *holding first place,* in reference to the early conception that the proteins were essential constituents of all animal tissues. Protein material is, after water, the most abundant constituent of the soft tissues, forming about 18% of the body weight. Proteins are concerned with the growth of cells, and as part of enzyme systems, they control many of the processes of metabolism. Animals, as a rule, do not build up the protein material of their bodies from simple inorganic compounds, but derive it from the protein in their food. Plants form protein directly from the material of the soil and air.

TABLE 3:1
Requirement and Amount of Essential Amino Acids in Some Common Foods

Amino Acid	*Recommended Daily Intake for Adults,*	*F.A.O.* [a] *Pattern*	*Grams in 100 Grams of Protein Derived from*					
			Milk	Meat	Eggs	Corn	Rice	Wheat
Tryptophan	0.5	1	1.44	1.4	1.5	0.7	1.2	1.2
Isoleucine	1.4	3	6.5	6.0	7.7	4.0	4.7	4.0
Leucine	2.2	3.4	10.0	8.0	9.2	12.7	8.4	7.0
Lysine	1.6	3	8.0	10.0	7.0	2.7	3.4	2.7
Methionine	2.2	3	3.4 [b]	4.4 [b]	6.4 [b]	3.4 [b]	2.9 [b]	4.3 [b]
Phenylalanine	2.2	2	5.0	5.0	6.3	4.5	4.7	5.1
Threonine	1.0	2	4.7	5.0	4.3	4.1	3.6	3.3
Valine	1.6	3	7.0	5.5	7.2	5.3	6.3	4.3

[a] Food and Agriculture Organization of the United Nations Pattern means that for efficient use by the body, the diet should contain three times as much Isoleucine, twice as much Threonine, etc., as of Tryptophan, which is given a value of one.

[b] Includes cysteine, which can serve as a partial substitute for methionine.

Proteins are composed of a large number of nitrogen-bearing compounds known as amino acids. The amino acids have been termed the building blocks for proteins and their place in the protein structure may be compared to the use of letters in the spelling of words. Thousands of combinations may occur just as countless combinations are possible with the letters of the alphabet in forming words. Eighteen different amino acids are concerned in nutrition, eight of which are considered indispensable to nutrition and must be supplied in adequate amount by the diet; see Table 3:1.[3,10] The others may be needed, but if not present in the diet, the body can form them from the other amino acids. One exception to this is the finding that histidine is essential to the nutrition of infants who in early life must have this amino acid supplied by the diet.[4] Recommended daily protein allowances are given in Chapter 6, Table 6:1.

In the digestive process, the proteins in food are split into their component amino acids, which the blood stream then brings to each growing cell, to form the specific protein needed. Should there be a deficiency in protein or a disturbed protein metabolism, resistance to disease is lowered and in case of illness, recovery is delayed. Aged persons tend toward a lack of protein in their diet. The recommended intake is about 0.9 gram of protein per kilogram of body weight. The United States Food and Drug Administration (Federal Register, Jan. 19, 1973) states that the Recommended Daily Allowance (RDA) of protein in a food product is 45 grams, if the protein efficiency ration (PER) of the total protein in the product is equal to or greater than that of casein. If the PER is less than that of casein, the RDA is 65 grams. Adding a few pounds of lysine to a ton of wheat is sufficient to make its protein quality practically equal to that of casein.

From Table 3:1 it may be seen that a person needs a little over three times as much lysine as tryptophan. Milk, meat, or eggs provide more than this ratio of lysine, but rice and wheat do not; therefore lysine is a limiting amino acid in these cereals; in milk, it is methionine. By combining milk and cereals, individual deficiencies in lysine and methionine are balanced and better use is made by the body of both proteins than would occur if they were present separately in the diet. Milk proteins rank next to those of meat as a source of high-quality protein in the American diet. The diets of adolescent girls and of older people are likely to be low in protein compared with the recommended amounts.

Foods high in protein, such as dry non-fat milk, cheese, and lean meat, permit a high intake of protein, without increasing the bulk of the food consumed. Although they are cheaper and more plentiful, proteins of vegetable origin are, in general, of less nutritional value than those of animal origin. The nutritional value of a protein is limited by the amount of essential amino acid present in the least required amount. Advances in the synthesis of amino acids permits the upgrading of diets.

Milk and milk products supply from 20 to 30% of the total protein in the diets used in the United States, Canada, and most of western Europe. In Italy, Greece, and Portugal it supplies from 6 to 18% while in most of Asia and Latin America, less than 10% comes from dairy products.

An example of the influence of a lack of protein in the diet is shown in the severe disease of malnutrition known as "kwashiorkor", found among native children in Africa and among Andean Indians. The name means "the disease of the first child after the second arrives". It occurs when the first child is weaned directly to a diet of low-protein, high-carbohydrate gruel and the new-born child is breast-fed.

The amino acid content of the principal proteins in milk is given in Table 3:2.

TABLE 3:2
Amino Acid Content of Major Milk Proteins [5,7]
(Grams per 100 grams of protein)

Amino acid	*Total milk proteins*	*Casein*	*Lactalbumin*	*β-Lacto-globulin*
Alanine	3.6	3.1	7.0	6.8
Arginine	3.5	4.2	3.1	2.9
Aspartic acid	7.5	6.5	11.1	11.7
Cystine	0.9	0.4	2.7	3.0
Glutamic acid	21.7	23.6	17.7	19.8
Glycine	2.1	2.1	2.5	1.7
Histidine	2.7	3.0	1.8	1.7
Isoleucine	6.5	6.6	6.7	7.4
Leucine	9.9	10.1	12.0	15.0
Lysine	8.0	8.2	9.7	11.9
Methionine	2.4	3.3	1.9	3.3
Phenylalanine	5.1	5.8	4.0	3.8
Proline	9.2	12.3	4.7	5.2
Serine	5.2	6.3	4.8	4.3
Threonine	4.7	4.5	5.2	5.2
Tryptophan	1.3	1.5	1.8	2.3
Tyrosine	4.9	6.3	3.2	4.0
Valine	6.7	7.4	5.3	5.8

Combinations of amino acids with the lactose in milk, especially human milk, is discussed in Chapter 13. The development of a brown color in heated milk, the Maillard reaction, and subsequent flavor deterioration is described in Chapter 4.

CASEIN

Casein, the principal protein of milk, is that fraction of raw skim milk which is precipitated by acidifying to *p*H 4.6-4.7. It comprises about 80% of the proteins of milk, the amount present in milk varying from about 2.6 to 3.4%, depending upon the breed of cow that produced the milk. Casein consists of a mixture of at least three protein components, termed alpha-, beta-, and gamma-casein.[10] It is present in milk as a colloidal suspension of complex particles called micelles.

The electron microscope shows that the casein particles in fresh, raw milk appear as individual spheres, whose diameter vary between 10 and 200 millimicrons, with some particles as large as 800 millimicrons. Pasteurization does not appear to alter the dispersion of the casein. The homogenizing of milk causes some of the particles of casein to agglomerate with the fat globules. The fat-casein agglomerates may reach a diameter of several microns.

The casein particles in milk may be separated by high-speed centrifuging or by the addition of acid. Souring of the milk by bacterial action also causes the casein to precipitate. When sufficient acid is present to change the *p*H of the milk to about 5.2-5.3, precipitation occurs, with a gradual solution of the calcium and phosphorus salts associated with the protein. At the iso-electric point, *p*H 4.6-4.7—at which point it is least soluble—the casein is essentially free of all assooiated inorganic salts.

Alpha and beta-casein are formed in the cow's udder, but gamma-casein, which forms 3 to 7% of the casein complex, is first found in the blood stream from which it passes into the udder by some manner not yet understood. Associated with the alpha-casein complex is a portion known as *kappa*-casein.[12] Kappa-casein makes up about 15% of the total casein complex and acts as a stabilizing factor in that it holds the entire casein complex in colloidal suspension, which gives milk its white color. When milk is acidified, either by the addition of acid or by bacterial action, the casein forms a clot or curd, known as paracasein. A portion of the alpha-casein is liberated and becomes part of the whey proteins.

Certain enzymes, especially rennin, can remove the kappa fraction from the casein complex. When this occurs, its stabilizing action is lost. Casein, free of the kappa component, is insoluble in the presence of calcium ions, such as are normally present in milk, and so it precipitates as a clot or curd. This is the basis of cheese manufacture. (See Chapter 19). If calcium ions are not present, no casein clot is formed, even though rennin action has split off the kappa component.

Casein obtained by coagulation by rennin more nearly represents the protein complex as it exists in milk than does acid-precipitated casein. The latter may be considered as free casein, whereas the rennet-casein is a calcium-caseinate complex. Because of their different chemical and physical properties, these caseins have different commercial uses.

Manufacture of Casein

Casein is made commercially from skim milk of as low a fat content as is practical. It is precipitated from the skim milk by the gradual addition of dilute acid, with gentle agitation, until the mixture has reached about *p*H4. Depending upon the acid used, the product is termed sulfuric, muriatic (hydrochloric), or lactic acid casein. The finished product may be further designated by the method of preparing the curd, such as *pressed-curd, cooked-curd, grain-curd,* and continuous-process casein. The same general characteristics apply to any of these preparations, except that owing to its heat treatment, cooked-curd casein is less soluble and contains more ash than the other caseins. The careful control given to the amount of acid used, the temperature of precipitation and the degree of washing the curd make grain-curd and continuous-process casein products of superior quality, low in ash and readily soluble.

If the skim milk is at a temperature below 95°F, a soft, finely divided curd is obtained, which generally is not desirable. At 100°, the curd is coarser and easier to wash. When the skim milk is heated to 130° or over, a cooked-curd casein is obtained, which is hard, high in acid and ash, and difficult to dissolve. After the whey is drained off, the curd is washed several times with cold water and then pressed to remove excess water. The pressed product is shredded in a curd mill and finally dried to a moisture content of about 8%. A yield of 2.5 to 2.9 pounds of casein is obtained from 100 pounds of skim milk.

Rennet-casein is prepared by adding 7 to 10 ml of rennet extract per 100 pounds of skim milk warmed to 95-100°F. The rennet extract is diluted with about 200 ml of water before it is added to the skim milk. Coagulation usually occurs within 20 minutes. The skim milk is then heated to about 150° while the curd is being stirred and cut, much as in the manufacture of cheese. After draining off the whey, the curd is washed with water, then drained, pressed, shredded, and dried. Rennet-casein is used in the manufacture of casein plastics (q.v.).

The manufacture of casein decreases when there is considerable world demand for non-fat dry milk, since the conversion into milk powder usually offers more profit. Comparatively little is made in the United States and exact statistics are not available. Imports are high, especially from New Zealand, Australia, France, Argentina, and Canada. In 1971, the United States imported 135.3 million pounds.

USES OF CASEIN

As Food

Casein, as such, is not used as a food product, except in experimental diets for animals and as a constituent of certain health foods. Casein also is used in the preparation of media for bacteriological work. Ammonium, sodium, and calcium caseinates are used in foodstuffs for special diets and are sold under various trade names; for example, Casec (a calcium caseinate), Sheftene C2 (a sodium caseinate), Ca-Sal (a preparation containing casein, lactalbumin and beta-lactalbumin). Products of this type have been recommended for enriching bread and cereal products because the essential amino acid content, especially that of lysine, is thereby increased.

Calcium caseinate is used to improve the whipping property of cream topping made from vegetable fats, and when legally permissible, to improve the body of sour cream and yoghurt. Sodium caseinate is used as a binder in processed meat products, such as sausages and loaves. It is also utilized as an ingredient of ice cream in order to minimize shrinkage and improve overrun. In a doughnut mix, it prevents excessive absorption of fat during deep-fat frying.

Sodium caseinate is made from edible-grade casein, solubilized with pure sodium hydroxide. The resulting solution is spray-dried to form a product that contains not over five precent moisture.

Increasing amounts of sodium caseinate are now used in the preparation of the many substitute dairy products, such as, imitation milk, coffee whiteners, dips, and toppings (see Chapter 21).

Industrial Uses of Casein

Casein Glues:

These preparations may be considered to be solutions of casein in an alkaline solvent. Casein glues made with the addition of calcium compounds become water-repellant upon drying. The alkaline solvent may be lime, sodium carbonate, borax or an organic base such as triethanolamine.

Paper Coating:

Casein is used in the manufacture of coating for paper. This, essentially, is a glue made from casein and mixed with clay and mineral pigments. The casein binds the clay and other materials to the paper. Coated paper has a superior printing surface which takes ink more evenly and gives much better reproduction of pictures and color reproduction than does uncoated paper.

Coatings made from starch compounds are more widely used than those made from casein, especially for machine-coated papers. In some applications, synthetic resins and soy-bean protein are displacing casein. The use of high-tack inks, which require a very adhesive paper coating, and the demand for better water-resistance tend to give the synthetic resins an advantage over casein and starch in coatings.

Plastics:

A plastic material may be made from casein. Usually rennet-casein is used, but sometimes acid-precipitated casein is employed in the manufacture of transparent and translucent plastics. The casein is ground and small amounts of water are added, together with any desired dye, pigment, or filler. The powdery mixture is fed to a screw-operated extruding machine where it is forced, at a temperature around 180°F, through resistance screens. Under the heat and pressure applied, the mass becomes plastic as it emerges from the nozzle of the machine in the form of rods. If sheets are desired, the extruded rods are placed in molds and formed into sheets by hydraulic pressure. The rods and pressed sheets are immersed in a weak solution of formaldehyde. This renders the plastic insoluble and hardens it. After the curing process the material is dried, the sheets are flattened and the rods are straightened and ground to the desired diameter.

The finished material is not moldable but is fabricated by machining, grinding, or carving. Casein plastic cannot be used where dimensional stability is needed, as it takes up or loses moisture, with an accompanying change in dimension. The principal use of casein plastic is in the manufacture of buttons, buckles, ornaments, and costume jewelry.

Textiles:

A fiber with many of the characteristics of wool may be made from casein. The first successful casein fiber was made in Italy about 1935, and it is still being manufactured there and in England. In Italy it is called Lanitol or Merinova; in England, Caslan or Fiberlane. A similar fiber was made in the United States between 1940 and 1950 under the name of Aralac. The introduction of a number of fully synthetic fibers made the manufacture of casein fiber uneconomical in the United States.

Casein fiber is resistant to moths and mildew, it does not shrink, and like wool, it has good wrinkle-resistance and warmth.

A composite fiber, made of 30% casein and 70% of acrylonitrile was made by the Toyobo Company in Japan. Known as K-6, it had the appearance and physical properties of silk.

Other Uses of Casein:

Casein is used as an adhesive in a number of insecticidal and fungicidal sprays. Ground casein and lime are combined to form calcium caseinate, which, after mixing with water and the insecticidal material, is sprayed on fruit or foliage. The casein compound favors a complete and even coverage of the material that is sprayed. A similar compound of casein is used in the preparation of emulsified oil sprays.

Compounds of casein with arsenic, iron, mercury, iodine, and silver have been used in medicine.

Iodinated casein, sometimes called *thyroprotein*, when included in dairy feed, acts like thyroxin, a hormone secreted by the thyroid gland. It is sometimes used to increase milk production and the fat content of the milk. [6]

PROTEINS IN WHEY

Lactalbumin

After the removal of the fat and casein from milk, the resulting liquid is known as *whey*. About 0.5 to 0.7% of soluble protein material is retained in the whey, namely the proteins, *lactalbumin* and *immunoglobulin*. Next to casein, lactalbumin or milk albumin, is the protein that occurs in the largest amount in milk. Like the albumin of the white of the egg, lactalbumin is coagulated easily by heat. The alpha-lactalbumin is the most heat-stable of the lactalbumins. A small amount probably is coagulated when milk is pasteurized and nearly one-third is coagulated when milk is heated to 71.1°C. (160°F)

TABLE 3:3
Some Properties of Proteins in Cow's Skim Milk,*

Approved Name	*Common Name*	*Percent in skim milk Protein*	*Percent of Total*	*Distinctive Characteristic*
	Casein	2-3.5	76-86	Precipitated by acid at pH 4.6
α-casein	1.4-2.3	1.4-2.3	45-55	Contains 1% phosphorus Is a mixture of proteins(a)
β-casein		0.5-1.0	19-28	0.6% phosphorus(a)
γ-casein		0.6-0.22	3-7	0.1% phosphorus (e)
	Whey protein	0.5-0.7	14-24	
β-Lactoglobulin A	Lactalbumin	-	-	(a)
β-Lactoglobulin B		0.29-0.42(b)	7-12 (b)	(a)
β-Lactoglobulin C	-	-	-	(a)
Blood serum albumin	-	0.2-0.5	0.7-1.3	(d)
IgG Immunoglobulins		0.05		
IgG1		1-2		(c)(e)
IgG2		0.2-0.5		(c) (e)
Proteose-peptone fraction		2-6		High in glutamic and aspartic acids and in amino sugars.

* Adapted from Nomenclature of the Proteins of Cow's Milk.[8]
(a) Formed in the Udder.
(b) Mixed A and B
(c) Preformed from Blood.
(d) Apparently identical to bovine blood serum.
(e) Carry antibodies present in high concentration in colostrum.

for thirty minutes. Unlike casein, lactalbumin is not coagulated by rennin.

Sometimes, after the manufacture of cheese, the whey is heated to coagulate the lactalbumin present. The coagulated material is then gathered and formed into a cheese-like product known variously as *ricotta, ziger,* and *mysost*. It is used especially by the Italians and Scandinavians (see Chapter 19).

Soluble lactalbumin may be prepared from whey by the method of Flanigan and Supplee [2] Dilute hydrochloric acid is added to the whey to bring it to *p*H 4.3-4.8, and then it is heated to above 160°F. The lactalbumin is precipitated and removed from the whey by passing through a filter press. The soluble product is obtained by dissolving the press-cake with dilute alkali, adjusting the *p*H of the solution to 6.9 with hydrochloric acid, and then spray-drying the solution.

Lactoglobulin and Other Milk Proteins

Milk contains about 0.05% of closely related proteins, which, because they are insoluble in water, but soluble in dilute salt solution, are known as globulins. Lactogloblin is a complex of closely related proteins which are known as beta-lactoglobins. They are a contaminant of commercially prepared lactalbumin.[13]

There are two types of beta-lactoglobulin. These are determined genetically. They differ in that several amino acids of one may be missing from the protein molecule of the other or one amino acid may be substituted for another. Some cows produce a mixture of both types, other cows will produce either one or the other. Beta-lactoglobulin contains sulfhydrol groups which appear to be associated with the development of a "cooked flavor" in heated milk (see Chapter 10).

The presence of a protein constituent adsorbed on the globules of fat in milk has already been mentioned. According to Dunkley and Sommer[1] this protein is a factor in causing the fat globules to cluster or adhere to each other, thus causing the formation of a cream layer in milk.

Colostrum, the secretion of the mammary gland for a few days after birth of the calf, is rich in globulins. Of special interest are the immunoglobulins, which have antibody activities. The term "*immunoglobulins*" refers to the protein substances previously called "euglobulin" and "pseudoglobulin". Three groups of immunoglobulins are known, designated IgM, IgA, and IgG.[1a] They occur in both the blood and the milk. As much as 50 to 75% of the protein in bovine colostrum consists of immunoglobulins, nearly all of which are classed as IgG.

The immunoglobulins serve as carriers for the antibodies that protect the calf against disease-producing organisms. The colostrum fed to the suckling calf on the first day of its life enables the antibodies to pass from the digestive tract into the blood stream. After the first day the calf loses the ability to transfer the antibodies, probably through changes in the tissue structure of the intestine.

Associated with the immunoglobulins is a protein fraction known as lactoferrin. About 2% is present in normal bovine colostrum, but in the globulins of the udder tissue of the dry cow about 50% may be present. The presence of a high lactoferrin concentration apparently increases the protective ability of the immunoglobulins to resist infection. Most bacteria need iron for growth and lactoferrin can prevent such growth by combining with available iron. When iron is bound to lactoferrin, a red color is imparted to the protein, giving rise to the term once used, "red protein of milk".

The protein components in the whey portion of 100 g of mature, human milk (11 to 60 days post-partum) was found to contain about .44 g casein, .21 g lactoferrin, .28 g alpha-lactalbumin, .06 g serum albumin and .05 g remaining proteins. Human colostrum and transitional milk contains about .46 g casein, .47 g lactoferrin, .42 g alpha-lactalbumin and .33 g remaining proteins (globulins)![15]

About 5% of the total nitrogen in milk is present as non-protein materials. They are probably by-products or residues formed during the nitrogen metabolism in the cow's body and in the formation of milk. When milk is boiled, the non-protein nitrogen content is increased because of changes in the protein portion. Pasteurization at 143°F does not change the distribution of nitrogen, but heating to 155° for 30 minutes does increase the non-protein nitrogen. A similar change that occurs in evaporated milk becomes more evident upon storage for several months.

In addition to the compounds named in Table 3:4, non-protein nitrogen is found in phospholipids and in certain vitamins and enzymes.

TABLE 3:4
Some Non-Protein Nitrogenous Components of Milk[5,11]

	Mg/100 ml milk	
Compound	Raw	Pasteurized
Ammonia	0.79	0.93
Urea	15.1	9.2
Creatinine	0.9	0.46
Creatine	3.1	4.0
Uric Acid	1.9	2.2

REFERENCES

1a. Bovine Immunoglobulins: A Review, Butler, J. E. *J. Dairy Sci.*, 52:1895-1909, (1968).
1. Dunkley, W. L. and Sommer, H. H., The Creaming of Milk, *Wisc. Agr. Exp. Sta. Res. Bull.* 151, (1944).
2. Flanigan, G. E. and Supplee, G. C., U.S. Patent 2023014 (1935).
3. Frost, D. V., Protein and Amino Acid Requirement of Mammals; Academic Press, New York (1950).
4. Holt, L. E. Jr., Gyorgy, P., Pratt, Snyderman, S. E., and Wallace, W. M. Protein and Amino Acid Requirements in Early Life, New York Univ. Press, (1960).
5. Kon, S. K. and Cowie, A. T. Milk: the Mammary Gland and its Secretion. Academic Press, New York and London, (1961).
6. Moore, L. A., Thyroprotein for dairy cattle, *J. Dairy Sci.*, 41:452-455 (1958).
7. National Research Council, Composition of Milk, Pub. 254, (1953).
8. Nomenclature of the Proteins of Cow's Milk: Third Revision, *J. Dairy Science*, 53:1-17 (1969).

9. Rose, W. C., Wixon, R. L., Lockhard, H. B., Lambert, G. F., Amino acid requirements of man, J. Biol. Chem., 217:987 (1955).
10. Samuelson, E. G., Proteins of Milk, International Dairy Fed., Ann. Bull., Part II (1962).
11. Shahari, K. M. and Sommer, H. H., The protein and non-protein nitrogen fractions in milk, *J. Dairy Sci.*, 34:1035 (1951).
12. Waugh, D. F. and von Hippel, P. H. Kappa-casein and the stabilization of casein micelles. *J. Am. Chem. Soc.*, 78:4576. (1956).
13. Whittier, E. C. and Webb, B. H., *Byproducts from Milk*; Reinhold Publishing Corporation, New York (1950).
14. Whitney, R. M., The minor proteins of bovine milk, *J. Dairy Sci.*, 41:1303 (1958).
15. Nagasawa, T., Kiyosawa, I., and Kuwahara, K., Amounts of Lactoferrin in Human Colostrum and Milk, *J. Dairy Sci.*, 55:165-1659 (1973).

chapter 4

Lactose

Lactose, or milk sugar, is the main carbohydrate found in milk; it is found in the milk of nearly all mammals and in no other natural foodstuff. In another definition, lactose is found in the milk of placental mammals, that is, those which nourish the fetus in the placenta. Milk produced by marine animals has a high content of total solids, protein, and fat but the lactose content is low and in some species may be entirely absent. The milk of the California sea lion, the hooded seal, and the harp seal is reported to contain no lactose.

Lactose is a disaccharide, that is, it can be hydrolyzed into two other sugars, *glucose* (dextrose or corn sugar) and *galactose*. Traces of glucose and of galactose are present in cow's milk, about 0.007% and 0.002%, respectively.[3] Minute amounts of compounds of lactose with other sugar-like and nitrogen containing substances, the oligosaccharides, are also present. Human milk contains about 0.4% of oligosaccharides compared to about 0.004% in cow's milk.[6] (See Chap. 12).

The milk of the platypus and the echidna, monotremes or egg-laying mammals, contain only small amounts of free lactose, about 0.1%, compared to 7% for human milk.[8]

Blood glucose is a precursor of lactose. Two protein substances function in its synthesis; the A protein is the enzyme galactosyl-transferase, the B protein is lactalbumin, the same lactalbumin present in whey.[1,2]

Lactose was first recognized as a constituent of milk in 1615, but it was not until about 1780 that the Swedish chemist, C. W. Scheele, showed that it is a sugar. The commercial manufacture of lactose was confined to Switzerland up to 1880 but now is made in most places with a dairy industry.

Usually it is prepared from the whey obtained as a by-product in the manufacture of cheese and, when available, from that of casein. The potential supply is over 500 million pounds. Of this large potential amount, about 87.3 million pounds of lactose was made in the United States in 1972. Wisconsin alone produced about 58.4 million pounds with Minnesota second with 16 million pounds. In 1972, 5 million pounds were imported.[5]

Its uses are limited and for many purposes the less costly sugars, such as sucrose and glucose, are available. Much of the available supply of whey is produced in such locally limited amounts and in so widely separated areas that it is not economical to use it for the manufacture of lactose. In areas where large amounts of whey are obtainable, the manufacture of lactose not only produces a valuable product, but also utilizes by-products of milk which otherwise could give rise to problems in their sanitary disposal.

The whey from the manufacture of hydrochloric (muriatic) acid casein is preferred for the manufacture of lactose. Whey from sulfuric acid casein contains metallic sulfates which are difficult to remove and would make a lactose solution cloudy. In self-sour and cheese wheys, considerable lactose is lost through its being converted to lactic acid but sweet cheese whey is now the chief source.

In one method of manufacture, the whey is heated almost to boiling and a slurry of lime is added until the mixture is only slightly acid—about 0.5% as lactic acid, or *p*H 6.2. The proteins and most of the mineral matter (calcium phospate) separate from the whey and are allowed to settle. The clear fluid is drawn off and evaporated to a concentration around 30% lactose. The liquid is then filtered and the sirup is further evaporated to about 50% content of solids. The hot sirup is placed in crystallizing vats, where it is cooled slowly and carefully agitated to induce crystallization. The crude lactose is separated from the adhering liquid in a sugar centrifuge and washed sparingly with cold water. The by-product liquid obtained after the separation of the lactose usually is concentrated or dried for use as feed, especially for poultry.[16,17]

Crude lactose is further refined for the preparation of the edible and United States Pharmacopoeia grades. For this purpose, the crude lactose crystals are washed in the centrifuge until the effluent water is clear. The crystals then are dissolved in hot water and treated with activated charcoal to remove coloring matter. The solution then is filtered, and treated with lime slurry to flocculate any remaining protein and insoluble salts; these are removed by filtration. The remaining solution may be crystallized and dried to make a product of *edible* grade. For the *pharmaceutical* (USP) grade, the lactose crystals are further washed and then dried at about 100°F, after which they are pulverized to pass a 100-mesh screen. This grade of

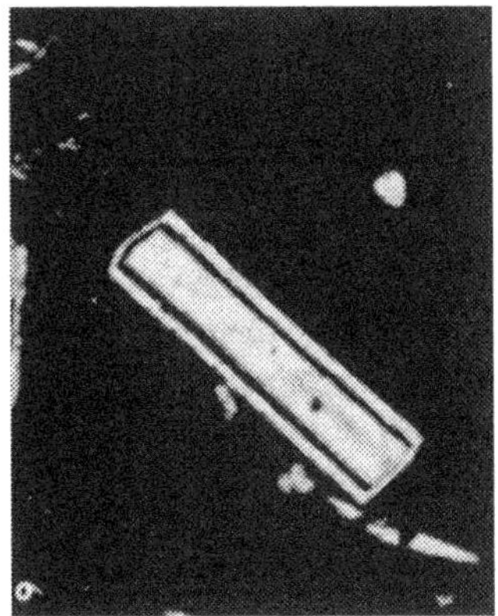

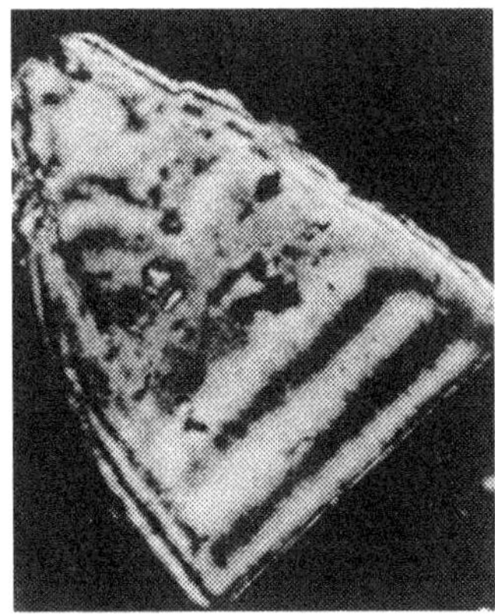

Fig. 4-1. Lactose Crystals (313X): *a* *Beta*-Lactose crystal *b* Typical axe-shaped crystal of *alpha*-lactose (Missouri Agr. Expt. Sta. Bull. 373)

lactose is also made by spray-drying a concentrated, purified solution of lactose.

Lactose made by the methods described is known as *alpha-lactose* hydrate. Its solubility in water is only 17.8% at 25°C (77°F) and in equal concentration it is only about one-third as sweet as cane sugar. (See Table 4:1). The other form of lactose, *beta-lactose*, is made by crystallizing the sugar at a temperature above 93.3°C (200°F) and drying a concentrated solution of lactose on a drum dryer. In another process, lactose crystals are heated in the presence of water vapor, in a closed container, under pressure, and at a temperature around 121°C (250°F). A solution is formed from which the sugar separates as *beta*-lactose.[13] At room temperature, *beta*-lactose is about seven times more soluble than the *alpha* form. The increased solubility

TABLE 4:1
Relative Sweetness of Sugars
Percent Concentration to yield Equivalent Sweetness
(R. M. Pangborn, J. Food Sci. 28, 726-1963)

Sucrose	*Glucose*	*Lactose*
0.5	0.9	1.9
1.0	1.8	3.5
2.0	3.6	6.5
5.0	8.3	15.7
10.0	13.9	25.9
15.0	20.0	34.6

makes it appear sweeter to the taste and favors the use of it in dietary and baby foods.

The low solubility of lactose in water and its tendency to form a supersaturated solution is of concern to the manufacturer of dairy products. In a supersaturated solution, more lactose is present than can normally remain dissolved. If such a solution is cooled, or if sucrose is added in the amounts used to make ice cream or sweetened condensed milk, the lactose may separate in the form of *alpha*-type crystals. This may also happen when milk is concentrated beyond the three-to-one ratio, as in the manufacture of concentrated milk. The formation of relatively large lactose crystals gives rise to a condition in the product known as *sandiness*, because the crystals are hard, slowly soluble, and give the sensation in the mouth of the presence of sand.

Crystallization of lactose in sweetened condensed milk cannot be avoided, so a small amount of very finely divided lactose, or whey powder, is added during manufacture (see Chapter 15). The added lactose forms a large number of centers or nuclei upon which the excess of lactose in the product will crystallize and so create very small crystals which are not noticed when eaten. The crystals should not be larger than about 15 microns (0.0006 in.) in order not to be detected in the mouth.

The lactose in dried milk, spray or roller dried, is present as a glass or concentrated sirup. The speed of drying does not allow time for crystal formation. In this form the lactose is hygroscopic and will absorb moisture from the air until enough is taken up to form the crystalline alpha-hydrate which is not hygroscopic.

Use of Lactose

Although a vast amount of lactose is consumed in the form of milk products, the sugar itself has only few industrial applications. Much milk sugar is used as a constituent of infant foods and medicinal products. Physicians prescribe lactose for the modification of cow's milk for infant feeding, in order to bring its composition closer to that of women's milk. It has been suggested that as galactose (which is present in lactose) is needed in the early stages of formation of the brain, milk sugar, rather than other sugars should be added to the infant's diet.[15] In this sense, galactose may be considered to be an indispensable carbohydrate.

Infant foods are probably the largest user of edible grades of lactose; about 35 million pounds are reported to be used per year.

In the pharmaceutical industry, lactose is used for the manufacture of tablets and capsules, and as a general filler. Confectioners use lactose in fondants and tablet-like candies. In the baking industry, lactose is used to produce a desirable brown color in pie crusts, cookies, and other baked goods. This is effected by a reaction during baking between the lactose and protein material in the dough, the so-called browning, or Maillard reaction. This reaction also occurs during the manufacture of evaporated milk, as described in Chapter 16. The addition of lactose to chocolate milk, buttermilk, and modified skim milk has been advocated in order to improve the flavor and body of the product.

Lactose is an important ingredient in dry mixtures used as substitutes for cream in coffee. Lactose is added to some foodstuffs in order to improve their dispersibility. The product is then spray-dried and given an "instantizing" treatment. (See Chapter 17).

The slow rate at which lactose is fermented favored the use of it in the original methods for the manufacture of penicillin. Lactose is still used as an ingredient of the medium in which the mold is grown, but other sugars, such as glucose, are now used in the manufacture of penicillin and the synthetic penicillins. This is done by continuously supplying the glucose solution to the medium at the same rate as it is being fermented by the growing molds.

Interesting, although minor, applications of lactose, include use of it in military technology for the making of smoke screens, and signal and target candles, where the slow burning of the sugar prolongs the burning time and deepens the color of the discharge. Lactose has been used as a reducing agent in the silvering of mirrors. The growth of its crystals forms a decorative, frost-like coating on the inside of some bottled liqueurs.

Lactic acid is made commercially to some extent by the fermentation of the lactose in whey, but generally it is made from other sources, such as starch and blackstrap molasses. Lactic acid is used in the manufacture of chemicals and medicinal products, and as a substitute for acetic and citric acids in foodstuffs, such as pickles. In the leather industry, the acid is used to treat hides after the liming operation. The function of lactic acid in the fermentation and souring of dairy products is described in Chapter 14.

Nutritional Value of Lactose

The function of lactose in nutrition has not been given the same study as has been given to sucrose and other carbohydrates in the daily diet. In grown animals, lactose does not have the fattening effect of most other sugars and there is some evidence that lactose

TABLE 4:2
Typical Analyses of Various Grades of Lactose[3]

Percent	*Crude*	*Edible*	*Spray Process U.S.P.*	*Lactose U.S.P.*
Lactose (by difference)	97.75	98.75	99.25	99.60
Free moisture	0.40	0.60	0.50	0.20
Acidity (as lactic)	0.03	0.04	0.03	0.03
Protein (NX6.25)	0.75	0.20	0.10	0.05
Ash	0.65	0.15	0.05	0.03
Fat (ether extract)	0.20	0.15	0.10	0.05
Bacteria, per gram (standard plate count)	—	1000	1000	1000

stimulates the growth of small animals. When it was fed to rats, they lived longer and grew more rapidly than those fed sucrose.[16] The higher content of lactose in human milk (7%), makes its caloric value equal to that of cow's milk, which has about 4.9% of lactose, but more fat than women's milk. (Table 1:2)

Some investigators have observed that fat may be needed for the efficient utilization of lactose. Rats fed no food other than whole milk do not eliminate lactose through the kidneys, but do so when put on a diet of skim milk.[12] The addition to skim milk of three or four percent of a fat such as butter or lard, is sufficient to enable the animal to make use of the lactose in its diet. An excess of lactose in the rat's diet may lead to the formation of cataracts, owing to their inability to utilize a large amount of lactose.[4] It should be emphasized that experiments on the feeding of animals add to our knowledge of dietary factors, but that the diets used are very abnormal and bear little relation to the ordinary balanced diet used by human beings.

As lactose is broken down slowly during the digestive process, some of it reaches the lower intestine unchanged. Lactose is an excellent food for acid-forming bacteria that may be present in the intestine. These organisms use the lactose to produce organic acids, such as lactic acid. These acids help to check the growth of less desirable organisms of the putrefactive type. The acid condition formed in the intestine also favors the absorption of calcium in the diet as well as the utilization of vitamin D. Lactose has found some use in the treatment of constipation, but excessive amounts may result in diarrhea. (See below)

LACTASE

Lactose is not utilized by the body until it is first hydrolized, that is, split into its two component simple sugars, glucose and galactose. This is accomplished by the action of the enzyme lactase, properly called beta-D-galactosidase.

Lactase is present in the digestive tract of many mammals, including human beings, calves, goats, sheep, and dogs. It is reported to be missing in the intestinal tract of cows and in the mammary glands of goats and cows. Small amounts of lactase have been found in apples, apricots, and almonds. It has not been found in normal milk. Lactase activity is greatest at birth, but after weaning about 90% is lost.

There is a relationship between a lack of lactase in the small intestine and the difficulty some people have to digest milk. This difficulty differs from the intolerance which may arise from allergic causes (see Chap. 14). Lactose intolerance may be attributed to undigested lactose that reaches the colon. Here fermentation by bacteria forms lactic acid which causes flatulence, cramps, or diarrhea.

Lactose intolerance has geographical and racial limitations. It is prevalent in areas where there are few dairy animals or where the adults use little or no milk. It is not significant in areas where dairy animals are used or lactose-rich foods are consumed. Among persons who have been given a lactose-tolerance test only about 2% of those in western Europe and their descendants have shown lactose-intolerance. The natives of eastern Africa, who keep herds of dairy cattle, show little lactose-intolerance. In contrast, the natives of west Africa and the many American Negroes who are their descendants, are lactose-intolerant. About 70% of adult American Negroes and up to 95% of the Asiatic peoples may be lactose-intolerant. This condition is not present at birth but develops at puberty.

Since lactase is rarely totally absent, two or three glasses of milk may be tolerated. The use of milk products such as, yogurt or cultured buttermilk, in which the lactose has been fermented, is recommended for persons with a lactose deficiency.

In order to promote the use of dairy products by the lactose-intolerant person an effort is being made to develop low-lactose products. A commercial preparation made from the mold *Saccharomyces lactis* has been used to reduce the lactose content of milk.[7] Such milk is not adversely affected in flavor, acidity, or color, although the sweetness is increased.

REFERENCES

1. Brew, K., Vanaman, T. C., and Hill, R. L., *Proc. Nat. Acad. Sci.* Washington, D.C. 50:491 (1968).
2. Brodbeck, U. et al., *J. Biol. Chem.*, 242:1391 (1967).
3. Call, A. O., Lactose Symposium: Utilization of Lactose, *J. Dairy Sci.*, 41:332 (1958).
4. DeGroot, A. P. and Hoogendoorn, P., Neth. Milk and Dairy J., 11, 290 (1957).
5. Foreign Agr. Circular, U.S. Depart. of Agriculture, FD-2-73, May 1973.
6. Kon, S. K. and Cowie, A. T., *Milk: the Mammary Gland and its Secretion.* Academic Press, New York (1961).
7. Kosikowski, F. V. and Wierzbiski, Lactose Hydrolysis of Raw and Pasteurized Milks by *Sacharomyces lactis* Lactase, J. Dairy Sci., 146-148; 56 (1973).
8. Messner, M. and Kerry, K. P., Milk Carbohydrates of the Echidna and the Platypus, Science 180:201-203 (1973).
9. Manufactured Dairy Products. U.S. Dept. of Agr. August 1968.
10. Patton, S., Browning and associated changes in milk and its products: a review, *J. Dairy Sci.*, 38:457 (1955).
11. Rosensweig, N. S., Adult Human Milk Intolerance and Intestinal Lactase Deficiency. A Review, *J. Dairy Sci.*, 52:585 (1969).
12. Schantz, E. J., Elvehjem, C. A., Hart, E. B., The relation of fat to the utilization of lactose in milk, *J. Biol. Chem.*, 122:381 (1938).
13. Sharp, P. F. and Hand, D. B., U.S. Patents 2,182,618 and 2,182,619, Dec. 5, 1939.
14. Tomarelli, R. M., et al., The Effect of Lactose Feeding on the Body Fat of the Rat, *J. Nutri.*, 71,221-228 (1960).
15. White, A., et al., *Principles of biochemistry*, McGraw-Hill Book Co., Inc., New York (1959).
16. Whittier, E. V., Lactose and its utilizations: a review, *J. Dairy Sci.*, 27:505 (1944).
17. Whittier, E. V. and Webb, B. H., *Byproducts From Milk*; Reinhold Publishing Corporation, New York (1950).
18. Regez, W., Experience with bacteria-free milk: Uperization process; *Proceedings*, Laboratory Section, Milk Industry Foundation, Washington, D.C. (1962).

chapter 5

The Minerals and Water in Milk

When the water of milk or other food is removed by evaporation and the dry residue is incinerated at a low red heat, there is left a white or nearly white ash which contains the mineral substances. Owing to chemical changes that occur during the ashing process, the ash contains carbonates, oxides, and phosphates which are not present as such in the original food. Some of the phosphorus and sulfur compounds in the ash of milk are derived from the milk proteins; other compounds, such as the citric acid in milk, are the source of the carbonates in the ash. The ash of milk is always alkaline in reaction. The major mineral constituents of milk are shown in Table 5:1, those in other milk products in Table 5:2. The average ash content of milk is 0.72%.

Mineral elements essential for human nutrition are calcium, magnesium, potassium, sodium, phosphorus, copper, iron, sulfur, iodine, and chlorine; and there is much evidence that manganese, zinc, fluorine, molybdenum, cobalt, and several others must be included. These elements generally are considered individually but it must be stressed that mineral metabolism is not simple, and the activity within the body of one element may depend upon the

TABLE 5:1
Average Mineral Content of Milk
And Ash of Milk

Element	*% in Milk*	*% in Ash of Milk*
Potassium	0.140	20.0
Calcium	0.125	17.4
Chlorine	0.103	14.5
Phosphorus	0.096	13.3
Sodium	0.056	7.8
Magnesium	0.012	1.45
Sulfur	0.025	3.6

TABLE 5:2
Mineral Content of Some Dairy Products
(Percentages)

Product	*Calcium*	*Phosphorus*	*Sodium*	*Potassium*	*Iron*
Butter (salted)	0.02	0.15	0.960	0.02	0.00
Buttermilk	0.11	0.09	0.13	0.14	0.00
CHEESE					
Blue	0.35	0.33	1.8	—	0.0005
Cheddar	0.75	0.50	0.70	0.09	0.001
Cheddar (Processed)	0.80	0.85	1.5	0.08	0.0008
Cottage (4% fat)	0.09	0.17	0.30	0.07	0.0002
Cream	0.65	0.95	0.25	0.07	0.0001
Edam	0.80	0.50	1.5	—	—
Limburger	0.52	0.36	1.4	—	—
Swiss	0.99	0.57	0.70	0.09	
Swiss (processed)	0.89	0.87	1.20	—	
Cream (20% fat)	0.10	0.07	0.045	0.07	
Cream (35% fat)	0.08	0.06	0.04	0.05	
Ice Cream (vanilla, 10%) fat)	0.13	0.10	0.10	0.09	
MILK, Cow's (3.8% fat)	0.12	0.09	0.05	0.14	
Skim	0.12	0.10	0.05	0.15	
Evaporated	0.25	0.20	0.10	0.27	
Dry, whole milk	0.90	1.02	0.38	0.97	
Dry, Non-fat	1.30	1.03	0.53	1.12	
Milk, Goat's	0.13	0.11	0.03	0.18	
Whey Powder	0.68	0.58	—	—	

presence of other elements and compounds. An example of this relationship is shown in the requirement for calcium, phosphorus, and fluorine in the formation of the teeth. Other examples are the influence of copper upon the utilization of iron by the body and the need for minute amounts of iodine and of cobalt for the maintenance of normal nutrition.

In addition to the minerals previously named, small amounts of copper, iron, manganese, zinc, and iodine are present in milk. Traces of many other elements, such as aluminum, barium, cobalt, chromium, germanium, lithium, rubidium, silver, strontium, tin, titanium, and vanadium have been found in milk. Arsenic, boron, and fluorine also have been detected in milk.[9]

The mineral content of milk is fairly constant and is influenced but little by the feed. To a considerable extent a deficiency of calcuim and phosphorus in the diet is not reflected in the composition of the milk because the cow will draw upon her own skeleton to furnish these elements. In this way, nature gives the calf sufficient minerals from its mother to insure normal bone growth. Experiments have shown that iron, copper, calcium, and phosphorus cannot be added to the milk by inclusion in the feed. The iodine content can be increased appreciably in the milk by feeding iodine compounds, but there is some evidence that this tends to lower the yield and the fat content of the milk. The citric acid in milk, 0.16%, is present in combination with the alkaline elements such as potassium and sodium.

Milk has the ability to dissolve small amounts of many metals with which it may come in contact. Copper, iron, nickel, and monel metal are slightly soluble in milk whereas aluminum and stainless steel are very resistant to corrosion by it. As the acidity of milk increases so does its ability to dissolve zinc, tin solder, and aluminum.

In addition to their importance to nutrition, some of the minerals in milk have an influence upon the processing operations in the manufacture of dairy products. The influence of calcium upon the action of rennin in the manufacture of cheese and upon the thickening of evaporated milk during manufacture is of special interest and is discussed in later chapters. Other minerals, such as copper and iron, may be associated with the development of undesirable flavors in milk products.

Bacterial action in the milk has an influence upon the salt balance as the milk sours, causing the milk to become unstable to heat and curdle easily. Lactic acid and lactates are formed, which originally are not present in milk. The citrates originally present may disappear owing to conversion to other products, such as diacetyl.

Calcium

About 1.6% of the weight of the adult human body is calcium. It forms a larger portion of the body weight than any other inorganic substance, except water. About 85% of the ash of the skeleton consists of tri-calcium phosphate, the form in which a great part of the calcium of in the body is found in the bones and teeth, the other 1% in the body fluids, nerves, heart and muscles, where it assists in the proper maintenance of body functions. It also has an influence upon the clotting power of the blood, and is necessary for normal regulation of the heart beat.

The calcium level of the blood (actually the blood serum) is controlled by the parathyroid gland. If disease brings about a low calcium level in the blood, this may result in irritability of muscles and nerves and even lead to tetany, a nervous disorder characterized by muscular spasms. Rickets, which usually is caused by a lack of vitamin D, also may result from insufficient intake of calcium and phosphorus or from an improper balance of these minerals.

The exact amount of calcium needed in the diet of the adult is not known, but generally it is held that the American diet is deficient in foods that are rich in calcium. A normal requirement is about one gram of calcium for 100 grams of protein, in a diet that contains about as much phosphorus as calcium. To be on the safe side, the daily diet should contain at least 0.8 grams of calcium. The requirement for pregnant and nursing women is from 1.5 to 2.0 grams per day. At birth, the child has no reserve of calcium and it is essential that it be supplied in an adequate amount to meet the needs of the period rapid growth.

The recommended calcium intakes are given in Table 6:1.

Milk contains more calcium per unit of dry matter than most other foodstuffs. With the exception of the leafy vegetables, no other food has sufficient calcium to meet the dietary requirements. In general, children do not utilize the calcium of vegetables as well as that from milk.

The calcium content of milk is remarkably constant. Removing the mineral from the diet of the cow does not alter the calcium content of her milk, as she will then transfer the element from her skeleton to the milk in order to maintain its normal calcium content. By this action, nature provides the suckling calf with a source of calcium needed to build its bones. It is impossible to increase the calcium content of the milk by feeding the cow a diet high in this mineral.

About one-third of the calcium in milk is in true solution, the rest in colloidal suspension combined with casein, phosphorus, and citrate.[9] When milk is boiled or pasteurized, ten to twenty percent of the calcium becomes less available to the body, as measured by feeding experiments with rats. As cow's milk contains more calcium than women's milk, this loss is not important so far as the nutrition of children is concerned. It is generally held that one quart of milk a day will supply the calcium required by a growing child.

The comparative values of dairy products as sources of calcium are given in Table 5:2.

The absorption of calcium from the intestinal tract is favored by an acid medium, such as is established by the lactose in milk. The presence of milk-fat also aids in the absorption of the calcium from milk, probably by favoring the use of lactose, as mentioned in the previous section on the nutritional value of lactose. Vitamin D also is essential for the utilization of calcium. The effect of chocolate in the diet upon calcium availability is discussed in Chapter 12.

Phosphorus

Phosphorus is an important constituent of all body cells. It has a place in the chemical reactions whereby fats, protein, and carbohydrates are metabolized in the body to supply energy and materials for growth and repair. It is involved with the activity of vitamins and enzymes and is important in helping the blood maintain the acid-base balance. The relationship with calcium in the formation of bone has been discussed. As in the case of calcium, the absorption of phosphorus is favored by an acid condition of the intestinal tract. About 85% of the total phosphorus in the body is combined with calcium in the skeleton.

It has been estimated that the adult human skeleton contains about 600 grams of phosphorus, the muscles about 56 grams, and the brain and nervous tissue about five grams. The minimum daily requirements of phosphorus are the same as the amounts required for calcium, except that for infants under two months of age the requirement is 0.2 gram, for those aged two to six months, 0.4 gram, and those six to twelve months of age, 0.5 gram.[14]

If the diet supplies sufficient calcium and protein, it most probably also supplies the phosphorus requirement, because such foods are good sources of phosphorus.

The comparative values of some dairy products as sources of phosphorus are given in Table 5:2.

Magnesium

Magnesium, which is essential to normal nutrition, has functions which relate it to calcium and phosphorus. Magnesium takes part in the reactions involved in the formation of protein from amino acids and in the metabolism of carbohydrates and fats. The adult body contains about 1.3 ounces of magnesium, about 50% of which is in the bones. Most of the rest is found in the soft tissues; comparatively little magnesium is in the blood, only 1 to 3 mg in 100 ml of blood.

The average daily requirement of magnesium is believed to be about 0.3 gram. A deficiency of magnesium is very unlikely to occur with the ordinary diet. In animals, experimental diets low in magnesium produce extreme nervousness, damage to the heart and finally death from convulsions.

Sodium, Potassium and Chlorine

Although sodium and potassium are similar in many of their chemical characteristics, one cannot substitute for the other in the body functions. Sodium, which makes up about 0.2% of the body weight, is found in the blood and fluids of the body whereas potassium is found principally within the cells. Sodium is especially important for its part in regulating the proper balance of acid and base in the blood.

Potassium, which makes up about 0.09% of the body weight, is in association with sodium, important in maintaining a normal balance between the water content of the cells and the fluid surrounding them. For example, an increase of sodium in a body fluid causes a compensating transfer of water from the cell to the fluid. In some diseases of the heart and kidneys, the ability to compensate for variations in sodium intake is decreased and a diet low in sodium may be prescribed. The manufacture of low-sodium milk is described in Chapter 12.

The normal daily intake of sodium varies from 3 to 20 grams, usually about 8 grams, but it depends upon the amount of table salt a person uses. The daily intake of potassium varies from about 1.5 to 4.5 grams, depending upon the diet. The requirement is estimated to be 0.8 to 1.3 grams, and it is related to the amount of protein and carbohydrate in the diet.

The chlorine content of milk is closely related to its sodium content, but it most probably is present as chloride ions, rather than in combination with sodium and other minerals. The average chloride content of normal cow's milk is about 0.14% with a range between 0.11 to 0.16%. Milk from cows with diseased udders usually has a high content of salt and tastes salty. The chloride content of milk from cows with mastitis may be as high as 0.3%. In such milk the lactose content is decreased. In order to compensate for the decrease of lactose and maintain the normal osmotic pressure of the milk, there is an increase in its chloride content. This characteristic has been used to distinguish normal from abnormal milk and is often expressed as the Koestler or *chloride-lactose number:* 100x%chloride/% lactose.[10] This value is generally between 1.5 and 3 for normal milk, but much higher values are obtained with mastitic milk.

The chloride content of the body is largely in combination with sodium. It serves as an enzyme activator and as a component of the hydrochloric acid of the stomach.

Sulfur

Sulfur is an important constituent of all body tissues. Milk contains about 0.2%, largely as part of the amino acids methionine and cystine. Methionine contains about 0.69% of sulfur and cystine contains about 0.09%. Inorganic sulfur, such as sodium sulfate may be converted in the cow's rumen to precursors of amino acids.

Trace Minerals

By sensitive methods of examination, such as atomic absorption analysis, traces of many elements may be found in milk. Contamination of the milk from an external source, such as the container, may introduce some unusual element, such as nickel, zinc or lead. Tin, aluminum, or chromium, when in contact with milk, are resistant to solution and rarely find their way into the product. Copper and iron are soluble in milk and because of their bad effects on the flavor they have been given special attention by investigators. Radioactive contamination of milk is discussed in Chapter 8.

Many of the elements found only in trace amounts are essential to normal nutrition and a lack of them leads to nutritional disturbances. The elements involved are found in the various enzymes, where they appear to act as the link between the protein portion of the enzyme and the substance in the body upon which the enzyme acts. Other elements have other functions, such as the iron in hemoglobin, the iodine in the thyroid gland, and cobalt in vitamin B_{12}. Chromium probably is needed to make insulin more effective in promoting glucose metabolism, especially in some diabetics.

Iron

Although iron makes up only about 0.004% of the body weight, life would be impossible without it. As a part of the hemoglobin molecule, iron is concerned with the transport of oxygen by the blood and with the various cellular oxidative processes. Enzymes that contain iron are *catalase* and *peroxidase*, discussed in Chapter 7. The utilization of iron to form hemoglobin depends upon the presence of an adequate amount of copper [5]

The form of iron in the diet is important because the body can assimilate only the soluble form. Some foodstuffs comparatively rich in iron are nevertheless poor sources of it for nutritional purposes as the iron is present in a non-soluble form. For example, in spinach only about one-fourth of its iron content is available for the body, whereas in eggs and milk all of the iron is in a readily utilizable form. Many simple inorganic salts of iron are assimilated by the body, and such salts may be used to supply the mineral if the diet needs a supplementary amount of iron.

Provided that the mother's diet has been adequate, the infant at birth has about a three months' supply of iron. A diet that includes an assimilable form of iron will provide the infant's requirement when the production of hemoglobin becomes active. Such iron is obtained from fortified cereals and strained meats.

The adult man needs about 10 mg. of iron daily, women about 18 mg. Children need 10 mg. with an increasing amount, up to 18 mg., between the ages of 12 to 20 years.

As the amount of iron in milk is small, milk is an inadequate source of supply of this mineral for the diet. The average amount present is about 0.1 mg in 100 grams. Higher amounts have been found, but any appreciable increase is due to contamination of the milk by contact with iron equipment. An increase in the iron content of the cow's feed does not increase that of the milk. Special compounds of iron are used in mineral-fortified milk since the usual iron salts cause milk to develop a disagreeable flavor either through oxidation of lipid material or by activation of lipase. Ferric iron salts, up to 20 mg. per liter, may be added to nonfat dry milk with no adverse flavor effects [20]

Copper

Copper is closely associated with iron in nutrition. A deficiency of copper results in an anemia, because without copper present, the body is unable to make use of its iron reserve to form hemoglobin. Foods that contain iron usually are also good sources of copper. A deficiency of copper is rarely found in the average diet. The infant is born with a reserve of copper in its liver for use during the first few months of life until other foods are added to the milk diet.

The copper content of milk varies between 0.04 and 0.2 mg. per liter, with an average of about 0.09 mg.[12]

Colostrum and milk in the early stages of lactation have slightly higher copper content than in the later stage of lactation. Women's milk, at all stages of lactation, contains slightly more copper than does cow's milk.[3] Milk may dissolve small amounts of copper should it come in contact with equipment made from or containing copper. Dairy equipment made from tinned copper, or with white metal fittings, is still in use, but it has been largely displaced by stainless steel. As little as one part per million of copper in milk is sometimes enough to cause the development of disagreeable "oxidized" flavor in the milk. The presence of copper also will cause an appreciable reduction in the ascorbic acid (vitamin C) content in milk.

An exception to the use of copper for dairy equipment is the traditional copper kettle used in the manufacture of Swiss cheese. Swiss cheese may contain about 15 ppm. of copper. When a copper kettle is not used, copper may be added to the milk used for making the cheese (See Swiss cheese, Chaper 19).

Iodine

Iodine is important in nutrition because it is essential to the production of thyroxin, a substance secreted by the thyroid gland. The adult body contains only about 25 mg of iodine, practically all of which is found in the thyroid gland. The daily requirement is estimated to be between 0.05 and 0.1 mg for adults and about three times this amount for children.

Foodstuffs vary greatly in their iodine content. The higher content is found in foodstuffs obtained from regions near the sea, or where the soil is rich in the element.

In a large part of the United States, reaching from the Appalachian Mountains in the East, along the Great Lakes and the northern states westward to the Pacific Coast, the soil and water are poor in iodine. Foodstuffs grown in this region contains very little iodine. If such foods were the sole diet of a person he would soon show symptoms of iodine deficiency, such as sensitiveness to infections, enlarged thyroid gland or goiter, and lowered mental activity. The addition of iodized salt is one means of combatting a deficiency of iodine in the diet.

Iodine is present in milk, but in very small and widely varying amounts. It is one of the few substances that are transmitted easily from the feed into the milk. Ordinarily, milk may contain from 0.04 to 0.06 part of iodine per million parts of milk, a very small amount. Cows in regions near the seashore give milk that is relatively rich in iodine, containing up to two parts per million parts of milk. Milk from cows that have been fed substances that contain iodine may contain up to fourteen parts of it per million of milk.[1] Colostrum may contain up to three times as much iodine as the normal milk. Radioactive iodine is discussed in chapter 8.

Bromine

Traces of bromide may be present in milk, ranging from nil to 8 ppm. Feeding a ration that contains up to 220 ppm of total bromide may raise the content in the milk to around 20 ppm.[11]

Manganese

Manganese is a normal constituent of milk, but only traces are present, usually between 0.03 to 0.2 mg per liter with an average of about 0.09 mg.[12] The mineral is believed to be essential, although symptoms of deficiency have not been demonstrated in man. The function of manganese in the diet is not known. In experimental work, it has been shown to be needed for the activation of the phosphatases in the formation of bone and also for the activation of other enzymes. In experiments on animals, manganese also has been shown to be necessary for normal reproduction and for the utilization of thiamine. Manganese forms part of the enzyme arginase found in the liver.

Cobalt

It is probable that the minute amount of cobalt in the body is all associated with its vitamin B_{12} content. It is therefore an essential element, which, like iodine, is needed only in very minute amount. A lack of cobalt in the diet of sheep and cattle, which may occur in areas where the soil is deficient in the element, causes serious nutritional disease in the animals. The average human diet, especially if it contains leafy vegetables, has been estimated to provide more than ten times the cobalt needed in relation to the requirement of vitamin B_{12}.

The cobalt content of cow's milk is reported to be between 0.002 and 0.02 part per million, and may be increased about four times by feeding cobalt supplements in the normal ration.[2] It also has been reported that feeding cobalt salts increased the vitamin B_{12} content of ewe's milk (see Vitamin B_{12}).

Fluorine

Besides iron and iodine, fluorine is the only other element in the diet, for which, according to present knowledge, man can show a deficiency. Fluorine is a constituent of the teeth and it is considered to be essential to health of the teeth. In areas where the drinking water contains about 1 ppm of fluorine, the incidence of dental caries in children is lower than in those where the water contains less fluorine. For this reason, many authorities believe that during the years when the teeth are formed, the child's diet should supply about 1.0 mg of fluoride daily. When the fluoride content of the water supply and daily diet exceeds about 1.5 ppm, objectionable mottling of the dental enamel may occur.

Fluorine is accumulative in the body, especially in the bones, the amount present depending upon the fluorine content of the diet. There is a very narrow margin between the amount considered beneficial and the toxic level. In large doses, fluorine is an acute poison. Studies have shown that when the food and water consumed daily does not furnish more than 4-5 mg of fluorine daily, very little is retained in the body. At higher levels of intake, the element accumulates rapidly in the bones. In cattle, if the fluorine content of the feed, calculated on a dry basis, exceeds about 0.003%, their bones thicken, become soft, and break easily.

Few foods, other than sea fish, contain more than 0.1 to 1 ppm of fluorine. Tea is an exceptionally high source of the element, containing up to 100 ppm. Cow's milk contains from 10 to 50 micrograms of fluorine in 100 grams (0.1 to 0.5ppm). The fluorine content of milk is not increased by addition of fluorides to the cow's ration. Butter may contain about 1.5 ppm; and cheddar cheese, about 1.6 ppm of fluorine. The fluoride content of various Italian type cheeses was found to vary between 4 and 7 parts per million of dry cheese solids.[8]

The suggestion has been made that fluoride be added to milk, rather than to the water supply, in order to provide dental protection to children. Workers at the Louisiana State University added sufficient sodium fluoride to the milk consumed by children to supply 1 mg of fluoride to the daily diet. Tests over a period of three and one-half years with 171 children, aged six to nine, showed a 75% drop in dental cavities.[18]

The addition of fluoride to a food stuff is not permitted by the United States Food and Drug Administration. Fluoridation is permitted only for use in public water supplies.[7]

Zinc

Zinc is one of the trace elements essential to growth and well-being. It is found in many foods, especially in grain, protein foodstuffs and in milk. No evidence has been reported of any zinc deficiency in human nutrition except in cases where the diet is deficient in protein and iron, as may happen in Egypt and Iran.[6] Zinc is needed to maintain a normal concentration of vitamin A in the body.

The amount of zinc in the milk of cows, ewes, and women varies from 3 to 6 mg per liter.[12] It may be increased to about 5 mg by feeding a zinc supplement to the cow.

The metal is dissolved readily by products from sour milk, such as sour whey and buttermilk. Instances of zinc poisoning of animals have occurred when they were fed sour milk held in galvanized containers.

Other Elements

Milk produced in Oregon was found to contain from 0.005 to 0.067 ppm of selenium. Human milk contained 0.013 to 0.053 ppm.

The feeding to rats of a diet in which essentially all of the protein came from dried skim milk caused the animals to die from a necrosis of the liver, especially if the milk were roller-dried. Feeding fresh skim milk, even when heated to 120°C, caused no injury to the rats. A daily dose of 0.4 microgram of sodium selenite prevented the onset of necrosis of the liver of rats that were on the diet of dried skim milk. Selenium has been establishd to be an essential nutrient, the requirement being 0.1 to 0.3 ppm. Selenium may not be added to a foodstuff because in high doses it can be carcinogenic, although some experiments have suggested that it may protect against cancer. [22]

Rubidium is present in trace amounts. In one survey, the average content of market milk varied between 0.57 and 3.39 parts per million. In general, milk in coastal areas of the United States had a higher rubidium content that milk produced in inland States.

The strontium content of milk varies between 0.04 and 0.48 mg per liter, within average value of 0.17 mg. [12]

Mercury is found to be present in some samples of milk. Amounts up to 0.027 ppm were found in dry whole milk, with an average value of 0.01. For fresh whole milk the mercury content varied up to 0.009 ppm, but the average amount was less that 0.001 ppm. [19]

There are little available data on the presence of cadmium in milk. Traces of cadmium are found in the air, water, and vegetation. The United States Public Health Service has set an upper limit of ten micrograms per liter for cadmium in drinking water. One report gives the cadmium content of freeze-dried whole milk as 0.7 to 2.9 ppb; of dried skim milk 2.4 ppb. [21]

The lead content of milk was found to vary between 0.023 and 0.079 parts per million. A higher lead content may occur if the cows lick objects containing lead or lick painted surfaces. [13]

The average lead content of evaporated milk is reported to be about 0.11 ppm. This may arise in part from the solder used to seal the can. The lead content of nonfat dry milk may reach 0.8 ppm. Human milk has been found to contain 0.02 ppm of lead.

Water

About 87% of the weight of milk is water. It is, therefore, the principal constituent of milk, just as it is the chief component of most living matter. The water carries in solution milk sugar, mineral salts, such as those of sodium and potassium, and the water-soluble vitamins. Suspended in the milk are the fat, casein, and certain mineral salts, such as calcium phosphate and magnesium phosphate. It once was popularly supposed that the water in milk had some special nutritive value, but this is not so.

TABLE 5:3
Water Content of Edible Portion of Some Foods
(Percentages)

Milk	87.3	Chicken (raw)	74.3
Cantaloupe	95.0	Roast Beef (cooked)	51.0
Watermelon	92.0	Spaghetti (cooked)	60.0
Carrots	88.2	Salmon	64.0
Lettuce	94.8	Oat Meal (cooked)	85.0
White of Egg	88.2	Cottage Cheese	78.5

Nutritionists recommend that adults consume one milliliter of water for each calorie in the diet. Ordinarily about 2.5 liters (2.6 quarts) is an appropriate allowance, much of which is contained in the daily prepared meals. The infant requires a large amount of water, about 2.5 ounces per pound of body weight, and during warm weather this amount may need to be increased.

Many solid foods contain more water than does milk as indicated in Table 5:3. The water content of various dairy products is given in the sections dealing with the specific product. The adulteration of milk by addition of water is discussed in Chapter 20.

REFERENCES

1. Archibald, J. G., Iodine in Cow's Milk, *J. Dairy Sci.*, 41:711 (1958).
2. —, Cobalt in cow's milk, *J. Dairy Sci.*, 30:293 (1947).
3. Beck, A. B., The copper content of the milk of sheep and of cows, *Australian J. Exptl. Biol. Med. Sci.*, 19:145 (1941).
4. Curran, G. L., Azarnoff, D. L., Bolinger, R. E., Vanadium inhibition of cholesterol synthesis in man, *J. Clin. Invest.*, 38:1251 (1959).
5. Elvehjem, C. A., Duckles, D., Mendenhall, D. R., Iron versus iron and copper in the treatment of anemia in infants, *Am. J. Diseases Children*, 53:787 (1937).
6. Eminians, J., et al., *Clin. Ped.* 6:603 (1967).
7. Federal Register, Wash. D.C., May 29, (1963).
8. Gutmanis, F. & Chen, S. L., Determination of Fluoride Content in Cheese, *J. Dairy Sci.*, 49, 1212 (1966).
9. Jenness, R. and Patton, S., *Principles of Dairy Chemistry*; John Wiley & Sons, Inc., New York, 1959.
10. Koestler, G. A., The detection of milk altered by secretion disturbances, *Mitt. Geb. Lebensm. Unters. Hyg.*, 11:154 (1920).
11. Lynn, G. E., Shrader, S. A., Lamter, G. A., Occurrence of bromide in the milk of cows fed sodium bromide and grain fumigated with methyl bromide, *J. Agr. Food Chem.*, 11:87 (1963).
12. Murthy, G. K., Rhea, U. S., and Peeler, J. T., Copper, Iron, Manganese, Strontium, and Zinc Content of Market Milk, *J. Dairy Sci.*, 55:1666-1674 (1972).
13. Murthy, G. K., Rhea, U. and Peeler, J. T., Rubidium and lead content of market milk. *J. Dairy Sci.*, 50:651 (1967).
14. National Academy of Science, Nat. Res. Council, *Recommended Daily Dietary Allowances*, Wash., D.C. 1974.
15. Parkash, S. & Jenness, R., Status of Zinc in Cow's Milk, *J. Dairy Sci.*, 50:127 (1967).

16. Porter, J. W. G., Nutritive value of milk and milk products: Part 2, *J. Dairy Res.*, 27:329 (1960).
17. Report of committee on nutrition, *Pediatrics*, 26:715 (1960).
18. Rusoff, L. L., et al., Fluoride Addition to Milk and Its Effect on Dental Caries in School Children, *Am. J. Clin. Nutrition*, 11:94 (1962).
19. Tanner, J. T., Friedman, M. H., and Lincoln, D. N., Mercury in Foods. *Science* 177:1102-1103 (1972).
20. Kurtz, F. E., Tamsma, A., and Pallansch, M. J., Effect of Fortification with Iron on Susceptibility of Skim Milk and Nonfat Dry Milk to Oxidation, *J. Dairy Sci.*, 56:1139-1143 (1973).
21. Cornell, D. G. and Pallansch, M. T., Cadmium Analysis of Dried Milk by Pulse Polarographic Techniques, *J. Dairy Sci.*, 56:1479-1485 (1974).
22. Scott, M. L., The Selenium Dilemma, *J. Nutrition*, 103-803-810; 1973.

chapter 6

The Vitamins in Milk

Long before vitamins were known, careful observers had noted the association of certain foodstuffs with the prevention or cure of some diseases. Early in the sixteenth century, a decoction of spruce needles was used in England as a cure for scurvy, a disease especially then prevalent among sailors that made long voyages. By 1600, oranges and lemons were made part of the sailor's diet as a preventive of the disease.[9] This important observation was little appreciated, and it was not until 1804 that the daily use of lemon (or lime) juice was adopted by the British Navy.

Many subsequent references were made to the use of fresh fruits and vegetables for the prevention of scurvy during the following years, but it was not until 1880 that another outstanding announcement, that could be related to the vitamins, was made. Then there was published in Germany a statement that there must be something in milk that was indispensable to nutrition because without milk mice could not be nourished upon a diet of purified protein, fat, carbohydrate, and salt.[20]

In 1882, Admiral Takaki of the Japanese Navy, found that beri-beri among the sailors could be prevented if their diet contained meat and vegetables instead of the customary rice. By 1905, enough evidence had been gathered to enable one investigator to write that "there is in a natural foodstuff, such as milk, a still unknown substance which even in very small amounts is of paramount importance in nutrition."[26] About the same time similar findings were made in England where it was announced that milk must contain *accessory food factors*, the very term we now often use for vitamins.[14]

In 1912, Casimir Funk[8] believed that there were four compounds needed in the diet to prevent beriberi, pellagra, rickets, and scurvy. He introduced the word "vitamine" upon the assumption that they were necessary to life, vita, and to denote that they contained a nitrogenous substance, amine. This assumption was shown to be wrong and the final *e* was dropped from the word in order to remove the association with nitrogen-containing amines. Later it was proposed that the vitamins should be designated by letters of the alphabet. This manner of identification generally was accepted but it has outgrown its usefulness because it is now possible to identify most of the known vitamins by their chemical composition.

In 1913 American investigators found that milk fat contains a fat-soluble substance which is necessary for growth.[24] Further reasearch showed that there were present in other foodstuffs substances of unknown composition that had similar properties.

A vitamin is a substance found in relatively very small amount in a foodstuff, which is needed for normal nutrition and whose absence from the diet causes a specific deficiency disease, *avitaminosis*. A specific vitamin cannot be synthesized by the organism needing it and must be supplied by the diet. Each vitamin exerts a definite control over specific chemical reactions in the body. A number of the water-soluble vitamins do this by being part of an enzyme system; that is, the vitamin functions as a coenzyme.

VITAMIN A

Pure vitamin A occurs as pale yellow crystals at room temperature. Chemically, it is a high-molecular-weight alcohol of known composition. It is widely distributed in the form of vitamin A_1, known as "retinol" in mammals and salt-water fish; in fresh-water fish, this vitamin, known as A_2, "dehydroretinol", has a slightly different chemical composition. When first studied, the vitamin was known as *fat-soluble A*. Pure vitamin A occurs only in animals.

The relationship between this vitamin and the yellow coloring material in plants was noted long before the chemical nature of the vitamin was established. It is now known that the vitamin A activity of plant materials is due to their content of carotene and related yellow pigments. The chemical name *carotene* is derived from the word *carrot*, a vegetable that is very rich in the pigment. There are several vegetable pigments, especially *alpha*-carotene, *beta*-carotene, *gamma*-carotene, and cryptoxanthine, that can be converted into vitamin A by the animal body and therefore are called *provitamins* A. Cryptoxanthine is found in egg yolk and to some extent in milk-fat. Animal products, such as fats, milk and butter may contain both carotene and vitamin A. Fish-liver oils contain only the vitamin.

The symptoms of vitamin A deficiency have been described by many investigators. One of the first to appear is night blindness. As the name implies, this is an inability to see in dim light, such as one may encounter upon entering a motion picture theatre. Vitamin A is a component of rhodopsin or visual purple, a red-colored substance found in the retina of the eye, which is needed for vision in dim light.

Another disease of the eye, known as *xerophthalmia* or extreme dryness of the mucous membrane of the eye, is caused by a lack of vitamin A. It is cured by the addition of the vitamin to the diet, provided that bacterial invasion of the tissue of the eye has not caused irreparable damage. This disease is rare in the United States, but at the time of World War I, it was prevalent in Denmark, especially among children. The consumption of butter had been decreased greatly, as it was being exported to England and elsewhere, and as a result the population suffered from a lack of vitamin A in the diet. In order to prevent such an occurrence, many butter substitutes today are fortified with vitamin A obtained from fish-liver oil or made synthetically.

The daily requirement for a vitamin in the diet is expressed in terms of international units. An international unit is the measure of the amount of a vitamin in a food which is available for use in the body. One mg of vitamin A contains 4.5 million international units. One I.U. of provitamin A is 0.6 microgram of pure *beta*-carotene, an amount so small that 50 million units weigh one ounce. One vitamin A_1 unit is biologically equivalent to a unit of provitamin A. The recommended daily requirements are given in Table 6:1. Any preparation containing more than 10,000 I.U. of vitamin A per dose is limited by FDA rules to prescription only.

Vitamin A in Milk

The blood of bovine animals, in contrast to that of sheep and goats, contains both carotene and vitamin A, both of which are passed into the milk by the mammary gland.

Practically all of the vitamin A in milk is associated with the unsaponifiable portion of the milk fat. Both the vitamin and the carotenoids of milk appear to be concentrated on the surface of the fat globules and in amounts related to the size of the globules.[36] There is no important difference in the vitamin A content of the milk of different breeds of cows when they are fed the same ration, and the measurement is made on the basis of the fat content of the milk. As Guernsey milk contains more fat, a quart of it, as a rule, contains more of the vitamins than does a quart of milk from Holstein cows, which has a lower content of fat.[31] The Guernsey cow secretes a large proportion of vitamin A into the milk as carotene, producing a pigmented milk. Other breeds, such as the Holstein-Fresian and the

Ayrshire, produce milk low in carotene but equally high in vitamin A on a fat basis.

The amount of vitamin A in the milk appears to be associated directly with the carotene content of the feed. In general, it appears that a maximum of about 2.5 grams of carotene per day is all that the cow may transfer to the milk.[34]

Milk produced during the summer months or pasture season is richer in the vitamin than milk produced at the end of the winter feeding season because the carotene that is present in green grass is transferred to the milk and so increases its vitamin A potency, whereas the hay fed during the winter season is low in its carotene content. The vitamin A content of milk can be increased ten to twenty times by feeding rations high in vitamin A or carotene.[29]

There is no loss of vitamin A when milk is pasteurized, evaporated, or dried.[12]

The addition of vitamin A to milk, as in the preparation of fortified or multi-vitamin milk, may cause an off-flavor in the product, usually described as "hay-like." The use of a vitamin concentrate that is free of flavor and odor usually will produce a product free of this defect.

The average vitamin A content of various milk products is given in Table 6:2.

THE VITAMIN B COMPLEX

The material first identified as vitamin B is now known to consist of a number of constituents, of which eleven have been identified. The vitamins of the B complex are important to health and nutrition. Five are known to form parts of coenzymes needed for the metabolism of carbohydrate, protein, or fat. The vitamins of the B complex sometimes are referred to as the *water-soluble* vitamins.

The cow produces all of the B Vitamins in the rumen, and so does not depend upon their presence in the feed.

Thiamine

Thiamine, first known as vitamin B_1, is also called the *antineuritic* factor. Its isolation began about 1890, when experiments in Java showed that chickens that were fed polished rice developed a disease similar to beri-beri in man.[6,9] When natural rice or polishings from rice were fed the disease could be cured, almost within an hour. This investigation probably was the first in which animals were used to study a disease of dietary origin.

In some countries, especially in the Orient, a serious lack of the vitamin exists in the diet of people who live on polished rice and other

TABLE 6:1
FOOD AND NUTRITION BOARD, NATIONAL ACADEMY OF SCIENCES-NATIONAL RESEARCH COUNCIL
RECOMMENDED DAILY DIETARY ALLOWANCES [a] Revised 1974

Designed for the maintenance of good nutrition of practically all healthy people in the U.S.A.

								Fat-Soluble Vitamins				Water-Soluble Vitamins							Minerals					
	Age (years)	Weight		Height		Energy	Protein	Vitamin A Activity		Vitamin D	Vitamin E Activity	Ascorbic Acid	Folacin [f]	Niacin [g]	Riboflavin (B_2)	Thiamine (B_1)	Vitamin B_6	Vitamin B_{12}	Calcium	Phosphorus	Iodine	Iron	Magnesium	Zinc
		(kg)	(lbs)	(cm)	(in)	(kcal)[b]	(g)	(RE)[c]	(IU)	(IU)	(IU)	(mg)	(μg)	(mg)	(mg)	(mg)	(mg)	(μg)	(mg)	(mg)	(μg)	(mg)	(mg)	(mg)
Infants	0.0-0.5	6	14	60	24	kgx117	kgx2.2	420[d]	1,400	400	4	35	50	5	0.4	0.3	0.3	0.3	360	240	35	10	60	3
	0.5-1.0	9	20	71	28	kgx108	kgx2.0	400	2,000	400	5	35	50	8	0.6	0.5	0.4	0.3	540	400	45	15	70	5
Children	1-3	13	28	86	34	1300	23	400	2,000	400	7	40	100	9	0.8	0.7	0.6	1.0	800	800	60	15	150	10
	4-6	20	44	110	44	1800	30	500	2,500	400	9	40	200	12	1.1	0.9	0.9	1.5	800	800	80	10	200	10
	7-10	30	66	135	54	2400	36	700	3,300	400	10	40	300	16	1.2	1.2	1.2	2.0	800	800	110	10	250	10
Males	11-14	44	97	158	63	2800	44	1,000	5,000	400	12	45	400	18	1.5	1.4	1.6	3.0	1200	1200	130	18	350	15
	15-18	61	134	172	69	3000	54	1,000	5,000	400	15	45	400	20	1.8	1.5	2.0	3.0	1200	1200	150	18	400	15
	19-22	67	147	172	69	3000	54	1,000	5,000	400	15	45	400	20	1.8	1.5	2.0	3.0	800	800	140	10	350	15
	23-50	70	154	172	69	2700	56	1,000	5,000		15	45	400	18	1.6	1.4	2.0	3.0	800	800	130	10	350	15
	51+	70	154	172	69	2400	56	1,000	5,000		15	45	400	16	1.5	1.2	2.0	3.0	800	800	110	10	350	15
Females	11-14	44	97	155	62	2400	44	800	4,000	400	12	45	400	16	1.3	1.2	1.6	3.0	1200	1200	115	18	300	15
	15-18	54	119	162	65	2100	48	800	4,000	400	12	45	400	14	1.4	1.1	2.0	3.0	1200	1200	115	18	300	15
	19-22	58	128	162	65	2100	46	800	4,000	400	12	45	400	14	1.4	1.1	2.0	3.0	800	800	100	18	300	15
	23-50	58	128	162	65	2000	46	800	4,000		12	45	400	13	1.2	1.0	2.0	3.0	800	800	100	18	300	15
	51+	58	128	162	65	1800	46	800	4,000		12	45	400	12	1.1	1.0	2.0	3.0	800	800	80	10	300	15
Pregnant						+300	+30	1,000	5,000	400	15	60	800	+2	+0.3	+0.3	2.5	4.0	1200	1200	125	18[h]+	450	20
Lactating						+500	+20	1,200	6,000	400	15	80	600	+4	+0.5	+0.3	2.5	4.0	1200	1200	150	18	450	25

[a] The allowances are intended to provide for individual variations among most normal persons as they live in the United States under usual environmental stresses. Diets should be based on a variety of common foods in order to provide other nutrients for which human requirements have been less well defined. See text for more detailed discussion of allowances and of nutrients not tabulated.

[b] Kilojoules (kJ) = 4.2 x kcal

[c] Retinol equivalents

[d] Assumed to be all as retinol in milk during the first six months of life. All subsequent intakes are assumed to be half as retinol and half as β-carotene when calculated from international units. As retinol equivalents, three fourths are as retinol and one fourth as β-carotene.

[e] Total vitamin E activity, estimated to be 80 percent as α-tocopherol and 20 percent other tocopherols. See text for variation in allowances.

[f] The folacin allowances refer to dietary sources as determined by *Lactobacillus case* assay. Pure forms of folacin may be effective in doses less than one fourth of the recommended dietary allowance.

[g] Although allowances are expressed as niacin, it is recognized that on the average 1 mg of niacin is derived from each 60 mg of dietary tryptophan.

[h] This increased requirement cannot be met by ordinary diets, therefore, the use of supplemental iron is recommended.

TABLE 6:2
Average Vitamin Content of Milk and Milk Products (100 gram Portions)

Product	*Vitamin A* I.U.	*Thiamine* mg	*Riboflavin* mg	*Niacin* mg	*Ascorbin Acid* mg	*Other Vitamins*
Milk, Cow's —, Fresh, raw yearly average)	160	0.035	0.17	0.08	2.0	Vitamin D, I.U. 0.5 to 4.4 (44 in Fortified Milk); Vitamin B_{12}, 0.5 microgram; Vitamin K. trace; Biotin, 0.3 to 0.5 micrograms; Carotene, Av. 30mg (Holstein-28 mg; Guernsey-60 mg); Choline (total), 15 mg; Folic Acid, 3 to 8 micrograms; Inositol, 18mg; Pantothenic Acid, 0.3-0.45 mg., Pyridoxine, 0.05-0.1 mg; Tocopherols 0.08 mg.
—, Non-Fat	4	0.04	0.17	0.8	1.0	
—, Evaporated	360	0.05	0.35	0.20	1.0	B_{12}, 4.5 microgram; Folic Acid, 0.7 microgram; Pyridoxine, 0.06.
—, Condensed	335	0.10	0.40	0.20	0.9	B_{12}, 0.2 microgram.
—, Dry, Whole	1400	0.32	1.5	0.7	5.8	Choline 80-100 mg. B_{12} 13 to 28 micrograms; Pyridoxine, 0.33-0.8 mg.
—, Dry, Non-Fat	40	0.35	20	1.25	7.0	Choline, 160 mg; Pyridoxine, 0.55 mg. Pantothenic Acid, 3.1, Inositol, 80 mg
—, Malted	1110	0.25	0.9	..	3.0	
Milk, Goat's	160	0.035	0.11	0.20	1.0	Pyridoxine, 0.01 mg, Folic Acid, 0.6 microgram; Pantothenic Acid, 0.38 mg; Biotin, 0.1 microgram.
Milk, Human	65	0.014	0.037	0.18	5.7	Vitamin D-I.U., 0.4-10, Vitamin E., 0.13 to 3.6 mg.; Carotene, 25 mg; Biotin, 0.8 microgram; Folic Acid, 3-8 micrograms, Inositol, 45 mg., Pantothenic Acid, 0.25 mg; Pyridoxine, 18 micrograms, Tocopherols, 0.2 mg.
Butter (year-around average)	3,300	..	..	.1	..	Vit. D Average, 90 I.U. Carotene, 10-17 micrograms; Choline, 8-13 mg.; Tocopherol, 2.4 mg.
Buttermilk, Dried,	(270)	(0.4)	(2.0)	(1.2)	(7.0)	Pyridoxine (3.0); Insitol (76).

TABLE 6:2 (*continued*)

Product	*Vitamin A* I.U.	*Thiamine* mg	*Riboflavin* mg	*Niacin* mg	*Ascorbic Acid* mg	Other Vitamins
Margarine, Fortified	3,300	..	..	..	..	Vit. D fortified; Tocopherols, total 54 mg; alpha 28 mg
Cheese						
—, Cheddar	1350	0.02	0.4	0.03	..	Folic Acid, 14 micrograms, Inositol, 25 micrograms;
—, Cottage (creamed)	190	0.03	0.29	0.09	..	Biotin, 2 mg; Pantothenic Acid, 0.35-1.0 mg; Pyridoxine, 0.1 mg
—, Cottage (uncreamed)	9	0.03	0.28	0.09	..	
—, Cream	1480	0.02	0.25	0.15	..	Folic Acid, 20-45 micrograms
—, Roquefort type	1250	0.04	0.06	0.55	..	
—, Swiss	1140	0.04	0.30	0.19	..	
Cream						
—, Half and Half (12% Fat)	492	0.028	0.16	0.04	..	
—, Table (20%)	890	0.029	0.14	0.04		
—, Medium (33%)	1340	0.025	0.12	0.04	..	Vit. D—av. 17 I.U.
—, Heavy (39%)	1600	0.021	0.10	0.04	..	
Ice Cream: Plain, 12% Fat	520	0.05	0.25	0.14	1.0	
Ice Milk (5% Fat)	210	0.05	0.2	0.1	1.0	Tocopherol 0.39.
Whey (dried)	50	0.4	2.5	0.9	..	

*This Table is based upon values given in Bulletin 72, *Nutritive Value of Foods*, U.S.D.A., 1960 references given in this chapter, and values furnished by the American Dry Milk Institute. Figures in parentheses are calculated values.

restricted diets. In the United States symptoms of a gross deficiency of thiamine in the diet are rare. The enrichment of flour and cereal products to restore their original content of thiamine and certain other food essentials is now an accepted procedure to alleviate dietary inadequacies. Thiamine is found to some extent in all the foodstuffs common to the American diet, it being absent only from fats, oils, and refined sugar. Lean pork, dry beans, and peas are among the best sources.

A deficiency of thiamine in the diet leads to such symptoms as moodiness, fear, and mental as well as physical fatigue. A more serious deficiency causes loss of appetite, muscular weakness, vague pains or neuritis. As with many of the other vitamins, a lack of thiamine results in retardation of growth. It appears that with increasing food consumption the bodily requirement for thiamine also increases.[3] The occasional administration of large amounts of thiamine, such as twenty times the normal requirement, has no harmful effect. Continued feeding of very large amounts leads to an abnormal production of free fat in the liver cells and other disorders.

Thiamine is synthesized by plants, yeasts, and bacteria but apparently not by higher animals. In the cow and other ruminants, however, a synthesis of the vitamins of the B complex occurs because of bacterial action in the first stomach or rumen. Thiamine must be supplied in the diet of man, and as the body has little ability to store thiamine, the supply must be renewed constantly. Of the body tissues, only the liver, kidney, heart, and brain contain appreciable amounts of this vitamin. As would be expected, only a small amount of thiamine is needed by the human body, the minimum requirement being about 0.23 milligram for every 1000 calories in the diet. The daily requirements for thiamine and other vitamins is shown in Table 6:1.

In the body, thiamine forms thiamine pyrophosphate, which also is known as *cocarboxylase*. This is a coenzyme which takes part in the chemical reactions by which glucose furnishes energy to the body tissues.

Thiamine Content of Milk:

Cows on a diet low in thiamine may secrete milk that contains a normal amount of the vitamin. This is due to the synthesis of the vitamin by the action of bacteria present in the rumen. Even though the feed is rich in thiamine there is comparatively little transfer of it from the feed to the milk and it is impossible to exceed a maximum level of thiamine in the milk. This level is about 0.4 milligram per quart of milk. Unlike the vitamin A content, the amount of thiamine in milk is very constant and except that the level is highest during early lactation, it does not vary with the season of the year.

Milk is a fair but not a rich source of thiamine. Pasteurization destroys about 10 to 20% of the thiamine originally present in the milk and the loss is somewhat greater in the manufacture of evaporated and dried milk.[10,13] A loss of only about three percent is reported to occur when milk is pasteurized by the short time-high temperature process. Dried milk exposed to gamma ray radiation (up to 5.58 Mrad) showed no loss of thiamine.[37] Sterilization by heat may destroy up to 40 percent of the thiamine originally present.

The average thiamine content of various milk products is given in Table 6:2

Riboflavin

Riboflavin is the name given to the vitamin that previously had been called vitamin B_2 or G. It is of historical interest that milk (or whey) was the material from which riboflavin-containing material was first studied, long before its nature as a vitamin was known. In 1879 a yellow pigment was found in whey, which later became known as *lactochrom* or *lactoflavin.* In Europe, riboflavin is often called lactoflavin.

Riboflavin is yellow in color. Its aqueous solution has a yellow-green fluorescence, to which much of the characteristic color of whey is due. It is widely distributed and usually is found with other members of the B complex. Milk is an outstanding source, other good sources being green-leafed vegetables, heart, kidney, and lean meat. Riboflavin is prepared commercially by a number of patented, synthetic processes, usually by microbiological fermentation methods.

The recommended daily allowances for riboflavin are given in Table 6:1. Unlike thiamine, it appears that the need for riboflavin is related to body weight, rather than to the caloric intake. It is unusual to find a person showing only a deficienty of riboflavin as a diet low in this usually is low in the other vitamins of the B-complex. In human beings, a deficiency of this vitamin causes an inflamed and scaly condition of the skin around the corners of the mouth, the base of the nose, and the ears. A disturbance of vision also may occur.

The riboflavin content of milk and various milk products is given in Table 6:2. The amount in milk is fairly constant for a given breed of cow. The Channel Island breeds produce milk that contains from 20 to 50 percent more riboflavin than do the Ayrshire and Holstein breeds. As riboflavin is water-soluble, it is not found in the milk fat. When skim milk or whey is dried, the powder is an especially rich source of riboflavin. As the vitamin is heat-stable, especially in acidic media, milk can be pasteurized, sterilized or boiled for a short time without any material effect upon its riboflavin content. The heat treatment given evaporated milk, or storing it for a year at room temperature has no effect upon its content of riboflavin.[12,13]

According to most investigators if milk is exposed to bright light, especially direct sunlight, a loss of riboflavin takes place, usually with the development of an off-flavor.[25] Homogenized milk, because it is more opaque, undergoes less riboflavin loss than ordinary "cream line" milk. The loss of riboflavin due to exposure to light may reach 80% or more.[30] Milk contained in amber glass bottles loses little riboflavin by exposure to light but fiber containers do permit some destruction of the vitamin. Ordinary storage in the household refrigerator has no effect upon the riboflavin content. Fluorescent light is less harmful than sunlight.

In the manufacture of cheese, such as cheddar, brick, and blue, from 20 to 30% of the riboflavin originally present in the milk is retained in the cheese.[15]

Niacin

Niacin is a white, crystalline compound, first prepared about 1870 by the oxidation of nicotine. It was not known to be a vitamin until about 1937. Between 1928 and 1937 various investigators found that in foods rich in "water-soluble B" there also was a distinct factor that promoted growth and prevented pellagra, and which was identical to nicotinic acid. With the establishment of the vitamin properties of nicotinic acid, its name in the United States was changed to *niacin* because it was considered that the general public might confuse or associate nicotinic acid with poisonous nicotine. In most other countries the term *nicotinic acid* is used.

In some part of the southern United States, and in tropical and sub-tropical areas where the nutritional disease, *pellagra*, is prevalent, diets are deficient in niacin, although other dietary deficiencies are also involved in the disease. The diet of low-income groups in these areas consists largely of corn, which is very low in niacin and tryptophan. Part of the body's need for niacin may be obtained from the amino acid tryptophan; from 10 to 20 parts of tryptophan are needed to take the place of one part of niacin.

Milk is a poor source of niacin but a good source of tryptophan from which the body can form niacin. In most cereals, the niacin present is not available to man. Beef, chicken, heart, kidney, liver, and peanut butter are among the more common foods relatively rich in niacin. In plants, the vitamin is present as niacin; in animal tissues as a related compound, *niacinamide*. In the body, niacin acts as a part of those coenzymes that activate repiration in the tissues and produce energy through the oxidation of glucose.

Over two million pounds of niacin and niacinamide are made synthetically in the United States every year for use in vitamin preparations and for the enrichment of bread and cereal products. The daily requirements for niacin are given in Table 6:1. The average niacin content of various milk products is given in Table 6:2.

There is no evidence that exposure to light, pasteurization, evaporation, or drying causes any loss of niacin in milk.[3]

Pantothenic Acid (Vitamin B_3)

Pantothenic acid, in free form, is a yellow, oily liquid. In nutritional experiments and as a pharmaceutical preparation, its calcium salt is used. Traces are present in many foodstuffs, in reference to which it derives its name from the Greek word meaning *from everywhere*. Its existence was first indicated in 1901, when a growth factor termed *bios* was isolated from yeast cells, but it was not until 1930 that its identity as a vitamin was established. At one time it received attention as the "anti-gray-hair" factor because the lack of it in the diet of rats caused them to develop gray hair on their bodies; however, this does not apply to man.

Pantothenic acid is a component of coenzyme A, which is needed for the reactions by which fatty acids are metabolized to provide energy to the body. On the basis of animal experiments, it is estimated that man needs not over 10 mg a day. The amount present in a normal diet is more than enough to supply this requirement.

The pantothenic acid content of milk is independent of the feed or breed of the cow. Heat treatment and storage produce little or no change in the pantothenic acid content of milk products.

The amount of pantothenic acid in various milk products is given in Table 6:2.

Folic Acid (Pteroylglutamic Acid)

Folic acid is a yellow, crystalline substance, slightly soluble in water. A related form is known as *folinic acid* or *citrovorum factor*. Folic acid is a growth factor for animals and many microorganisms. It exerts beneficial results in the treatment of some forms of anemia, but it is not the only factor concerned with the disease. In man and animals, both folic acid and vitamin B_{12} are needed for red blood cell development.

Folic acid is present in liver, green vegetables, meat, yeast, and to some extent in bananas and strawberries. Milk is a poor source. Cow's milk and human milk have about the same folic acid activity, but goat's milk has appreciably less. On the day of parturition the folic acid content of bovine colostrum is up to 30 times the level of normal milk, but by the second or third day it lowers to the normal level.

Pasteurization has little effect upon the folic acid content. Heat sterilization, in the presence of air, destroys about 50% of the folic acid activity in milk. Cow's milk appears to have enough folic acid to provide the daily needs of infants.[39] The amount of folic acid in some milk products is given in Table 6:2.

Biotin

In 1916, it was found that a diet in which the protein came from raw egg white, was toxic to rats. This condition was prevented if the diet also contained liver or yeast. The protective factor, first called "vitamin H" was later identified as *biotin.*[33] This vitamin was isolated and synthesized in 1943. It is needed for growth and well-being by microorganisms and animals, including man. Liver, eggs, and legumes as well as milk are good sources of the vitamin. Microorganisms in the digestive tract appear to synthesize enough biotin to supply human needs.

Neither the exposure to light or the heat treatment given to milk during pasteurization materially alter its biotin content. The amount found in some milk products is given in Table 6:2.

Pyridoxine (Vitamins B_6)

Vitamin B_6 is found naturally in three forms, *pyridoxine, pyridoxal,* and *pyridoxamine*, any of which may be used interchangeably by the body. The phosphates of pyridoxal and pyridoxamine serve as coenzymes which function in the utilization of protein by the body and for the metabolism of certain amino acids. The vitamin also is needed for the metabolism of unsaturated fatty acids, especially linoleic and arachidonic acids.

Vitamin B_6 is widely distributed in foodstuffs; good sources are liver, yeast, eggs, green vegetables, and the bran from cereal grains. Milk is a fair source of the vitamin, but treatment by heat, especially sterilization, causes some reduction in the vitamin content.[10] As children have a definite need for this vitamin, the loss during the processing of milk may at times be of importance. There is one report of children fed a sterilized milk formula who developed convulsive seizures, which were cured by administering therapeutic doses of pyridoxine.

Because of the several natural forms of the vitamin which react differently to various test procedures, analytical figures from different sources may appear inconsistent. Table 6:2 gives an estimate of the Vitamin B_6 content of various milk products.

Vitamin B_{12}

Like the folic acid and Vitamin B_6 group, vitamin B_{12} includes a number of related factors. Each of these factors contains about 4% of cobalt; vitamin B_{12} itself is *cyanocobalamin*. This group is important to the formation of red blood cells, and in the prevention and cure of pernicious anemia. Absence of an "intrinsic factor" in the gastric juice leads to failure of absorption of the vitamin and to the development of the anemia, even in the presence of an adequate amount of vitamin B_{12} in the diet.

Vitamin B_{12} is found in animal tissues such as liver, kidney, muscle meats, and to some extent in corn, wheat, and milk. Most animals synthesize the vitamin in their intestines, but man must obtain it from his diet. Commercially, the vitamin is obtained from the mother liquors left over after the microbiological fermentation used to obtain antibiotics such as streptomycin. In addition to the purified vitamin used in pharmaceuticals, preparations of the vitamin are used in animal feeds to promote growth and well-being. Such feeds may contain up to 0.003% of vitamin B_{12}.

As indicated in Chapter 5, the vitamin B_{12} content of milk may be affected by the cobalt content of the diet. A ten-fold increase (0.08 to 0.7 mg) of cobalt a day increased the B_{12} content of ewes' milk from 1 millimicrogram to 2.6 millimicrograms per ml of milk. The vitamin B_{12} content of goats milk is discussed in Chapter 12.

Pasteurization does not alter the vitamin B_{12} content of milk. The vitamin content of various milk products is given in Table 6:2.

Other Factors in the Vitamin B Complex

A number of compounds other than those mentioned in the previous paragraphs are present in the B complex. Among these are choline, inositol, and *para*-aminobenzoic acid, all of which have been found present in milk. Choline has been discussed, as it is part of the lecithin molecule; it can be synthesized by the body. Neither inositol nor *para*-aminobenzoic acid has been found to be essential to human nutrition. Table 6:2 gives the amounts present in milk.

VITAMIN C—ASCORBIC ACID

In the first part of this chapter, mention is made of the historical record concerning certain foodstuffs in the diet and their association with the prevention of scurvy. Over three hundred years passed before the essential dietary factor was identified as vitamin C. By 1933, vitamin C had been synthesized and named *ascorbic acid.* Today large amounts of ascorbic acid are made commercially; in one process, corn sugar is oxidized to make *sorbose*, a related sugar, which then is converted into ascorbic acid.

The richest natural sources of ascorbic acid are rose hips, pine needles, liver, fruits and green vegetables. The vitamin exists in two forms, D- and L- ascorbic acid, but only the L-ascorbic acid is biologically active. Ascorbic acid is practically non-toxic and large doses may be taken without harm.

In addition to its anti-scorbutic properties, ascorbic acid has other biological functions. It is needed for the formation of teeth and bones, the healing of wounds, and, in general, it has a part in the bodily defense against infections. Ascorbic acid acts as a coenzyme for the metabolism of the essential amino acid, tyrosine.[28] A biological relationship exists between ascorbic acid, folic acid, and vitamin B_{12}, in that it assists in the conversion of folic acid into the *citrovorum factor.*

At least 10 mg of ascorbic acid are needed daily to prevent scurvy. Mild or sub-clinical scurvy may occur in infants between the ages of six and twelve months, who have been nursed on processed milk formulas that are not supplemented with a source of ascorbic acid. Such scurvy is rare in infants under four months or over eighteen months of age. A very early sign of scurvy in infants is pain or tenderness of the legs with the child is handled.[32] Breast-fed infants receive sufficient ascorbic acid if the mothers are well nourished, because women's milk may contain up to six times as much ascorbic acid as does cow's milk.

Ascorbic acid is unique among the vitamins in that it is both an acid and a chemically active reducing agent. In a slightly acid solution, ascorbic acid is stable, but it quickly oxidizes in the presence of copper or iron ions. The first oxidation product is dehydroascorbic acid, which has the same biological activity as the original ascorbic acid. It is the form in which practically all of the vitamin is present in fresh, raw milk, shortly after it is drawn from the udder.

The ascorbic acid content of cow's milk appears to be fairly uniform, and independent of the season of the year and the animal's diet. The average amount in fresh, raw milk is about 20 mg to the quart, which is less than the recommended dietary requirement. During pasteurization by the holding method most of the dehydroascorbic acid present in milk is further oxidized into a biologically inactive product (diketogulonic acid). Any ascorbic acid left in the milk continues to be oxidized, the rate depending upon many factors, such as the temperature and time of heating, the presence of metallic ions, and the amount of exposure to air and light. Pasteurization by the HTST method is much less destructive of dehydroascorbic acid. Milk held under refrigeration at 10°C shows a progressive loss of ascorbic acid.

Milk exposed to light undergoes a gradual loss of its ascorbic acid content, especially if held in clear glass bottles. Brown glass containers or plastic-coated cartons offer protection for the vitamin. Fluorescent light is less harmful but causes a loss of the vitamin. Even after storage in the dark, the loss of ascorbic acid continues after it is once started by exposure to light.

In the manufacture of evaporated milk, the loss of ascorbic acid may reach 60% and a further loss—up to 40% more—may occur upon storage for a year. In spray-dried whole milk, the loss may reach 20%; in the roller-dried product, about 30%. Fruit ice cream may contain an appreciable content of ascorbic acid because the acid is used as an antioxidant in the preparation of the frozen fruit used in the ice cream. This additional ascorbic acid also serves to retard the development of "oxidized" flavor in the ice cream.

That ascorbic acid may have an influence upon the development of off-flavor milk is indicated by the finding that the development of a tallowy (oxidized) flavor is inhibited if the ascorbic acid is all oxidized; that is, in the form of dehydroascorbic acid, before the milk is pasteurized and stored.[17] The normal amount of ascorbic acid in milk tends to favor the production of the oxidized flavor, but if the milk is fortified with an additional 50-100 mg to the liter, the acid acts as an effective antioxidant.[16]

The recommended daily requirements of ascorbic acid are given in Table 6:1; the amounts found in various milk products are in Table 6:2.

THE VITAMINS D

The existence of a substance that prevents rickets was first reported in 1921, and recognized as a vitamin in 1922.[21,29] It is now known that rickets, which may start in the second or third month of an infant's life, is due to a lack of, or improper balance of, calcium, phosphorus, and vitamin D in the diet. The disease is characterized by softness of the bones and delayed dentition. In adults, a lack of vitamin D may cause softening of the bones, especially in nursing mothers. An important function of the vitamins D is to increase absorption of calcium from the intestine. Apparently the alkaline phosphatase enzyme present in the body is activated with a resulting deposition of phosphate in the bone structure.

Although the chemical composition of the different forms of vitamin D is known, there is as yet no generally satisfactory chemical method for estimating them. The usual method employed to measure the vitamin D activity of a foodstuff is to feed definite amounts of it to test animals, usually rats or chicks that have been raised on a vitamin D-free diet, and to note the rate of formation of the bones. Many vitamin D assays are made by means of the *line test*, in which the animal is killed and the amount of new bone in the leg is observed. The first indication of the formation of new bone is a more or less continuous line of band-like deposition of calcium phosphate in the bone, hence the name, line-test. As healing in chicks does not show itself as a line, usually the ash in the bone in the center toe is measured. See Fig. 6:1.

According to recent investigation, the body converts vitamin D to a metabolite, 25-hydroxycholecalciferol, which is the active form that cures and prevents rickets and allied bone diseases.[4]

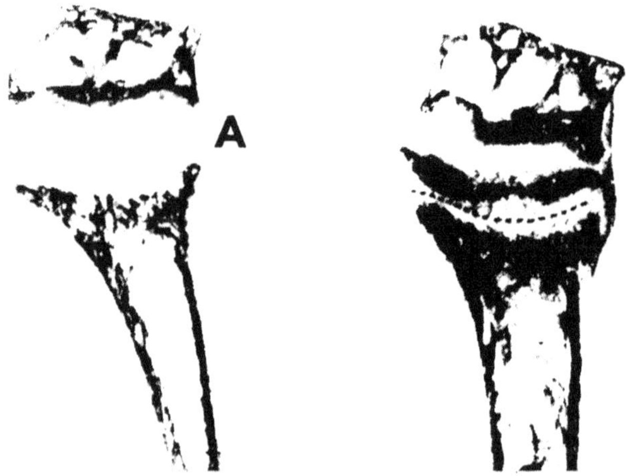

Fig. 6-1. Line Test for Vitamin D: Left: Leg bone (radius) of rat fed a diet deficient in vitamin D. The white band (A) shows growing portion of bone, devoid of calcium, between the tip of the bone and the older calcified bone below. Right: Let bone of rat (radius) showing calcification after diet contained vitamin D. Note that calcium (dark band above dotted line) is being deposited in the area formerly devoid of the mineral.

Early investigators noted that many foods acquire the antirachitic activity of vitamin D when exposed to ultraviolet light. This was found to be due to activation of certain sterols. Of the many substances that may have vitamin D activity, two are of importance to human nutrition, D_2 and D_3.

TABLE 6:2
The Provitamins and Vitamins D [5]

Provitamin	*Vitamin*
Ergosterol	D_2 - Ergocalciferol
7-Dehydrocholesterol	D_3 - Cholocalciferol
22,23-Dihydroergosterol	D_4 -
7-Dehydrositosterol	D_5 -
7-Dehydrostigmasterol	D_6
7-Dehydrocampesterol	D_7

There is no vitamin D_1 since it was shown to be a compound of D_2 and another sterol. Vitamin D_3 is more active than vitamin D_2; vitamin D_4 is of intermediate activity. The other vitamins are relatively inactive.

Vitamin D_2:

Vitamin D_2, or ergocalciferol, does not occur naturally, but is formed when ergosterol, derived from yeast, is irradiated with ultraviolet light. D_2 is the form in which the vitamin is found in many pharmaceutical products and in metabolized vitamin D milk, that is, from cows fed irradiated yeast. It is the form of vitamin D added to milk, margarine and non-poultry feed.

Vitamin D_3:

The other vitamin D, D_3, is formed when the cholesterol derivative, 7-dehydrocholesterol, is irradiated with ultraviolet light. It is more expensive than vitamin D_2 but is the only form of vitamin D that can be assimilated by poultry. Vitamin D_2 is the form found in natural sources, such as fish-liver oils, animal fats, egg yolk, and irradiated milk. Exposure of the skin to sunlight or ultraviolet rays results in the formation of vitamin D_3, which then is utilized by the body just as if it had been supplied in the diet.

Any preparation of vitamin D with more than 400 I.U. per dosage unit is limited to prescription only under new FDA rules.

The International Unit (I.U.) for vitamin D, adopted in 1951, is the activity of 0.025 microgram of pure vitamin D_3. The U.S. Pharmacopeia Unit has the same value. Vitamin D_3 is used as a standard, rather than D_2 because it is effective for all test animals, whereas D_2 is much less effective when chicks are used. For man and mammals in general, either vitamin is equally effective. If taken in massive doses over a period of time, vitamin D produces severe toxic reactions due to a general calcification of the tissues.

The recommended daily requirements and the amount of vitamin D in milk products are given in Tables 6:1 and 6:2.

Vitamin D in Normal Milk:

Milk, butter, eggs, and liver are the only foods that contain appreciable amounts of natural vitamin D. Most of the vitamin is found in the fat. Compared to certain fish-liver oils, milk fat is a poor source of the vitamin. When measured on the basis of their fat content, there is no difference in the vitamin D content of the milk of the different breeds. The content of vitamin D in milk is highest during the summer months and lowest in winter, which is due to variation in the cow's diet and exposure to sunlight. If the cow or her udder is exposed to sunlight or ultraviolet radiation, the vitamin D potency of the milk is increased to some extent, but the results of this procedure are too variable to be considered satisfactory or of commercial importance. As the diet of most children is deficient in vitamin D, it has become a commercial procedure to add the vitamin to milk.

Some authorities believe that in the United States, Canada, and Western Europe, infants may be given too much vitamin D, rather than too little. They believe that only fluid milk and infant milk foods should be fortified and that the daily intake from all sources should not exceed 400 I.U.[1] France does not permit the addition of vitamin D to milk.

Where unfortified fluid whole milk is used, cases of rickets are often reported. In Greece where less than one percent of the fluid milk is fortified, rickets is not uncommon, but not in the case of infants using fortified milk products. Even in Canada a similar situation is found. Universal fortification of milk would appear to be desirable, with due regard to the possibility of excessive amount of vitamin D being present in the whole diet.

Metabolized Vitamin D Milk:

Vitamin D is one of the few substances important to nutrition which when fed to the cow appears in the milk. Materials rich in vitamin D, such as fish-liver oils, irradiated yeast, or irradiated ergosterol, may be added to the cow's feed in order to increase the vitamin D content of her milk. The use of fish-liver oil is not recommended because it may impart an undesirable flavor to the milk or decrease the amount of milk given by the cow.

The cow is very inefficient in the transfer of the vitamin to the milk. Measurements show that less than 2% of the amount fed appears in the milk.[27] This small amount, nevertheless, is enough to increase the vitamin D potency of the milk about fifteen times, so that the milk will contain about 430 International Units to the quart. This potency coincides with the minimum daily requirement recommended for children. The use of irradiated yeast in the feed is no longer practiced in the United States.

Irradiated Milk:

Foodstuffs, including milk, may be irradiated with ultraviolet light to impart vitamin D activity.[35] Ultraviolet light cannot penetrate to any depth in a milk film and about 75% of the effective irradiation is absorbed in the first 0.02 millimeter (about 0.0008 inch) of the milk film. As long exposure to ultra-violet light imparts an unpleasant flavor to milk, the time of exposure is short, usually about three seconds, and the milk flow is a very thin film during the irradiation. Mechanical control of the irradiation makes it possible to produce a milk product with a definite vitamin D potency, usually not over 200 units per quart. In 1945, much evaporated milk was irradiated but this procedure was not found sufficiently practical to add 400 units, as now recommended. Irradiation is still used in parts of Germany and in Italy to produce milk with up to 320 units.

Fortified Vitamin D Milk:

Concentrates made from certain fish-liver oils, or from irradiated ergosterol mixed with vegetable oil, milk or cream, or dissolved in propylene glycol, may be added directly to milk in order to increase its vitamin D content and yet not affect the flavor of the milk. Usually sufficient concentrate is added, prior to pasteurization, to bring the potency up to at least 400 units of vitamin D per quart of milk. Most canned evaporated milk is so fortified that when it is diluted with an equal volume of water, the resulting solution contains 400 units of vitamin D per quart. Most homogenized milk used in the home has 400 units of vitamin D added per quart. The direct addition of a vitamin concentrate to milk has the advantage that the potency can be controlled by adding known amounts of the vitamin. See Chapter 12.

THE VITAMINS E

In 1922, it was noted that rats on experimental diets failed to reproduce even though they apparently were otherwise normal. This observation led to the discovery of the fat-soluble vitamin E. This later was found to consist of a number of related compounds, known as tocopherols. These are identified by Greek letter prefixes.

Alpha-tocopherol and beta-tocopherol are found in wheat-germ oil, gamma-tocopherol is found in cottonseed oil, and delta-tocopherol in soya-bean oil.

In most animals a deficiency of vitamin E results in muscular degeneration, sterility in rats and cardiac muscle degeneration in cattle. A vitamin E deficiency disease is not known in man, but by analogy to other animals, it is believed to be needed for man. The vitamin stabilizes certain fatty acids, sulfur-containing amino acids and vitamin A.

Some alpha and a little gamma-tocopherol is present in milk, but milk is a poor source of the vitamin. Vitamin E may have some effect in retarding the development of oxidized flavor in milk. Vitamin E values for milk products are given in Table 6:2.[19] Human milk contains about twice the vitamin E content of cow's milk.[38]

As mentioned in Chapter 2, the need for vitamin E is increased when the amount of polyunsaturated fatty acids in the diet is increased. The presence of more than a normal amount of tocopherol in butter is an indication of adulteration with a vegetable oil.[22]

REFERENCES

1. American Acad. of Pediatrics, Committee of Nutrition, Pediatrics, 198:96 (1963).
2. Bunnell, R. H., et al, Alpha-Tocopherol Content of Foods, Amer. *J. Clinical Nutrition*, 17:1-10 (1965).
3. Cowgill, G. R., *The Vitamin B Requirement of Man*, Yale University Press, New Haven, 1934.
4. DeLucca, H., et al, *Chemical Communications*, No. 14:801 (1968).
5. Dyke, S. F., *The Chemistry of the Vitamins*, Interscience Publishers, London, 1965.
6. Eijkman, C., *Geneesk. Tijdschr. Nederland-Indie*, 30:259 (1890).
7. Ford, J. E. and Scott, K. J., The Folic Acid Activity of Some Milk Foods for Babies, *J. Dairy Res.*, 35:85 (1968).
8. Funk, C., *State Med.*, 20:341 (1912).
9. Harris, L. J., History of vitamins, in *Biochemistry and Physiology Nutrition*, Vol. 1; Academic Press, Inc., New York (1953).
10. Hassinen, J. B., Durbin, G. T. and Bernhart, F. W., Vitamin B_6 content of milk products, *J. Nutrition*, 53:246 (1954).
11. Henry, K. M., Houston, J., Kon, S. K., Oshorne, L. W., The effect of commercial drying and evaporation of the nutritive properties of milk, *J. Dairy Res.*, 10:272 (1939).
12. Henry, K. M., Houston, J., Kon, S. K., Thompson, S. Y., The effect of commercial processing and of storage on some nutritive properties of milk, *J. Dairy Res.*, 13:329 (1944).
13. Holmes, A. D., Lindquist, H. G., Jones, C. P., Wertz, A. W., Effect of high temperature, short time pasteurization on the ascorbic acid, riboflavin and thiamin content of milk, *J. Dairy Sci.*, 28:29 (1945).
14. Hopkins, F. G., The analyst and the medical man, *Analyst*, 31:385 (1906).
15. Irvine, O. R., et al., Retention of nutrients in cheese-making, *Sci. Agr.*, 25:817 25:817 (1945).

16. Jenness, R. and Patton, S., *Principles of Dairy Chemistry*, p. 378, John Wiley & Sons, Inc., New York, 1959.
17. Krukovsky, V. N. and Guthrie, E. E., Ascorbic acid oxidation as a factor in the inhibition or promotion of the tallowy flavor in milk, *J. Dairy Sci.*, 28:565 (1945).
18. Krukovsky, V. N. and Loosli, J. K., Influence of tocopherol supplementation in the vitamin content of milk fat, stability of milk and milk and fat production, *J. Dairy Sci.*, 35:834 (1952).
19. Kanno, C., Yamauchi, K., and Tsuga, T., Occurrence of Tocopherol in Bovine Milk Fat, *J. Dairy Sci.*, 51:17131719 (1968).
20. Lunin, N., *Z. Physiol. Chem.*, 5:31 (1881).
21. McCollum, E. V., Simmonds, N., Becker, J. E., Shipley, P. G., Experimental demonstration of a vitamin which promotes calcium deposition, *Biol. Chem.*, 53:293 (1922).
22. Mahon, J. H. and Chapman, R. A., Detection of adulteration of butter with vegetable oils by means of the tocopherol content, *Anal. Chem.*, 26:1195 (1954).
23. Mellanby, E., *Medical Research Council (British) Special Report Ser.* 61 (1921).
24. Osborne, T. B. and Mendel, L. B., The influence of butter fat on growth, *J. Biol. Chem.*, 16:423 (1913).
25. Patton, S., The mechanism of sunlight-flavor formation in milk with special reference to methionine and riboflavin, *J. Dairy Sci.*, 37:446 (1954).
26. Pekelharing, C. A., *Ned. Tijdschr. Geneesk.* 70, 11 (1905).
27. Russell, W. C., Wilcox, D. E., Waddell, J., Wilson, L. T., The relative value of irradiated yeast and irradiated ergosterol in the production of vitamin D milk, *J. Dairy Sci.*, 17:445 (1934).
28. Sealock, R. R. and Goodland, R. L., Ascorbic acid, a coenzyme in tyrosine oxidation, *Science*, 114:645 (1951).
29. Sherman, H., *Vitamins and Hormones;* Academic Press New York, 1950.
30. Shetlar, M. R., C. L. Shetlar, J. F. Lyman, Determination of riboflavin in chocolate milk and the comparative photochemical losses in chocolate and whole milk, *J. Dairy Sci.*, 28:873 (1945).
31. Sutton, T. S. and Krause, W. E., Vitamin A in milk, *Ohio Agr. Expt. Sta. Bi-Monthly Bull. No.* 164 (1933).
32. Tisdale, F. F., Vitamins in infancy and childhood, in *Nutrition in Everyday Practice*; Canadian Med. Assoc. (1939).
33. Vigneaud, du V., Structure of biotin, *Science*, 96:455 (1942).
34. Webb, B. H. and Johnson, A. H., *Fundamentals of Dairy Chemistry*, The Avi Pub. Co., Inc., Westport, Conn., 1965.
35. Weckel, K. S. and Jackson, H. C., The irradiation of milk, *Wis. Univ. Agr. Expt. Sta. Res. Bull. No.* 136 (1939).
36. White, R. F., Easton, H. D. and Patton, S., Relationship between fat globule surface area and carotenoid and vitamin A content of milk, *J. Dairy Sci.*, 37:147 (1954).
37. Ziporin, Z. Z., Kraybill, H. F., Thack, H. J., Vitamin content of foods exposed to ionizing radiations, *J. Nutrition*, 63:201 (1957).
38. Thompson, S. Y., Fat soluble vitamins in milk and milk products, *J. Dairy Res.*, 35:149 (1968).
39. Ghittes, J. and Tripathy, K., Amer. *J. Clin. Nutr.* 23:141 (1970).

chapter 7

Dairy Microbiology and Sanitation

BACTERIA — FUNGI — ENZYMES

The composition, and therefore the food value, of milk as it comes from the cow is an established fact, but the cleanliness and safety of the product are under human control. The importance of obtaining a milk of good sanitary quality and low bacterial content cannot be overemphasized. To be of good quality, milk must come from healthy cows that are fed wholesome food and kept in clean surroundings. To be of low bacterial content, the milk must be obtained with clean equipment and stored at a low temperature until it is processed and consumed.

A common procedure to judge the sanitary quality of milk is to determine its bacterial content. Bacteria are tiny organisms, belonging to the vegetable kingdom and they are among the simplest forms of life. They are invisible to the unaided eye and some are so small that they hardly can be seen under the microscope. Even the largest are so small that about 500 of them placed end to end would make a line one inch long. Most bacteria are less than one-tenth of this size, being from one to five microns in length and one-half to one micron in diameter. (A micron is 0.001 mm, or 1/25,400 inch). A drop of sour milk may contain more than 50 million bacteria.

Although invisible to the naked eye, the bacteria that may be present in a milk product become evident through their activities. Undesirable organisms in milk may cause it to sour or develop a bad odor and flavor. The daily industry makes use of certain desirable microorganisms in the manufacture of fermented milk products, butter, and cheese. Many persons associate something unpleasant or dangerous with bacteria, but fortunately very few of the large number of known bacteria are harmful to man. Most foods contain bacteria and practically everything with which one comes in contact is covered

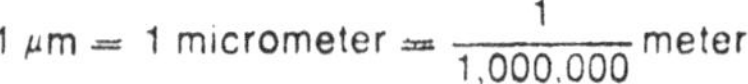

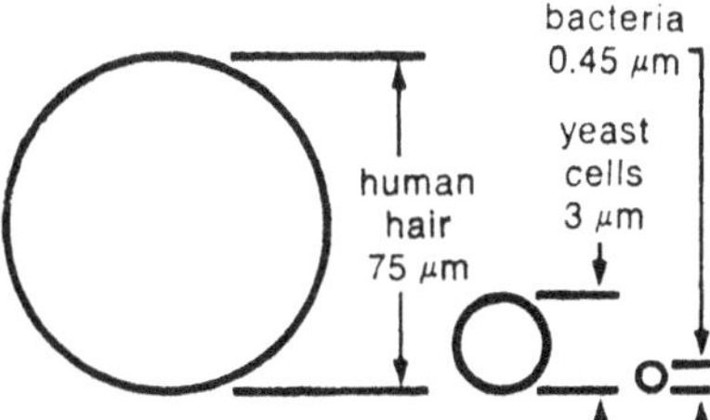

Fig. 7-1 Comparative Size

with them or contains them. They are present in and on our bodies and although they may be removed temporarily from the skin by special chemicals, it is impossible to remove them completely from the body.

The normal life cycle of a bacterium, like that of all living beings, includes growth, reproduction, and death. Bacteria reproduce by cell division; that is, the organism actually divides into parts, each of which then becomes a separate, living bacterium. In many types a new generation may be formed in this matter every 20 minutes. Thus, at the end of this time, one bacterium has become two; in about 40 minutes, these have become four and at the end of an hour the original organism has given rise to eight bacteria. As bacteria usually are present in fairly large numbers, reproduction at this rate can produce a very large population within a short time. Reproduction however cannot continue indefinitely because some of the bacteria die, the available food supply may become exhausted, the accumulated byproducts discourage growth, and may eventually result in death.

When conditions for growth are not favorable, certain bacteria, especially organisms belonging to the genera *bacillus* and *clostridium*, have the ability to transform themselves into small bodies called spores. The word *spore* comes from the Greek word for *seed.* With very few exceptions, only one spore is formed in each cell. The bacteriological spore is not a form in the reproduction cycle; it is only a means by which the organism protects itself against conditions that otherwise would be fatal. The spore, for example, can often withstand drying, the temperature of boiling water, and the action of some germicides, which would destroy the organism in its original growing or vegetative state. When suitable conditions return, the spore resumes its vegetative form and the bacterium again returns to the usual activities of its normal life cycle.

CLASSIFICATION OF SOME BACTERIA FOUND IN MILK

So many different kinds of bacteria are known that it is necessary to have some method of classification in order that they can be identified. One important means of classifying bacteria is to place them in groups according to their shape as seen under the microscope. Organisms that are longer than they are wide—rod-shaped—are called *Bacilli* (singular, *bacillus*, rod). Bacteria that are round or ball-shaped are known as *cocci*, (singular, *coccus*, seed or grain). Unlike many of the bacilli, the cocci do not form spores. The manner in which the cocci divide at reproduction furnishes a further means of classification.

If the cells remain connected end-to-end after division, a bead-like chain is formed. Such organisms are known as *streptococci*. Many streptococci form a polysaccharide which is different and specific for each particular organism. Identification of the polysaccharide enables the bacteriologist to type or classify the organism, the procedure being known as Lancefield typing. It is used to identify streptococci that are pathogenic to man. Cocci that form grape-like clusters are called *staphylococci*. Others that divide in such a manner that they form a flat or sheet-like mass of cells are known as *micrococci*. Some of the micrococci are able to survive pasteurization, whether 145°F or at high-temperature for a short-time. The micrococci do not form spores.

The bacilli and the cocci are the most common forms of bacteria found in milk. The streptococci often are present in milk which was not cooled properly after milking. The micrococci frequently are found on dairy equipment which was not cleaned properly or had not been sterilized or dried after cleaning. This is a reason why micrococci are common contaminants in dairy products.

Naming of Bacteria

The bacteriologist has a system of naming the bacteria. The first part of the name gives the *genus* or family which have characteristics in common; the second part indicates the *species* or variety within the genus. The name of the genus is spelled out when first mentioned, using a capital letter; thereafter the initial is used. The species is nearly always spelled with a small letter. Thus, *Streptococcus lactis* or *S. lactis* means that the organism belongs to the streptococci or chain-formers, and to the lactis variety. Occasionally when a characteristic difference occurs within the species, the species is subdivided to name the variety, as *S. cremoris* var. *hollandicus*.

Some organisms can live only in the presence of free oxygen; they are called *aerobes*. Other bacteria, called *anaerobes*, can live without air and get the oxygen they need from the food they consume. Another group, known as *facultative* organisms, have the ability to adapt themselves to changes in environment so that they can live in contact with the air or without it.

When living in liquid surroundings, many propel themselves through it by means of hair-like arms called *flagellae*. This word comes from the Latin word for whip and the bacterium actually moves itself by thrashing the liquid with its flagellae.

Some bacteria have their bodies covered with a viscous coating or envelope called the *capsule*. Capsulated organisms are associated with certain defects in milk, as will be mentioned later.(Chap. 10). It is of interest to note that with pathogenic organisms, the formation of a capsule is associated with their virulance. When the capsule is removed, the pathogeny is reduced.

The Significance of Bacteria in Milk

If milk is to be of good quality and safe to use, it must be obtained from healthy cows milked in clean surroundings. The equipment used to handle and process the milk must be of approved design and kept in a clean and sanitary condition. When care is taken in production, milk contains relatively few bacteria and, generally, it may be regarded as a better and safer product than milk with a high bacterial content. The presence of many bacteria or as it is commonly expressed, a *high count*, does not necessarily indicate that the product is unsafe to use as the organisms usually found in milk do not cause disease.

A high count indicates that the milk was handled in dirty equipment or produced under undesirable conditions, that the cows were diseased or that the milk was held in a warm place.

Although a low count indicates that the milk probably was produced under good conditions, this does not give the assurance that the milk was not obtained from a diseased cow, or was not otherwise subjected to contamination. Milk may be contaminated by a diseased milker or by contact with infected water.

The care given to the cows on the modern dairy farm, the use of milking machines, and of pipe lines that convey the milk directly to

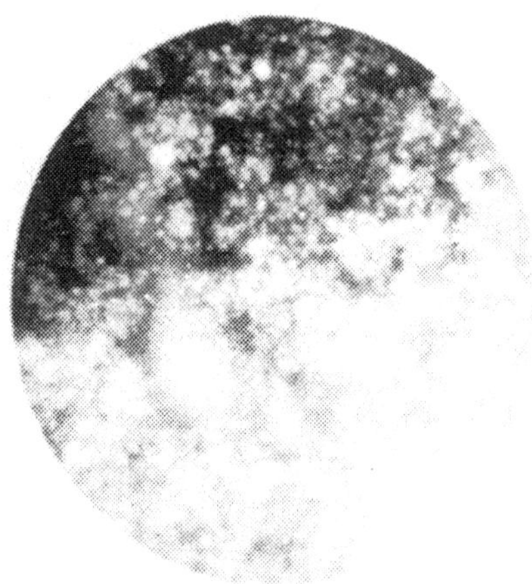

Fig. 7-2 Microscopic Appearance of High-grade milk. In this picture no bacteria are seen. The spaces left after the fat globules are dissolved from the smear appear as white areas. Leucocytes or epithelial cells may be present in high grade milk, but few bacteria may be found on microscopic examination. (New York State Agr. Expt. St., Geneva, Circ. 93.)

the refrigerated holding tank, have greatly minimized the possibility of contamination of the milk from external sources, such as air, dust, and manure. The potential danger of such contamination is not ignored and every care is taken to produce milk under sanitary conditions.

The knowledge and the appreciation of the precautions necessary to prevent contamination, and the realization by the milker and handler of their responsibility, contribute greatly to the production of a clean milk supply. Laxity in any one step in the production of a clean milk may cause other employees to attach less importance to regular sanitary procedures. Owing to the care taken in every step of milk production, from cow to consumer, milk is one of the cleanest foods in our diet.

Bacteria in the Cow's Udder

Even though it may be produced under the cleanest of conditions, milk usually contains some bacteria at the time it is drawn from the cow's udder. The contamination of milk begins while it is in the udder because bacteria present in the blood stream may find their way to the udder, or contaminating organisms may enter from the outside through the teats and milk ducts. Harmless micrococci are the ones most commonly found in aseptically drawn milk but streptococci also

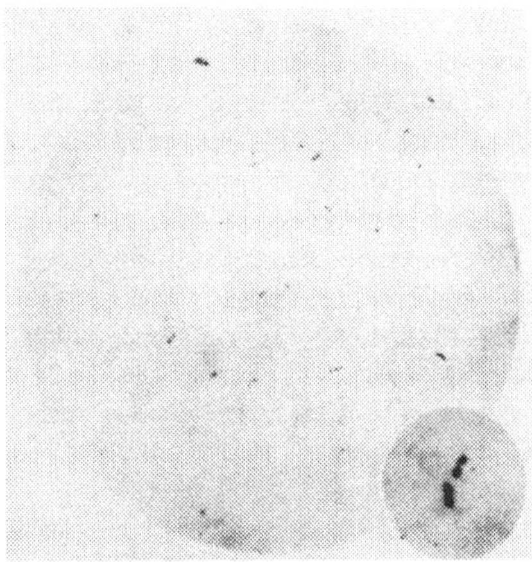

Fig. 7-3 Microscopic Appearance of Improperly Cooled Milk, Souring Normally: The predominant organism is Streptococcus lactis, which often is present in pairs (diplococci) and in threes and double pairs. The insert shows two pairs enlarged 2000 times. The lower pair shows a faint indication of division into four cells. (New York State Agr. Expt. Sta. Geneva, Circ. 93.)

occur very often. Organisms of brucellosis, tuberculosis, or other diseases may be found if the cow is infected.

The number of bacteria present in milk drawn aseptically from the udder of a healthy cow may vary greatly among different cows and even for each quarter of the udder. The number usually is low, one investigation showed a range of from 50 to 500 bacteria per ml, with an average of 240 per ml.[27] As a rule, the count lies between 1000 and less than 10,000 per ml.

The *fore-milk*, or the first few streams drawn from the udder, usually contains a larger proportion of bacteria than the last milk, called the *strippings*. The manipulation of the udder at the end of the milking, however, may dislodge bacteria and cause the strippings to contain more organisms than the middle portion of the milk drawn. The discarding of the fore-milk by the milker has no practical influence upon the total bacterial count of the main bulk of the milk, but its elimination is advisable, as the fore-milk often may be of inferior flavor. Usually the first two or three streams of milk are discarded. This portion of the milk is of relatively low fat content.

Germicidal Action of Milk

Many observers have noted that bacteria do not grow well in freshly drawn milk. The intensity of this germicidal action varies much among different samples of milk. It usually is present for several hours after the milk is drawn and may still be active for 24 hours if the milk is held at a low temperature. An inhibitory factor, named *lactenin*, has been isolated from milk.[25,26] Lactenin contains two symbiotic components; lactenin I which dominates in colostrum, and lactenin II, found principally in normal milk.[2]

Lactenin is not active under anaerobic conditions, so its activity in milk while in the udder must be limited. The germicidal action of milk has no practical importance. Any action it may have in restraining the growth of organisms normally present in milk may have some significance in places where freshly drawn milk is held without refrigeration, but this situation does not apply to modern dairy practice.

A substance, other than lactenin, that restrains the growth of certain strains of coliform organisms has been shown to be present in milk.[37] This factor is destroyed when milk is heated for 30 minutes at 53°C, whereas lactenin may still be active in milk that has been heated to 65°C for 30 minutes.

Bacteria in Air and Dust

The number and kind of organisms in the air may vary considerably according to the season of the year. The number of bacteria is usually smaller in the cooler and rainy months. Comparatively few acid-forming bacteria are found in the air, but yeasts and mold spores occur frequently. Under ordinary conditions, probably not more than twenty bacteria per milliliter of milk may be derived from the air.

The dust in milking barns, especially at feeding times or when the cows are being brushed, may carry enough bacteria to increase materially the bacterial count should the dust fall into the milk. Manure may enter the milk, and the bacteria contained therein grow rapidly in milk. Living organisms of bovine tuberculosis may enter milk in this way from tuberculous cows. Ordinarily, such care is taken in milking operations that actual contamination from this source may be considered to be insignificant. The number of organisms that might be added to milk from sources such as those mentioned is not nearly so large as is popularly supposed and even under dirty conditions usually less than 1,000 organisms per milliliter get into the milk.[23,40] Screening tests for airborne contamination are given in the *Standard Methods*.[43]

Counting the Bacteria in Milk

A commonly used procedure to measure the sanitary quality of milk is to estimate its bacterial content. There is no known method by

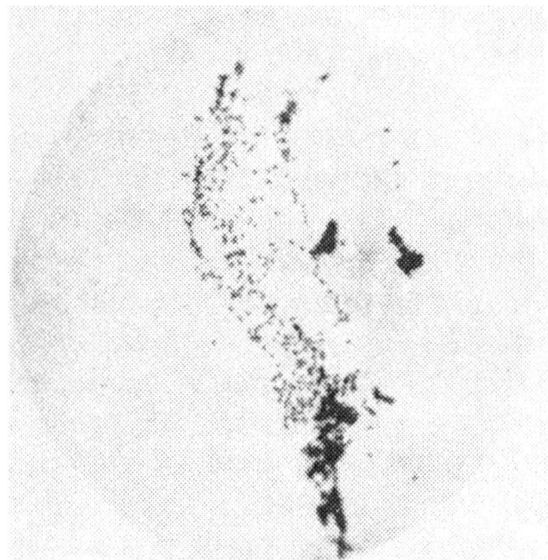

Fig. 7-4 Microscopic Appearance of Milk Handled in Unclean Utensils: The organisms present cannot be identified by microscopic examination alone. Masses of micrococci often are present, sometimes non-spore forming rods and occasional large spore-forming rods. X600. (*New York State Agr. Expt. Sta. Geneva, Circ. 93.*)

which the exact count of the number of bacteria present can be obtained. Furthermore, none of the methods in common use is capable of making known the presence or absence of disease-producing organisms. The bacterial count usually is expressed as the number of bacteria found per milliliter (*ml*), approximately 1/29th ounce. Often the term *cubic centimeter* (*cc*) is used instead of milliliter. The volumes* are practically the same but *milliliter* is the correct term. In viscous and non-liquid products the count is expressed on a weight basis, that is, the number of organisms in one gram of product. The various procedures used to estimate the bacterial count in dairy products are described in Chapter 20 but the *Standard Methods* must be consulted for important, necessary details.[43]

Somatic Cells

In the process of maintaining the functions of the body, including that of the udder, blood serum passes through the walls of the various capillaries. The red corpuscles do not pierce the capillary walls but some of the leucocytes or white blood cells do. As a result, their presence in milk is to be expected. In all bodily activities, gland cells and their constituent parts wear down and are eliminated in various

*One ml. = 1.000028 cc (U.S. Bureau of Standards, Cir. (434.)

ways. Accordingly, through the functioning of the udder, cells of various kinds are normal constituents of milk. These are the *somatic cells.*

The cells in milk, which may outnumber the bacteria present, have no special significance, unless present in excessive numbers. In contrast to the bacteria, the leucocytes are lifeless and do not increase in number after the milk is drawn. If there is blood in the milk, red cells or erythrocytes are present and a diseased condition in the cow's udder is indicated. When an infection takes place, the leucocytes migrate to the infected area in an attempt to combat the infection.

Various methods are used to count the cells in the milk. The method for the direct microscopic count of bacteria may be employed as the stained cells are readily seen under the microscope (Chapter 20). In another method, a sample of milk is placed in a small tube and whirled at high speed in a centrifuge. The sediment deposited on the bottom of the tube is removed, placed on a glass slide, then is stained with methylene blue, and examined under the microscope. Other methods, including electronic counting are described in Chapter 20 and in the Standard Methods for the examination of Dairy Products.[43]

There is no definite agreement as to the number of cells that may be found in normal milk. Early investigators thought that large numbers, many millions in fact, were normal. Today it is believed that the cell content of normal milk ranges between 50,000 and 100,000 per milliliter.[20] Milk from cows with diseased udders shows an increased number of somatic cells. A count around 150,000 warrants checking the condition of the cow and a count of 500,000 or more indicates an abnormal condition of the udder. In many areas of the United States a legal maximum count of 1.5 million somatic cells is permitted in raw milk.

Mastitis

An inflammation of the udder commonly is called *mastitis.*[54] Any injury to the udder, the use of too high a vacuum with milking machines, or leaving the machine on after the flow of milk has stopped tends to favor mastitis. The chronic form of mastitis is common among dairy cattle but it is not always readily recognized. The acute form, often termed *garget*, is more easily identified because the udder appears abnormal; secretion of milk is reduced and it may be tinged with blood.

A common form of mastitis is caused by *Streptococcus agalactiae.* This organism is stated to be a normal inhabitant of the udder.

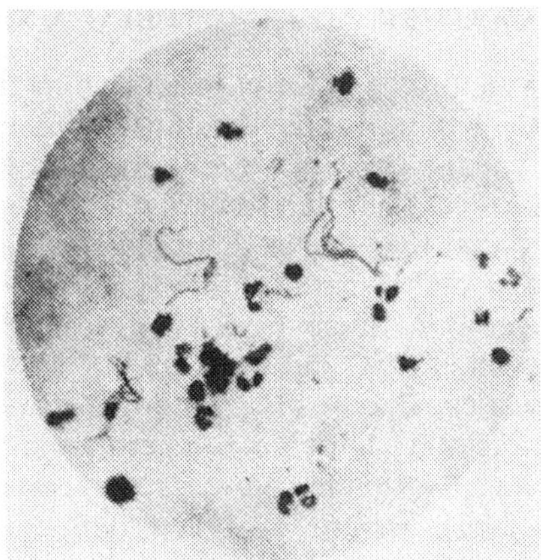

Fig. 7-5 Microscopic Appearance of Milk from Cow with Mastitis: Numerous leucocytes are present as well as long-chain streptococci. X600. (*New York State Agr. Expt. Sta., Geneva, Circ. 93.*)

Frequent incomplete milking appears to promote a rapid increase in the number of these organisms in the udder with subsequent development of mastitis. The organism is not pathogenic to man, and is destroyed when milk is pasteurized. Owing to the relative ease with which *Streptococcus agalactiae* is destroyed by antibiotics, other more resistant organisms, especially the pathogens *Streptococcus pyogenes* and *Staphylococcus aureus* have become important causes of infection. These organisms will invade injured tissue and bring about a serious attack of mastitis. At times, coliform organisms and bacteria found in soil and stagnant water are associated with mastitis.

As previously mentioned, large numbers of body cells usually appear in the milk during the early stages of mastitis. The milk may become more alkaline, with *p*H above 6.8. When the milker uses a *strip cup,** the first few streams of milk may show curdy particles or the milk may even be stringy. These observations are helpful, but bacteriological examination of the milk is the most trustworthy test for mastitis.[20] As this examination is time-consuming a number of rapid tests have been devised.

Mastitis has important significance in the economy of milk production. It is the most serious disease problem confronting the dairy farmer. In terms of animal and production loss, mastitis costs the American dairy industry about one-half billion dollars per year.[54]

*A cup covered with a black screen or gauze, through which the milk passes but upon which curdy particles are retained.

TEMPERATURE AND BACTERIAL GROWTH

Temperature is an important factor in controlling the rate of growth of bacteria. Most of them prefer a temperature between 20° and 36.7°, the range between that of a fairly warm room and that of the human body. These organisms, sometimes are called *mesophilic* bacteria, meaning that they prefer the middle or moderate range of temperature. Many of the pathogenic organisms of public health concern are mesophiles. Some organisms are able to grow at a temperature lower than 20°C and these are called *"psychrotrophic"* meaning organisms capable of significant growth in milk and milk products at commercial refrigeration temperatures, 2-7°C (35-45°F), irrespective of their optimum growth temperature. Species of *Pseudomonas, Achromobacter, Flavobacterium* and *Alcaligenes* are often encountered among psychrotrophic bacteria.

Psychrotrophic organisms, formerly termed *"psychrophilic"* often are present in raw milk, having as their source unsterilized utensils, the water supply, and dust from the barn and feed. They grow slowly in milk held at 1.7°C, but growth may be rapid as the temperature rises to and above 10°C. The modern methods of rapidly cooling milk as it is produced, the use of farm tanks and improved refrigerated storage favors the growth of psychrotrophic organisms and retards the growth of the more usual organisms, such as *S. Lactis*. Psychrotrophic organisms are not pathogenic and those found in milk are destroyed at temperatures of pasteurization. The majority are gram-negative rods. If present in properly pasteurized milk products, the source can be attributed to post-pasteurization contamination.

Psychrotrophic organisms may cause flavor defects and odors, especially fruity ones. These defects may occur in pasteurized milk after about a week of storage, especially if taken in and out of the refrigerator during this time. The psychrotrophic organism, *Pseudomonas fragi*, frequently is associated with spoilage in refrigerated milk.[5] The special procedure needed to count these organisms is mentioned in Chapter 21 and given in detail in the *Standard Methods*.[43]

Also of importance to the dairyman are the *thermoduric* and *thermophilic* bacteria. Thermoduric bacteria are not destroyed at the temperature used to pasteurize milk at 63°C (145°F) and they do not reproduce at this temperature. They are not pathogenic and generally do not form spores. The presence of thermoduric organisms in pasteurized milk may so increase its bacterial count that the product may not meet the standard for its grade. A high count of thermoduric or thermophilic organisms in milk indicates that it was produced or processed in unclean surroundings or under poor sanitary procedures.

Thermophilic organisms are those that not only withstand heat and like warm surroundings but are able to reproduce at the temperature used in the holding method for the pasteurization of milk, 63°C. The presence of these organisms in milk indicates contamination derived from feed, barn dust, soil, or the water supply. As a rule, thermophiles do not increase materially while the milk is on the dairy farm, but may do so during the heat treatment given the product at the processing plant. With the increased use of high-temperature, short-time pasteurization, thermophilic bacteria have ceased to be a serious problem in the processing of milk. Thermophilic bacteria found in pasteurized milk usually are rod-shaped and may form spores; they may readily be observed by direct microscopic examination (see Chapter 20). A commonly found thermophile is *Lactobacillus thermophilus*.

The effect of temperature upon the rate of growth of bacteria in milk was demonstrated in an investigation made in 1930 at the University of Vermont.[16] It was found that at 80°F (26.7°C) the number of organisms in a sample of milk doubled in about one and one-half hours; at 60°F (15.6°C) more than four hours were needed to double the number originally present; at 50°F (10°C) about eight hours, and if the milk was held at 40°F (4.4°C), about 39 hours. When held at room temperature, raw milk sours much more quickly than when held in a refrigerator, for the bacteria that cause the development of acid grow much faster around 70°F (21°C) than at a lower temperature. This also explains a popular superstition that milk will sour or curdle during a thunderstorm. The actual reason for this is that the warm weather preceding the storm favors the growth of bacteria. If the milk is kept cold nothing unusual happens.

Low temperatures slow down the growth of bacteria and lower the rate of reproduction. The number of organisms in a product generally decreases gradually when it is held under refrigeration. However, even extreme cold does not kill certain bacteria, and spores may withstand the temperature of liquid air, about −190°C. A high temperature, especially in the presence of moisture, is destructive to bacterial life. The exact temperature at which an organism is destroyed is rather indefinite, because it is influenced by the age of the cell, the medium in which it is present and the length of time for which it is exposed to heat. Young cells, for example, are killed more readily than older ones; dry heat is not as destructive as live steam.

Fig. 7:6. Time and Storage Temperatures for Milk Products.

The presence of sugar, as in ice cream, gives some protection to certain bacteria.

Another factor that influences the temperature at which bacteria may be destroyed is the temperature at which growth may have occurred. Organisms that grow at relatively low temperatures are more easily destroyed than those that grow at a higher temperature. This fact is of some importance when milk is pasteurized, since organisms that may have grown in cold raw milk are more susceptible to heat treatment than those that may have grown at a higher temperature.

Even though the manner in which milk is handled has changed for the better over the years, there has been no significant change in the types of microorganisms that might be present. Holding the milk under modern refrigeration has favored the growth of psychrophilic organisms. The use of antibiotics has brought about an increase in the number of penicillin-resistant organisms, especially of staphylococci that may be present in the raw milk. Holding the milk below 50°F (10°C) inhibits an increase in their number.

Differences in the thermal resistance of bacteria increases the difficulty in measuring the quality of raw milk to be pasteurized. A milk which has a high count of psychrophiles may be more desirable than one containing mesophiles and thermophiles since the former will be destroyed and many of the latter organisms may survive pasteurization.

Bacteria That Survive Pasteurization

As has been shown, all disease-producing organisms that may be present in milk are destroyed when it is pasteurized, but some entirely harmless bacteria do survive. When pasteurization was introduced it was believed that only spores and those organisms that were able to decompose milk protein could survive the heat treatment. It has since been shown that the relative proportion of the different groups of bacteria that may survive in pasteurized milk is about the same as in the raw product.

The kind of bacteria found in pasteurized milk depends upon the number and kind originally present in the raw product and also upon the time and temperature of pasteurization. In general, the higher the temperature, and the larger the number of bacteria present in the raw milk, the greater the precentage of destroyed bacteria. The age of the organisms has an influence upon the ease with which they are destroyed by heat. Milk that has been held at a temperature favorable to bacterial growth contains many young cells which are killed more easily by heat than are older organisms. A treatment at higher temperature or a longer exposure to heat is needed to kill the older cells.

The degree of resistance of micro-organisms to destruction by heat is measured by their "thermal death time". This is the shortest period of time needed to kill all of a suspension of the organisms under definite conditions of time and temperature. The medium in which the organisms are suspended, whether milk cr ice cream mix and the pH of the medium has an influence upon the thermal death time. Inasmuch as the organisms do not all die at a given moment, the term "thermal death time" is preferable to the older term "thermal death point".

One of the earliest studies on the bacteria that may survive pasteurization was made in 1913 by Ayers and Johnson of the United

States Department of Agriculture. They classified the bacteria into groups: *acid-forming, alkali-forming, peptonizing,* and *inert bacteria.* The peptonizing group consists of bacteria that decompose and liquefy milk protein. The inert group, as the name signifies, consists of bacteria that have no peptonizing, acid- or alkali-forming properties. It was found that when milk of high bacterial count, for example, with one million bacteria per milliliter, was pasteurized, about 99% of them were destroyed. Other investigators have shown that when milk of low bacterial count is pasteurized, the number of bacteria destroyed may be not more than 86 to 90% of the original number.[47] In general, about 96% of all harmless bacteria present in raw milk are destroyed when it is pasteurized, and all of any pathogenic organisms that may be present are completely destroyed.

When milk of rather high bacterial content is pasteurized at about 145°F (63°C), the largest proportion of surviving bacteria belong to the acid-forming groups. The proportion of the peptonizing group that survives is small. Pasteurization at a higher temperature causes more complete destruction of the acid-forming bacteria and so tends to increase the relative number of peptonizing organisms that survive. This difference in the kind and number of bacteria left living explains the tendency of milk pasteurized around 145°F (63°C) to become sour on standing because of the survival of some acid-forming bacteria. Milk pasteurized at a higher temperature for the same length of time may decompose on standing because of the survival of a greater proportion of the peptonizing bacteria.

The ability of certain strains of coliform organisms to survive pasteurization has been made the subject of much study.[4] Although they are harmless, these organisms are undesirable in milk. Some cells may manage to survive the heat treatment and their presence does not necessarily indicate that the milk had not been pasteurized properly. It does indicate that the milk was not produced under the best sanitary conditions. See below under coliform group.

Provided that no recontamination after pasteurization has occurred, an increase in the number of bacteria in the pasteurized milk over that originally present in the raw milk is an indication that thermophilic bacteria are present. Some investigators have found that thermoduric bacteria are not so readily destroyed by H.T.S.T. (high-temperature—short-time) pasteurization as they are by the holder method and, if they are present in raw milk, a considerable number of the thermoduric organisms may still be found in the milk if it is pasteurized by the H.T.S.T. method.[33,47] On the other hand, thermophilic bacteria have practically no opportunity to develop in the short time the milk is held at an elevated temperature in the H.T.S.T. process.

DESIRABLE AND HARMFUL BACTERIA

The bacteriology of dairy products would be comparatively simple if it were possible to divide the bacteria found in milk into two groups; one containing the organisms that are beneficial or desirable, and the other, those that are detrimental. Such a division is not practicable as the same organism may act differently under different conditions. In other words, certain bacteria may be useful in one product and yet be very undesirable in another. The bacteria that produce lactic acid by fermenting lactose illustrate this difference. Of special interest are the *Streptococcus lactis* group and some of the *Lactobacilli*. These organisms are not desirable in fresh milk because they cause the milk to sour. As described in Chapter 14, their growth is essential in the preparation of buttermilk, cheese, and some other dairy products.

Streptococcus lactis is harmless, but some of the other members of the genus *Streptococcus* are pathogenic to man; for example, *Streptococcus pyogenes,* may be excreted in the milk from an infected udder. It is associated with septic sore throat and scarlet fever.

Some defects in milk that are of bacterial origin are described in Chapter 10.

The Coliform Group

A group of non-pathogenic bacteria that are undesirable in milk are the organisms that inhabit the intestinal tract of animals and human beings. Among these is the organism *Escherichia coli*, named after Theodore Escherich (1857-1911), a pioneer bacteriologist; this organism often is called by an older name, *Bacterium coli*. Closely related to the *Escherichia* is the genus *Enterobacter*, formerly called *Aerobacter*. Organisms of this genus often are associated with decaying organic matter and are found in water and soil; they are not of animal origin but of plant origin. An important member of this genus is *Enterobacter aerogenes*. Organisms of the genera *Escherichia* and *Enterobacter* are called *coliform* organisms. At the Pennsylvania State College there is a National Coliform Reference Center, to aid in the identification of strains of the organism.

In the bacteriological examination of milk, the *Standard Methods* [43] defines as *coliform*

> ...all aerobic and facultatively anaerobic, gram-negative, non-spore-forming bacteria capable of fermenting lactose with the production of acid and gas at 32°C within 48 hours.

The term gram-negative refers to a method of classifying bacteria according to the manner in which they react to a special staining procedure, named after its originator, H. C. Gram, a Danish physician, in 1884. The bacteria are stained with a solution of crystal violet dye, then treated with a solution of iodine, and afterwards washed with alcohol. The alcohol removes the violet dye from some kinds of bacteria while others retain the color. Bacteria that retain the violent color are said to be *gram-positive;* those that lose it are said to be *gram-negative.* The difference may be rendered more distinct by counterstaining with another dye of a color contrasting to the original one—safranin, which imparts a red color, for example. No counterstain is taken up in gram-positive organisms but it displaces the original dye in gram-negative cells.

Many of the organisms associated with the spoilage of dairy products are gram-negative. These gram-negative organisms also show resistance to penicillin.

The presence of coliform organisms in water from wells, rivers, and reservoirs, may indicate contamination from sewage and intestinal waste of human origin. Water contaminated in this manner may also be infected with typhoid organisms or other disease-producing bacteria of intestinal origin, and therefore may be dangerous to use. The coliform organisms found in milk, however, usually are of animal origin and gain entrance into the milk from dust, soil, manure, feed, water or contact with dirty utensils. It is not practical to produce milk that always will be free of these organisms and their presence in raw milk may be tolerated. If present in large number, say over ten coliform organisms to the milliliter of pasteurized milk, they indicate that the milk was produced under improper procedures. With very few exceptions, the coliform bacteria do not cause disease in man but their presence in milk is undesirable because of their relationship to other organisms of intestinal origin, such as those that cause typhoid fever and dysentery.

Lactic acid bacteria, such as *S. lactis*, can ferment milk sugar and form lactic acid, but during the fermentation only acid and no gas is formed. A characteristic of the coliform organisms is that they form both acid and gas when they ferment milk sugar or certain other kinds of sugars. Special tests exist by which it is possible to distinguish between the individual members of the coliform group. These tests are important because if bacteria of the aerobacter group are present in milk, they indicate undesirable contamination from soil or plants. On the other hand, if *Escherichia coli* is found, it may mean contamination of intestinal origin, possibly carried to the raw milk by insanitary conditions among those handling the milk.

These tests are used to measure the efficiency of procedures used to minimize bacterial contamination of processed dairy products. When present in raw milk a few may be found under normal production procedures.

Some individuals of the coliform group are quite hardy and upon occasion a few may survive pasteurization if they were present in the raw milk. They multiply slowly in milk held around 40°F, so the presence of coliform organisms in a sample of milk a day old may be of little significance as an indication of the extent of possible contamination after pasteurization. Their presence in pasteurized milk, however, is an indication that the milk was improperly pasteurized or, most likely, contaminated after pasteurization, either from dirty equipment or by personal contamination from those processing the milk.

Salmonella

The *Salmonellae* are a group of organisms first identified by Dr. W. E. Salmon, in 1885. Many of them, such as *S.typhi*, the causative organism of typhoid fever and *S. paratyphi* and *S. schottmuelleri*, associated with paratyphoid fever, are pathogenic to man. Other salmonellae are the cause of outbreaks of food poisoning. Over 1200 types are known. Man can be infected by contact with any of a variety of animals and household pets, which may carry salmonella in their intestinal tract without showing any outward sign of infection. The term salmonellosis usually refers to cases of acute gastroenteritis and diarrhea caused by salmonella.[53] Symptoms of infection begin within eight to forty-eight hours. There is nausea, vomiting, stomach pains, and persistent diarrhea. Although considered an intestinal disease, it is difficult to demonstrate growth of the organisms in the intestine. Apparently after ingestion, the bacteria enter the lymph system, and then enter the blood stream.

A person, after apparently recovering from an infection with salmonella, may become a carrier of the organism without showing any symptoms of disease. Ordinarily the carrier stage may not last more than a few months, but in the case of typhoid fever it may persist much longer. A carrier is a public health menace because food handled by one might transmit the disease to others.

Many authorities believe that relatively large numbers of salmonellae (about 500,000) must be present to cause illness. A food product contaminated with a few cells may not in itself be infectious, but their detection is important in order to eliminate a potential source of illness. For example, a contaminated dry milk, after reconstitution, may remain for hours at a temperature favorable for rapid bacterial growth and make the product a serious source of infection. Compared to foodstuffs such as eggs, poultry and meat, milk products have not been a serious source of salmonellosis. The presence of salmonella in dry milk products is discussed in Chapter 17.

The salmonella are rod-shaped bacteria, gram-negative and usually have flagellae. They are facultative anaerobes. They grow best at body temperature, but can endure a wide range of temperature. Growth ceases at about 65.6°C (150°F) and below 6.7°C (44°F).

The same sources that may contaminate milk with coliform organisms can also carry salmonellae. As a result, all efforts made to eliminate the salmonella will also help to eliminate contamination with other pathogenic organisms. Pasteurization will destroy the organism, but contamination after pasteurization must be avoided, since subsequent processing does not provide for destruction of the contamination. Infection may originate from equipment contacted by contaminated water, dust, flies, birds or the dairy plant worker. Most foods are not pasteurized or sterilized in their final container and so may be exposed to contamination. Too much stress cannot be placed upon cleanliness of equipment and the surrounding structures. The work of regulatory agencies and the inspection of milk plants to insure sanitary conditions serves to eliminate sources of infection. Continuing emphasis must be placed upon sanitary practices in production, pasteurization, and the prevention of contamination after pasteurization.

Shigella

The *Shigella* are organisms which differ from the Salmonella mainly in not being motile. In some places, for example Great Britain, the Shigella are a common source of dysentery. The organism is spread by human carriers, with contaminated milk as a common source of infection, although other food products have been implicated most often. It is often transmitted by water, especially contaminated well water. A bloody-stool is a symptom which often differentiates shigellosis from salmonellosis.

Staphylococci

Some species of micrococci (staphylococci) may be associated with food poisoning. These organisms, especially *Staphylococcus aureus*, may be present in the milk of cows suffering from acute mastitis. It is the most common pathogenic organism found in raw milk but most cases of infection may be traced to a human carrier. *S. aureus*, which forms a yellow pigment, may be found in boils and suppurative sores. Man is an important source of S. aureus contamination in foods. It is reported that 40%of normal adults carry the organism in the nose and throat, from which they may be carried by the fingers. S. aureus may be present in large numbers in a food without causing any change in its odor, flavor, or appearance. The staphylococci grow best in the presence of oxygen, but they also are facultative aerobes. They grow best in neutral or nearly neutral foods, and will not grow in acid foods.

The significance of staphylococci in foods differ in some ways from that of the salmonella.[19] The organism is easily destroyed by heat, but the toxin produced is not destroyed by ordinary cooking temperatures; 121°C for thirty minutes is needed to destroy the toxin in milk. It is generally assumed that a large number of organisms must be present (perhaps 500,000) in order to yield enough toxin to produce illness. Consumption of food containing the toxin may cause vomiting within one-half to five hours. Often this is followed by diarrhea, which, while violent, is not prolonged. The attack rarely results in death. Often the victim may have recovered by the time an investigation is made. The rapid onset of vomiting is characteristic of this type of infection, but it does not exclude the salmonella as a cause.

Some organisims, especially the lactobacilli, inhibit the growth of staphylococci. Staphylococcic food poisoning has occurred through consumption of cheese which did not develop acidity in a normal manner. Evidently, the acid-producing bacteria such as the lactobacilli, did not develop and so were not available to inhibit the growth of contaminating staphylococci.

Often the deciding factor of whether or not a particular lot of contaminated milk may or may not be toxic depends upon the temperature at which it was held. It appears that formation of toxin may be increased if contaminated milk is held at room temperature for a time, then placed in a refrigerator, and again warmed to room temperature before consumption. This procedure is like that followed in the home with foods left over from one mealtime to another. The widespread occurrence of several strains of micrococci may result in their being isolated from suspected foodstuffs, but the formation of toxin must be demonstrated to incriminate the organism. In an epidemic of food poisoning any suspected food that is grossly contaminated with micrococci should be considered as a contributing factor.

The coagulase test is frequently used to distinguish between those micrococci that do not produce enterotoxin, and those that are potentially causative agents of food poisoning. Micrococci that do not produce coagulase do not form enterotoxin, but not all coagulase-positive micrococci produce enterotoxin.[43] Coagulase is an enzyme that can coagulate blood plasma, usually forming a clot within two hours.

The development of a serological procedure to measure the amount of any toxin that may be present has provided a means to detect a contaminated food before its consumption may have caused illness or death.[39] Prior to this development a food had to be fed to human volunteers or to cats or monkeys in order to observe the appearance of poison symptoms.

Diseases Transmissible by Milk

The marked decrease in milk-borne disease during recent years may be related to three major factors, (1) the care given to the health of the dairy cow; (2) the effectiveness of environmental sanitation and preventive medicine in maintaining the health of the populace; and (3) the almost universal use of pasteurized milk. Modern sanitary measures, milk inspection, medical examination of milk handlers, pasteurization, and the use of Certified Milk are accepted means for preventing the spread of diseases through milk. These preventive measures are conducted so efficiently that, compared with even ten years ago, today relatively few cases of milk-borne diseases are reported.

Contamination of milk or milk products by an infected milk-handler is perhaps the most common source of milk-borne disease. Often there is but a single source of contamination and the disease usually is then confined to those that consume the product of one dairy or milk plant, or are patrons of the milk distributor upon whose premises the milk was infected. Most cases of milk-borne disease originate through the use of raw milk. Intestinal complaints and diarrhea, especially among children, appear to be caused by milk that contains very large numbers of bacteria, rather than by any specific infection carried by the milk.

A milk-borne infection of diphtheria is today virtually unknown even among users of raw milk. This may be attributed largely to the widespread use of immunization to the disease. The causative organism, *Corynebacterium diphtheriae* can be transmitted to milk from a human source but it is readily destroyed by pasteurization.

The organisms of some uncommon diseases of cattle, such as anthrax, foot and mouth disease, and actinomycosis, may gain entry to the milk but their transmission to man through consuming the milk is very remote. Although it is not of bacterial origin, the rare disease of cattle called *trembles* may be transmitted to human beings; it is then called *milk sickness*. A century ago this was common in the Mississippi Valley, but today it is practically unknown. In animals, trembles is caused by poisoning from eating white snakeroot, rayless golden rod, or jimson weed.[7,36] The poison is eliminated in the milk.

The U.S. Public Health Service Reports indicate that between 1961 and 1966 there were 16 outbreaks of foodborne disease traced to milk and milk products. In 1970 there were four outbreaks associated with fluid milk, two with cheese and nine with other milk products; in 1971, there were two outbreaks traced to milk, three to cheese and two to other milk products.

Pathogenic Organisms in Milk

The pathogenic organisms that may gain entry to milk may be placed into two groups;

1) Those eliminated in the milk by a diseased cow;

2) Those that may contaminate the milk through some human agency after it has left the cow's udder.

Diseases from Milk Infected while in the Udder

Tuberculosis:

In many countries, tuberculosis is widespread in cattle. In the United States, owing to the success of the nation-wide program of testing for the disease, eradication of the diseased animals, and the care taken to maintain the health of the dairy herds, only 0.08 percent of the cow population is tubercular, compared to around 0.5 percent in 1950.

The causative organism, in cattle, called *Mycobacterium bovis*, is one of the most heat-resistant of the non-spore-forming pathogenic bacteria, but fortunately it is destroyed by pasteurization. A cow with pulmonary tuberculosis may swallow her own saliva and this, with the infected material coughed up from the lungs, then passes through the whole digestive tract, and remains an active source of infection. Particles of infected dust or manure may contaminate the milk, or it may be infected directly from the tubercular udder. The tubercle bacilli do not multiply in milk, and their presence in large numbers is an indication of serious contamination.

Infected raw milk is the chief means by which milk-borne tuberculosis is transmitted to man. The feeding of skim milk and cheese whey derived from infected raw milk has been a source of tuberculosis in animals. The belief that goats are immune to tuberculosis in incorrect; a number of cases of the disease among them has been reported.[8] Some strains of mycobacteria, similar to those that are associated with tuberculosis, have been found to survive pasteurization.[54]

In the United States today, milk-borne cases of tuberculosis are rare compared to the incidence of the disease during the years when milk was not pasteurized and when tubercular cows were not eliminated from the herds.

Brucellosis:

Brucellosis is the name applied to the disease caused by members of the group or organisms known as the *Brucella*. They first were isolated in 1886 from cases of fever on the Island of Malta by Dr. Bruce, a British army officer. In man, brucellosis is also called Malta Fever, Mediterranean Fever, and Undulant Fever. In cattle the disease is often called Bang's Disease, or Contagious Abortion. Usually it is spread by contact with infected material, or by consumption of raw milk from diseased animals.[42]

In the United States, especially in the Southwest, goats infected with *brucella melitensis*, have been carriers of brucellosis. *Brucella abortus*, the organism that causes the disease in cows can be a source of brucellosis in man in the United States, but B. suis is the most common cause in this country, especially among slaughter-house workers. Brucellosis is not a medical problem since modern therapy, using antibiotics, is effective in human beings. *Brucella suis* is more infective to man than *Brucella abortus* and the infection may be transferred by milk containing relatively few of the porcine organisms, whereas milk more heavily contaminated with *Brucella abortus* may not transmit the infection.[6]

Since 1951, only four outbreaks, involving fifteen cases, of milk-borne brucellosis have been reported in the United States. The disease may be of wider occurrence in man than commonly is recognized.[24] The possibility of infecting man is increased by the occurrence of the disease in hogs and cattle.[28] Errors in diagnosis may conceal the infection because the symptoms include fatigue, headache, arthritic pains, and night sweats. The disease normally persists for one to three months, but it may last for several years with intervals of normal health between attacks. As often occurs with milkborne disease, brucellosis most commonly is found in rural areas where raw milk often is used. The causative organisms are destroyed easily when milk is pasteurized.

The Brucella Ring Test,[43] is a rapid screening test to detect the presence of brucellosis in a dairy herd. Milk from animals with the disease contains antibodies for the infecting organism, which are adsorbed on the fat globules. A specially prepared, dyed suspension of *Brucella abortus* is added to a sample of the milk to be tested, and the mixture is incubated for one hour. If antibodies are present for *Brucella abortus*, the stained cells are agglutinated and rise with the fat globules, giving a colored cream layer. If no antibodies are present, the stained cells remain suspended in the milk and the cream layer is not colored. This test is so sensitive that the presence of a few infected animals may be detected by testing the mixed milk of a herd of 150 cows. A positive test does not always indicate the presence of infection, because animals vaccinated against brucellosis may give more or less positive reactions and the presence in the milk to be tested of colostrum from a healthy cow will give a positive test. Severe mastitis may also cause a suspicious test.

As with tuberculosis, programs for eradication by areas are in effect to eliminate brucellosis in dairy herds. Infected cows are branded and slaughtered. In the testing program, if not more than one percent of the cattle nor more than five percent of the herds is infected, the area may be declared a modified, certified, brucellosis-free area for a period of three years. After three years the area must be retested in order to be recertified.[47] Vaccination of calves with a special strain of brucella vaccine is also done to control the spread of brucellosis.

The *whey test* will give a positive reaction in the case of a virulent brucellosis infection that has localized in the udder or the adjacent lymph nodes. This test, done in the manner of the plate agglutination test, is made on the whey obtained by coagulating the sample of milk by the addition of rennet.[43]

Leptospirosis:

Leptospirosis is a spirochetal infection which is found throughout the world, and which now is being detected more frequently by modern diagnostic methods. In man it is known as Weil's disease, or infectious or hemorrhagic jaundice.[6] Mild attacks resemble influenza and the illness also has been confused with typhoid fever. The fatility is low. Various species of the genus *Leptospira* have been associated with the disease and are found widespread in cattle, dogs and rats.[14,32] A common source of infection for man is believed to be through abrasion of the skin upon contact with stagnant water that has been contaminated with the urine of infected animals. The *Leptospira* have been found in the milk of diseased cows. There is some doubt that such milk is a source of the disease in man. As milk has a lytic action on the organisms, they are gradually disintegrated.[29]

Q Fever (*Coxiellosis*):

Q fever is a pneumonia-like disease of rickettsial origin. The rate of mortality is negligible. The causative organism is *Coxiella burnetii*, formerly called ***Rickettsi burnetii***, which is similar to the ones that cause typhus and Rocky Mountain spotted fever. The disease first was reported in Australia in 1937, but since then it has been found to be world-wide. In recent years over 1000 cases have been reported in California, where in 1968, it was found in raw Certified Milk. The dairies concerned were not allowed to sell the milk unless it was properly pasteurized. In 1951 it was reported that ten percent of the dairy cows in Los Angeles county were infected with C. burnetii and butter made from unpasteurized milk from an infected herd was itself infected. In 1962, the milk of 84% of the cows in 50 seroreactive herds in Wisconsin were infected with C. burnetii. Human cases of Q fever usually arise through exposure to infected livestock or contaminated premises. Human volunteers who consumed infected milk did not experience illness.

A sensitive skin test has been developed to detect possible infections of Q fever. In animals, the disease is not evident because they show practically no clinical symptoms. Coxiella burnetii is one of the organisms most resistant to pasteurization, and proper relationships of time and temperature must be carefully observed to ensure its destruction. Pasteurization at 60°C (140°F) was found to be inadequate and the requirements have been raised to 62.8°C (145°F) for 30 minutes or to 71.1°C (161°F) for at least 15 seconds. An additional 2.7°C (4.9°) has been recommended for pasteurization products such as ice cream mix and chocolate milk.

Diseases from Contamination of Milk by an External Source

The contamination of milk, milk products, or milk-handling equipment by persons recovering from an infectious disease, or acting as "carriers" of the disease, is perhaps the most frequent cause of milk-borne disease. In such cases infection by salmonella and staphylococci is most common.

FOOD POISONING FROM DAIRY PRODUCTS

Food poisoning, which must be distinguished from *foodborne infection*, is the term generally applied to the comparatively rapid onset of gastrointestinal disturbances caused by the consumption of food contaminated with a poisonous substance. This may be of bacterial origin or due to the presence of toxic substances of other origin. Thus certain kinds of mushrooms are poisonous, and at times some shellfish contain toxic substances. The allergic reaction or intolerance to milk which occasionally is found among individuals, especially children and infants, should not be confused with food poisoning as discussed here. Some persons may not be able to tolerate milk or any preparation made from it, such as cheese, butter, or ice cream. This allergy often is associated with intolerance to lactalbumin and lactoglobulin.[55] These milk proteins are rendered insoluble when milk is heated sufficiently, as when evaporated milk is made. The insoluble proteins are inert and relatively inactive in producing an allergic reaction.

Persons allergic to cow's milk may be able to use goat's milk for a time, but within a month or so, the milk allergy often again becomes evident. Lactose intolerance is discussed in Chapter 4.

One authority states that the gastrointestinal symptoms in infants and children, which often are attributed to an allergy to cow's milk, actually have their origin in a gluten-induced intestinal disturbance.[12]

Food poisoning may occur from the consumption of contaminated milk or milk product, but not infrequently, this poisoning may have its real source in other foods consumed at the same time as the milk. An outbreak of food poisoning often is associated with the use of foods contaminated with organisms of the *Salmonella* group or the staphylococci which have been discussed in previous sections of this chapter.

Botulism, caused by consumption of food containing the toxin formed by the organism *Clostridium botulinum*, has rarely, if ever, been associated with fluid milk; but it has a very few times been caused by infected cheese and canned milk.[9]

In 1971 only seven cases of food-borne illness were associated with dairy products in the United States. One case was attributed to E. coli in cheese, one to a staphylococcus contamination, and one to a streptococcus infection. The others were of unknown origin.

MICROORGANISMS OTHER THAN BACTERIA

Molds

Unlike the single-celled bacteria and yeasts, the molds are much more complicated in their structure. Usually their growth starts from a spore, which in this case is a true means of reproduction, unlike the spore of a bacterium. When the spore germinates it sends out a sprout that grows into a threadlike filament known as a *hypha* (plural, *hyphae*) from the Greek word for *web*. The hypha branches out to form other hyphae, so that finally a tangled, fluffy mass is formed, known as the *mycelium*, to which old and new cells usually are attached. Mycelium is derived from the Greek work for mushroom. Botanists classify molds and mushrooms in the same family, the

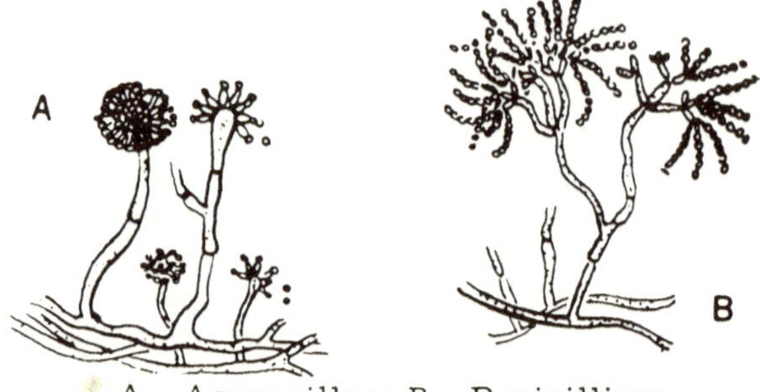

Fig. 7-7 Mold Forms about X2000. (Courtesy Univ. of Minnesota)

fungi. Molds produce both sexual spores, called *conidiospores*, and asexual spores, known as *sporangiospores*. The mature conidiospores, growing on the hyphae, often give the mold a distinctive color. *Penicillium* molds often are blue-green; *Alteneria* and *Cladosporia* are darkgreen to brown or black; *Rhizopus* species are black; and the *Aspergillae* produce black, brown, orange, or green growth.

Some oriental food products, such as soy sauce are made by the use of molds. Rhizopus is a common mold on bread. The Aspergillae are wide spread and some species are used in industrial fermentations to produce citric and gluconic acids.

Practically all molds are aerobic and can grown over a wide range of acidity and temperature. Mildew is a familiar form of moldy growth. Molds may grow on the surface of dairy products, sometimes in a rounded, button-like form, and frequently they cover the surface with a more or less complete layer. As the mold develops, it pushes some of its hyphae into the mass of the product on which it is growing, much as a plant extends its root system in order to obtain food. Other, external, hyphae develop spores, which are the reproductive bodies or fruits. The spores are microscopic in size and are found in the air, soil, manure, and on feedstuffs and unsterilized equipment.

Molds need a considerable supply of moisture for growth and, like the yeasts, they prefer to grow in a slightly acid medium. The acid formed in milk by the action of bacteria may reach a concentration unfavorable to the further growth of bacteria. This acid may serve as food for molds and yeasts. As they consume the acid, the decrease in acidity again favors the growth of bacteria. The bacteria that predominate in milk at this time are not acid-formers but are of the type that favor decomposition. This change from acid-forming bacteria to molds and yeasts and then to putrefactive bacteria is known as the *fermentation cycle*. Raw milk held for some time at room temperature usually undergoes this type of spoilage.

Molds are undesirable in most dairy products because they injure flavor and produce a musty odor. When growing on the surface of butter and cheese, they make the product unsightly and indicate that it is old or has not been stored in a clean place. Moldy cheese usually is edible; the growth is on the surface and when scraped off the product is still wholesome. Sometimes the growth of mold travels along cracks into the interior of ordinary cheese and so may ruin the product.

Certain kinds of molds are essential to the manufacture of Roquefort, Blue, Gorgonzola, Stilton, and other special types of cheese. The mold is encouraged to grow in these cheeses because it contributes to their characteristic flavor. Often a mold culture is added during the manufacture of Roquefort and Gorgonzola type of cheese and holes are made in the body of the cheese in order to admit enough air to enable the mold to grow. Camembert and Brie cheeses are covered with a layer of mold which grows during the ripening of these cheeses.

A mold that differs from many other types often is found growing on the surface of old cream and sour milk. This mold, *Geotrichum candidum* (formerly known as *Oospora lactis or Oidium lactis*), after prolonged growth, covers the surface upon which it grows with a wrinkled, velvety, and almost colorless mat or film such as is found on the surface of Camembert cheese. As it is destroyed by pasteurization, like other molds, the presence of *Geotrichum candidum* in pasteurized cream or butter may indicate improper processing.

Aflatoxin

Some molds can produce substances which are very toxic to most animals. In 1960, the feeding, in England, of moldy peanut meal to ducks and turkeys resulted in the death of many of them. It was found that the toxic substances, named *aflatoxin*, was produced by some strains of *Aspergillus flavus.*[1] Five aflatoxins are known, of which B_1 is most toxic; 0.5 to 1.0 mg per kilogram of body weight is sufficient to kill some animals. Moldy feed fed to cattle may result in their death. Low concentrations of the toxin may result in the formation of liver lesions. A food containing the toxin is considered to be potentially hazardous to the consumer.[34]

The milk from cows fed feed containing aflatoxin was found to contain the toxin in amounts sufficient to induce liver lesions in ducklings, as well as a metabolite, aflatoxin, M_1 and low levels of B_1. A comprehensive review of the mycotoxins in feedstuffs has been published.[15,34] A method that may detect as little as 0.5 ppm of aflatoxin has been devised.[61] The toxins are associated with the curd of milk, rather than in the whey. Even though aflatoxin is carcinogenic, there is no evidence to show that any trace that may be present in milk constitutes a health hazard. A survey of commercial milk samples throughout the United States detected no presence of aflatoxin M.[58] Out of 320 samples of cottage cheese curd, nonfat dry milk, and evaporated milk, 7.5% showed that they were made from milk that contained between 0.05 and 0.4 parts per billion of aflatoxin. The Food and Drug Administration has set 20 ppb as the action guide line for aflatoxin in food.[59]

A study has been made to note the development of aflatoxin on cheddar cheese inoculated with different strains of toxigenic aspergillus. Under favorable conditions of growth it was found that the molds formed substantial amounts of aflatoxins. The concentration in the upper layer of cheese was considered to be hazardous to human health if the cheese were consumed. Aflatoxins were highest in the mold mycelium and lowest in the portion below the upper layer of cheese.[15,31,34]

Inasmuch as cheese is not usually held at room temperature for a period of many weeks, it is believed that the growth of toxigenic aspergilli, if present, would not be encountered. The possibility of toxin development in packaged grated cheese may exist, if the product should contain sufficient moisture to support mold growth.

Yeasts

Besides bacteria, other microorganisms may be found in milk and its products. Among these are the yeasts. They are single-celled organisms, members of the fungi. They are oval, spherical or rounded in shape. They are larger than bacteria, averaging about five microns in diameter. They generally show a structure under the microscope. Except for one group of yeasts, the Schio-saccharomyces, they do not reproduce by fissure or division as do bacteria, but usually by a process known as *budding*. When the yeast cell buds, a swelling or lump forms on its surface and gradually grows larger, finally reaching the size and shape of the parent cell. Under certain conditions, reproduction occurs by the formation of three or four spores, known as *ascospores* within the yeast cell. In time the cell bursts and each spore develops into a yeast. During its lifespan, a cell may bud some 20 to 30 times, each time from a different area on its surface.

Yeasts are microorganisms that can ferment a solution of sugar into alcohol and carbon dioxide gas. Such a fermentation is the basis of the rising of bread during the baking process. It is also essential to the manufacture of beer and wine. Yeasts are undesirable organisms in milk products except in cases in which they help to ripen certain types of cheese and fermented milk.

As yeasts are present on feedstuff, in silage, and in the air, their introduction into milk is almost unavoidable. They grow well in milk or cream held at comparatively high temperature (24°-37.8°C), and can withstand an acid environment (*p*H 3.5), but do not grow well in an alkaline medium. The yeasts are readily destroyed by pasteurization. Various members of the *Saccharomyces* genus of yeasts may be present in fermented milk products and certain types of Italian cheeses. Old cream often has a characteristic yeasty odor. Yeasts that form pink colonies on the surface of milk or cream and on agar plates are rather common.

A group of yeasts that do not form ascospores are the so-called *wild yeasts*, formerly known as *Torulae*. Of this group, *Candida pseudotropicalis* (*Torula cremoris*) and *Torulopsis sphaerica* are of special interest because they often are present in yeasty and gassy cream, in which they ferment the lactose with the evolution of carbon dioxide and hydrogen gas. Other varieties may cause bitter flavor in milk, cream, butter, and cheese.

ENZYMES

Enzymes are proteins or proteins combined with other chemical groups, often associated with a metal which serves to "activate" the enzyme. The chemical group combined with the protein is called the *coenzyme* or prosthetic group. For example, a vitamin often forms the integral part of some coenzymes. Enzymes are formed in the living

Fig. 7-8 Typical Yeast Forms, about X2000.

body in which they specifically accelerate a particular chemical action.[45] Life would not be possible were it not for the various chemical transformations, such as the conversion of foodstuffs into the various substances that compose the body cells, brought about within the cells by enzymes. In the laboratory, strong chemicals and high temperatures and pressures are needed to bring about the reactions which small amounts of enzymes effect at body temperature. The enzyme, itself, remains essentially unchanged and it is capable of repeating its work almost indefinitely.

Each enzyme is generally very specific in its action. The compound upon which it acts is known as the substrate. The name of the enzyme usually is derived from that of the substrate together with the suffix *-ase*. Thus, *phosphatase* acts upon organic phosphates; *lipase* upon lipids (fats). Like other proteinaceous material, enzymes are denatured or inactivated when heated above some critical temperature, usually about 80°C or more. Their activity normally is confined to a narrow range of *p*H.

A number of enzymes normally are present in milk. As such, the milk enzymes have no known function in nutrition. The enzymes in raw milk must be considered in handling the product, as they may influence undesirable changes in flavor. A summation of the enzymes in milk is given by K. M. Shahani.[41]

Amylase

Two forms of amylase, *alpha* and *beta*, are known. The *alpha* form is found in milk, the *beta* form in plants, microorganisms, some milks, saliva, and in various body organs.[18] Amylase hydrolyzes starches, producing a sugar. The amount found in milk is variable and is highest in mastitic milk. It is most active at *p*H 7.4 and is inactivated when milk is heated at 45° to 60°C for 30 minutes. At one time it was proposed to use the amylase in milk as an index of the efficiency of pasteurization, but the indefinite temperature of inactivation makes this test of no value.[17]

An enzyme preparation containing an amylase of bacterial origin and a protease has been used to prevent the formation of oxidized flavor in milk in regions where such additions are legally permitted.

Catalase

Catalase accelerates the decomposition of hydrogen peroxide into oxygen and water. It is found in various body tissues, especially in liver and red blood cells. Commercial preparations are made from the liver of animals. It is most active at *p*H 7 and 0-10°C. When heated above 45°C, it loses its activity rapidly and becomes inactive at 65°C.

Catalase is used to remove hydrogen peroxide from milk to which it had been added to destroy bacteria (see Chapter 19). The normal content of catalase in milk is low, but increased amounts are present in colostrum and mastitic milk as well as in milk of high bacterial content.

Upon separation, much of the catalase appears in the cream. Separator slime has much catalase activity, probably due to its high somatic cell content.

The Catalase Test is used to help detect abnormal milk based upon the somatic cell content.

Lipase

Enzymes that hydrolyze fats are called *lipases*. They are present in many organs, microorganisms, and plant seeds. Under certain conditions, the lipases present in milk will liberate the fatty acids from milk fat.[45] The characteristic rancid flavor and odor of some of the fatty acids is desirable in some kinds of cheese but very undesirable in milk and cream.

Milk, usually condensed or dried in which part of the fat has been hydrolyzed by lipase, is used in the manufacture of milk chocolate. The free fatty acids impart a desirable and distinctive flavor to chocolate confections as well as tending to reduce the "sugar burn" (excessive sweetness) in these products. Depending upon its concentration, varying from about 1 to 5%, lipase-modified milk may impart a butter or cheese-like flavor to a food product.

The activity of lipase in milk decreases slowly at a low temperature (5°C) and the enzyme is destroyed at the temperature of pasteurization. The greatly increased number of fat globules in homogenized milk and their much greater surface area make homogenized milk readily susceptible to the action of lipase with subsequent development of rancidity. To prevent this, milk must be pasteurized before or immediately after homogenization.

Certain milk-processing operations make raw milk susceptible to induced lipolysis. Homogenization has been mentioned as a cause, but agitation and foaming, variations in temperature—such as cooling, warming to 30°C, and again cooling to 5°C—also induces this reaction. Warm milk should not be added to cold milk, as may happen if improperly cooled milk is added to a farm tank. The agitation of milk when pumped through pipe lines may accelerate activity of lipase. This occurs especially in the risers, the ascending parts of the pipe lines, where air may pass through the milk and cause foaming.

Certain bacteria and molds produce a lipase that can cause rancidity in milk and its products. Lipolytic organisms that may be present in milk include *Pseudamonas fragi, Ps. fluorescens, Candida lipolytica, Geotrichum candidum*, and *Penicillium roqueforti*. The last named mold is used in the manufacture of Roquefort and related types of cheese. If lipolytic organisms are present in milk they are destroyed when the product is pasteurized and rancidity is prevented. Pasteurization will not remove a rancid flavor if it is present in the raw product.

Phosphatase

Phosphatases constitute a group of enzymes that hydrolyze certain organic phosphates. Several phosphatases are present in milk. The principal one is an alkaline agent, most active at *p*H 9.65; to a much lesser extent, an acid phosphatase is present, active at *p*H 4. In milk, most of the alkaline phosphatase is concentrated on the surface of the fat globules; the acid phosphatase predominates in the skim milk.

As the acid phosphatase is relatively stable to heat it must be heated to 88°C for 30 minutes for inactivation. The alkaline phosphatase is inactivated at temperatures used to pasteurize milk and a number of procedures have been devised to correlate the destruction of phosphatase with proper pasteurization. The presence of more than a certain minimum amount of phosphatase after treatment by heat is regarded as evidence of improper pasteurization, or of contamination with raw product. Under some conditions, a reactivation of phosphatase occurs in milk and cream that has been pasteurized by a high-temperature short-time method. A description of the phosphatase test is given in Chapter 21.

When present in very large number, some microorganisms, such as *Lactobacillus enzymothermophilus*, produce a phosphatase in milk when it is heated to 143°F for 30 minutes. Bacterial phosphatase is more heat-stable than milk phosphatase and may be so distinguished in tests for pasteurization.

Peroxidase

Peroxidase, which is found in a number of natural substances, is one of the more heat-resistant enzymes in milk. In the presence of hydrogen peroxide, it catalyzes the oxidation of many organic substances. The Storch Test (see Chapter 21) is used in Europe as a test for heated milk. Peroxidase is destroyed when milk is heated to 75°C for five minutes. As milk must be heated to 70°C for about two and one-half hours to destroy the enzyme, it is not usable as a test for pasteurization.[3]

Other Enzymes

Among the enzymes found in milk are *xanthine oxidase, protease,* and *aldolase.* Xanthine oxidase, sometimes called *Schardinger enzyme,* is an oxidizing enzyme. It is also known as *reductase* owing to its ability to reduce or decolorize methylene blue in milk in the presence of formaldehyde. Milk must be heated to about 80°C to destroy this enzyme. Usual pasteurization temperatures inactivate only about 60% of the enzyme.

The enzyme is associated with the fat globules in the milk of ruminants (cows, ewes, and goats) but not in human, mare, or sow's milk. Homogenization, by decreasing the size of the fat globule and increasing their surface area, favors the ability of xanthine oxidase to pass the intestinal mucosa and reach the bloodstream. The hypothesis is advanced that the biochemical changes caused by xanthine oxidase activity favor the development of atherosclerosis. Non-homogenized milk is assumed to be inactive in this regard.[56] Xanthine oxidase is reported to attack aldehydes that are essential to maintenance of artery wall elasticity. In raw milk and non-homogenized milk the enzyme is digested in the intestinal track and is regarded to be inactive.[56]

Protease, which slowly decomposes milk proteins, was at one time named *galactase.* It was believed to be important in the ripening of cheese but modern investigations minimize its role, because bacterial activity is much more important in the ripening of cheese.

Aldolase is an enzyme involved in the metabolism of carbohydrates.[38]

Two important enzymes, *pepsin* and *rennin,* are discussed in Chapter 20.

CLEANING AND SANITIZING OF DAIRY EQUIPMENT

In order that the bacterial count of the product be as low as possible it is essential that all dairy equipment be kept clean, sanitized, and dry after being used. In the dairy industry, sanitizing means the cleaning of the equipment and, as completely as possible, the destruction of the microorganisms that may be present. To meet this requirement, dairy equipment is constructed to facilitate cleaning and sanitizing operations.

Specifications for various items such as cans, tanks, pumps, homogenizers, piping, and equipment for pasteurization are formulated by a group of public health officials and representatives of the U.S. Public Health Service and the Dairy Industry known as the 3A *Sanitary Standards Committee.*

Detailed tests to check the sanitation of equipment and containers are given in the *Standard Methods.*[43] The steps followed in a cleaning procedure are as follows:

1. A rise with water at 49°C. or higher in order to remove as much milk residue as possible.
2. Actual cleaning with an alkaline detergent or acid detergent.
3. Another rinse with warm water to remove loosened soil.
4. Application of a sanitizing solution.
5. A potable-water rinse, if required.

The effectiveness of any cleansing operation depends upon the surface to be cleaned, and the type of soil to be removed. Hot water, at a temperature of at least 82°C, and applied for not less than five minutes may be used to sanitize equipment. Whenever it is possible, the equipment should be immersed in such water. Live steam is a very satisfactory sanitizing agent. If applied in a closed container, such as a covered vat, steam sterilization will destroy non-spore-forming bacteria, yeasts, and molds. After this treatment, the equipment should be left to dry because moisture favors the germination of spores and the growth of surviving bacteria.

Pipe lines made of stainless steel or heat-resistant glass are widely used in dairy houses and processing plants to carry milk to storage or to processing equipment. These lines are so constructed that they need not be dismantled for cleaning and sanitizing, but are *cleaned-in-place*, constituting the CIP system, sometimes called "circulation cleaning." In this procedure, after the day's use, the pipes and all pieces of equipment are rinsed with warm water, and then a hot (60°-71°C) detergent solution is pumped continuously through them for at least 30 minutes. This time is needed to ensure removal of all traces of milk residue. All areas that come in contact with the product are cleaned and sanitized by programmed, automatic controls. Equipment used for cold milk is much easier to clean than that in which milk is heated.

The surfaces of milk-processing equipment may gradually acquire a deposit of milk solids. This material, known as *milkstone*, forms at temperatures favorable for the coagulation of milk albumin, but denatured casein, calcium, and phosphorus compounds together with small amounts of fat usually are also present.

In CIP systems, tanks are cleaned with permanently installed spray-nozzle devices (see Fig., 11:2). Plate-type pasteurizing equipment is cleaned by continuous re-circulation of the detergent solution, flowing at least four times the velocity used for milk. Separators and clarifiers are dismantled for cleaning manually.

The detergent solution may be reused several times with fresh detergent added to maintain the proper concentration. The cleanser should have wetting action, that is, its surface tension must be low enough to enable it to penetrate between the soil and the surface to be cleaned. For this purpose cleansers usually contain a non-foaming biodegradable anionic detergent, such as one of the ethylene oxide condensates.

Sodium hydroxide, sodium carbonate, sodium metasilicate, and tri-sodium phosphate are among the ingredients of most alkaline washing compounds. These compounds saponify and emulsify any fat present and also disperse and suspend the soil so that is does not redeposit. The silicates also provide corrosion inhibition. Some alkaline compounds tend to form deposits on equipment, especially if the water supply is hard. In order to keep scale-forming calcium and magnesium salts in solution, cleaning compounds often contain a chelating or sequestering agent, such as tetrasodium pyrophosphate, sodium tripolyphosphate, sodium hexametaphosphate, or sodium ethylenediamine tetraacetate. Sodium gluconate is used to prevent iron salts from forming deposits when highly alkaline cleaners are used.

Every fourth day the equipment is cleaned with an acid type cleanser in order to remove any adhering residues of alkaline detergent and to ensure the removal of firmly adhering traces of heated milk or milkstone. Phosphoric acid, gluconic, hydroxyacetic, and citric acid are used in the acidic cleaners.

Chemical Sanitization

Even though no single chemical compound is completely serviceable for every application, satisfactory and economical results may be obtained with chemical sanitizing agents, especially in places where it is not practical to apply heat, such as on the surface of large vats. As with any chemical reaction, chemical sanitization is influenced by the temperature, pH, and the length of time of contact.

Alkaline solutions, such as those made with lye or caustic soda, have definite germicidal properties. Gram-negative rods are destroyed readily by strong alkaline solutions, while Gram-positive organisms are somewhat more resistant. Alkaline solutions are corrosive to many metals used in dairy equipment and are hard to rinse off completely. A 0.5 per cent solution of sodium hydroxide (caustic soda) commonly is used for cleaning and sanitizing the rubber parts of milking machines.

Sanitizing solutions that contain chlorine compounds are used widely, because in the customary dilutions they are non-poisonous and have no objectional odor. In general, if hot water and steam are available, it is not advisable to depend solely on chemical sterilizers because certain factors detract from their effectiveness. The principal criticism against chemical sanitizers, and the chlorine compounds especially, is that in the presence of milk residues and other organic matter their efficiency is decreased.

Alkaline solutions of sodium hypochlorite, marketed under various trade names, are the most commonly used sanitzers. Organic compounds of chlorine, such as chloramine-T, chloramine-B, dichlorodimethylhydantoin, trichloroisocyanuric acid, and trichloromelamine are constituents of proprietary sanitizers. Preparations containing any of the last three compounds may be used in slightly acidic solutions. A solution of *n*-chlorosulfamate, made by the reaction of chlorine and sulfamic acid, is an effective sanitizing agent when used in neutral or slightly alkaline solution. The more alkaline any of these sanitizing solutions may be, the more strongly the chlorine is held in chemical combination, and the less opportunity it will have to exert its full germicidal action. A neutral or slightly acid solution is much more effective, but it loses strength rapidly and it may prove corrosive to some types of equipment.

A solution that contains 200 ppm of available chlorine will destroy bacteria on a cleaned surface within two minutes. In general, milk bottles and dairy equipment should be rinsed with a solution that contains 50 to 100 ppm of available chlorine. The use of rinsing solutions that contain up to 500 ppm will not impart undesirable odor or flavor to milk subsequently placed in the sterilized container. The amount of chlorine that ordinarily would be retained has no practical effect in reducing the bacterial content of milk. If sufficient chlorine is present to reduce the bacterial count, the milk will develop a distinctive odor and flavor. Traces of sodium chlorate are present in most hypochlorite solutions and tests for the presence of available chlorine in milk actually detect the chlorate. In England, hypochlorite solutions used for sanitation in dairies must contain not less than 0.7% of sodium chlorate, in order to facilitate detection of the sanitizing agent should it be present in a milk supply.

Cleaning-in-place may in time leave a very thin film of proteinaceous material on the surface of the equipment. Alkaline cleaning solutions that contain between 25 and 100 ppm of available chlorine have been found useful in removing this film. The chlorine probably degrades the protein residue and makes it more soluble in the cleaning solution. Chlorinated cleansing compounds also help to brighten the surface of stainless steel equipment by removing any oxide film that may be present.

Iodine-containing compounds, known as *iodophors*, have not attained the popularity of the chlorine sanitizers. These compounds contain an acid, usually phosphoric, iodine, and a nonionic surface-active agent which increases the solubility of the iodine in water and stabilizes the solution. About 80% of the iodine present is germicidally active. Solutions containing up to 50 ppm of available iodine are used in the same manner as the chlorine sanitizers. Solutions that contain less than ten ppm of iodine are ineffective Such solutions are detected easily by lack of color. As little as 8 ppm of an iodine compound has been reported to give an off-flavor to milk. Some observations indicate that the use of iodophors over a period of time may produce a medicinal odor in the milk house. White tile and paints may become discolored, probably through vaporization of iodine from the sanitizing solution as it is used.

A mixture of sodium hypochlorite and hypobromite, formed when tri-sodium phosphate, sodium hypochlorite, and potassium bromide are dissolved together has been found to be an efficient sanitizing agent.

Quaternary Ammonium Sanitizers

The quaternary ammonium compounds that are used as sanitizing agents in the dairy industry may be considered as derivatives of ammonium chloride in which the four hydrogen atoms are replaced by organic groups, one of which usually has 8 to 18 carbon atoms. The active portion of the molecule consists of the ammonium nitrogen atom with the attached organic groups and this constitutes the cationic—positively charged—part of the molecule. In contrast to the quaternary ammonium compounds, the active portion of a soap, or synthetic detergents such as the surface-active sulfonates and sulfated alcohols, is anionic and carries a negative charge. The quaternary ammonium compounds or "quats" are therefore not compatible with anionic compounds and only nonionic detergents can be used with them.

A solution that contains at least 200 ppm of quaternary ammonium compound is generally used for sanitization in dairies. Such a solution is non-toxic, odorless, stable to heat, and to the presence of milk residues. As a rule, gram-negative bacteria are more resistant to the "quats" than are gram-positive organisms. The cocci and clostridia are among the least resistant. The efficiency of the various quaternary ammonium bactericides varies widely and is affected by the mineral content of the water in which they are used, especially if it contains more than 400 grains of hardness. Only those quarternary compounds whose sanitizing properties are not materially reduced in the presence of hard water should be used for sanitization in dairies. Most of the alkyl dimethylbenzylammonium chlorides may be used.

The quaternary ammonium sanitizers may be used in alkaline solution in combination with nonionic detergents, in which case they are more effective than the sodium hypochlorite sanitizers in the destruction of Gram-positive bacteria.

If enough of a quaternary ammonium solution should be added to milk to act as a preservative or to reduce the bacterial count, its presence may be detected by the bitter flavor produced in the milk. There has been some objection to the use of quaternary ammonium sanitizers because traces may find their way into milk through improper rinsing of dairy equipment. As little as five ppm have been found to interfere with the development of flavor, the aroma, and the production of acid in cultures used for the preparation of fermented milk products and in the manufacture of cheese.

Acidic Sanitizers

Dairy equipment can be effectively sanitized with solutions that contain phosphoric acid and an anionic surfactant. A mixture that contains about 30% phosphoric acid and 5% of an alkylaryl surfactant, such as dodecylbenzene sulfonic acid, is representative of this type of sanitizer. A solution containing 200 ppm of sanitizer is commonly employed. Such a solution is not affected by the hardness of the water used for dilution and is relatively stable in the presence of organic matter. The acid content helps to control the formation of milkstones.

Checking Sanitary Conditions of Equipment

The Standard Methods [43] describes a number of procedures to determine the sanitary condition of dairy equipment. One modification makes use of samples of milk taken progressively at all access points along the lines. The points of entry of specific organisms can be determined by this method, employing appropriate differential media and incubation temperatures. Samples may be plated directly on nutrient media to determine where significant changes in total counts begin to occur.

REFERENCES

1. Allcroft, R. and Carnaghan, R. B. A., Ground-nut Toxicity. *Aspergillus flavus* toxin (aflatoxin) in animal products. *Vet. Record*, 74:863 (1962).
2. Auclair, J. E. and Berridge, N. J., The inhibition of micro-organisms by raw milk, *J. Dairy Res.*, 21:45-59; 370-374 (1954).
3. Aurand, L. W., Roberts, W. M., Cardwell, J. T., A method for the estimation of peroxidase activity in milk, *J. Dairy Sci.*, 39:568 (1956).

4. Buchbinder, L. and Alff, E. O., Studies on coliform bacteria in dairy products, *J. Milk Food Technol.*, 10:137 (1947).
5. Burgwald, L. H. and Josephson, D. V., The effect of refrigerated storage on the keeping qualities of pasteurized milk, *J. Dairy Sci.*, 30:371 (1947).
6. Cecil, R. L., *Textbook of Medicine*; W. B. Saunders, Philadelphia, 1945.
7. Crouch, F., Trembles (or Milk Sickness), *U.S. Dept. Agr. Circ.*, 306 (1933).
8. Cunningham, O. C. and Addington, L. H., Tuberculosis in goats, *J. Dairy Sci.*, 19:435 (1936).
9. Dack, G. M., *Food Poisoning*; Univ. Chicago Press, Chicago, (1959).
10. Dauer, C. C. and Sylvester, G., 1954 Summary of disease outbreaks, *Public Health Repts.*, 70:536, (1955).
11. Dept. of Health, Education & Welfare, Food & Drug Administration, Cincinnati, Ohio, Screening and Confirmatory Tests For The Detection of Abnormal Milk (1970).
12. Di Sant' Agnese, P. A., and Jones, W. O., Malabsorption (celiac disease) in pediatrics, *Borden's Review of Nutrition Research*, 22:No. 3 (1961).
13. Enright, J. B., Sadler, W. W., Thomas, R. C., Thermal inactivation of coxiella burnetii and its relation to pasteurization of milk, *Public Health Monograph* 47; U.S. Public Health Service, Washington, D.C., 1957.
14. Ferguson, L. C. and Bohl, The Leptospiral diseases of animals. *Proceeding* of *90th Annual Meeting*, Amer. Vet. Med. Assoc. (1953).
15. Frank, H. K., Diffusion of Aflatoxins in Foodstuffs, *J. Food Science*, 33:98 (1968).
16. Frayer, J. M. The influence of delayed cooling upon the quality of milk, Vermont Agr. Exp. Sta. Bull. 3.3 (1930).
17. Gould, B. S., The detection of inefficiently pasteurized milk based on a modification of the new Rothenfusser test. *J. Dairy Sci.*, 15:230 (1932).
18. Guy, E. J. and Jenness, R., The separation, concentration and properties of alpha amylase from cow's milk. *J. Dairy Sci.*, 41:13 (1958).
19. Harmon, L. G. Staphylococci: Its Significance to the Dairy Industry. *Amer. Dairy Rev.* April, 1967.
20. Current concepts of bovine mastitis, The National Mastitis Council, Inc., Washington, D.C., 1965.
21. Harrington, R., Jr. and Karlson. A. G., Destruction of various kinds of mycobacteria in milk by pasteurization. *Appl. Microbiol.* 13:494-495 (1965).
22. Herrington, B. L. Lipase; a review. *J. Dairy Sci.*, 37:775 (1954).
23. Hird, E. W., Weckel, K. G., Allen, N. N., The effect of clipping the udders of cows on the quality of milk, *J. Dairy Sci.*, 31:523 (1948).
24. Huddelson, J. F., *Brucellosis in Man and Animal* ; The Commonwealth Fund, New York, 1943.
25. Jones, F. S. and Little, R. B. The bactericidal property of cow's milk, *J. Exptl. Med.*, 45:319 (1927).
26. Jones, F. S. and Simms, H. S., Bacterial growth inhibitor (lactenin) in milk, *J. Exptl. Med.*, 51:327 (1930).
27. Jones, K., and Stevens, S. O., Bacteria in aseptically drawn milk, *Proc. Utah Acad. Sci.*, 18:9 (1941).
28. Kaplan, M. M. et al. Diseases transmitted through milk. Milk Hygiene, World Health Organization, Geneva (1962).
29. Kirscher, L. and Maguire, T., Antileptospiral effect of milk, *Brit. J. Exp. Path.*, 38:357 (1957).
30. Lewis, K. H. and Angellotti, R. Examination of foods for enteropathogenic and indicator bacteria, U.S. Public Service, Pub. No. 1142, Washington, D.C. (1964).
31. Lie, Jennie and Marth, E. H., Formation of aflatoxin in cheddar cheese, *J. Dairy Sci.*, 50:1708 (1967).

32. Little, R. B. and Baker, J. A., Leptospirosis in cattle, *J. Am. Vet. Med. Assoc.*, 116-105 (1950).
33. Macy, H. and Erekson, J. A., Seasonal variations in thermoduric organisms and methods of control, *Assoc. Bull.*, 34:127; International Assoc. Milk Dealers (1941).
34. Marth, E. H., Aflatoxin and other mycotoxins in agricultural products, *J. Milk and Food Technol.*, 30:192 (1967).
35. Marth, E. H., Salmonella and salmonellosis associated with milk and milk products: A review. *J. Dairy Sci.*, 55:283-315 (1969).
36. Milk sickness; a retrospect, *J. Am. Med. Assoc.*, 128:734 (1945).
37. Morris, C. S., The presence in raw milk of bacterial substance specific for certain strains of coliform organisms and the comparative rate of growth of bacteria in raw and pasteurized milk, *Dairy Ind.*, 10:180 (1945).
38. Polis, B. D. and Shmulker, H. W., Aldoase in bovine milk, *J. Dairy Sci.*, 33:619 (1950).
39. Read, R. B., Prichard, W. L., Bradshaw, J. and Black, L. A., In vitro assay of staphylococcic enterotoxins A and B in milk, *J. Dairy Sci.*, 18:411 (1965).
40. Ruehle, G. L. A. and Kulp, U. L., Germ content of stable air and its effect upon the germ content of milk, *New York State Agr. Expt. Sta. (Geneva), Bull.* 409 (1915).
41. Shahani, K. M. et al, Enzymes in Bovine Milk: A Review, *J. Dairy Sci.*, 56: 531-543 (1973).
42. Spink, W. W., *The Nature of Brucellosis;* Univ. Minnesota Press, Minneapolis (1956).
43. *Standard Methods for the Examination of Dairy Products*, 13th Ed.; Am. Public Health Assoc., Washington, D.C. (1972).
44. *Standard Methods for the Examination of Water and Wastewater*, 11th Ed., American Public Health Assoc., New York, (1960).
45. Summer, J. B. and Nyrback, K., *The Enzymes, Chemistry and Mechanisms of Action*; Academic Press, New York, (1951).
46. Thomas, S. D. et al, Incidence and significance of thermoduric bacteria in farm milk supplies: A reappraisal and review, *J. Appl. Bact.* 30:265-268 (1967).
47. *Brucellosis Eradication, Recommended Uniform Methods and Rules,* Animals and Plant Inspection Service, U.S. Dept. of Agriculture, Hyattsville, Maryland, April 1972.
48. *Summary of Milk-borne Disease Outbreaks During the Years* 1923-1959, *Inclusive*, Dept. Health, Education and Welfare, U.S. Public Health Service, Washington, April, 1961.
49. *Uniform Methods and Rules for the Establishment and Maintenance of Tuberculosis-Free Accredited Herds of Cattle and Modified Accredited Areas*, Bur. Animal Ind., U.S. Dept. Agr., Washington (1950); and *Recommendations for Brucellosis-Free Areas, Ibid.* (1952).
50. Wentworth, Bertinna B., Historical review of literature of Q fever, *Bacteriol. Rev.*, 19:129-149 (1955).
51. Witter, L.D., Psychrophilic bacteria—A review, *J. Dairy Sci.*, 44:983-1015 (1961).
52. Wuethrich, S., Richterich, R. and Hostettler, H. Investigations on Milk Enzymes—Enzymes in Cow's Milk and Human Milk. *Z. Lebensm. Untersuch. Forsch.* 124:336-344 (1964).
53. Zottola, E. A., Salmonellosis, Ext. Bul. 339, Agr. Ext. Service, Univ. Minnesota (1967).
54. Current concepts of bovine mastitis, The National Mastitis Council, Inc., Washington, D.C., 1970.
55. Spies, J. R., Milk Allergy, *Journal of Milk and Food Technology*, 36:225-231 (1973).

56. Zikakis, J. P., Homogenized Milk and Atherosclerosis, *Science* 183:472-473 (1974).
57. *Dairy Bacteriology*, Foster, Nelson, Speck, Doetsch and Olson, Prentice-Hall, 1957.
58. Brewington, C. R. and Weihrauch, J. L., Survey of Commercial Milk Samples for Aflatoxin M. *J. Dairy Sci.* 53:1509-1510 (1970).
59. *Consumer*, Food and Drug Administration, Feb. 1974.
60. Hileman, J. L., Leber, H. and Speck, M. L., Thermoduric bacteria in pasteurized milk. *J. Dairy Sci.*, 24:305 (1941).
61. Jacobson, W. C., et al, Determination of Aflatoxin B_1 and M_1 in Milk. *J. Dairy Sci.*, 54, 21 (1971).

chapter 8

Antibiotics—Pesticides Radioactivity

The dairy farmer and veterinarian use many drugs to treat diseases in dairy cattle. It is claimed that over 75 tons of antibiotic drugs are used each year to treat or prevent mastitis in the dairy cow. Penicillin is most generally used, but other antibiotics such as bacitracin, chlortetracycline, oxytetracycline, tetracycline, and streptomycin are employed. (Aureomycin is the trademarked name for a brand of chlortetracycline). Vast amounts of antibiotics also are used in medicated feeds to increase the efficiency of feed utilization and also to shorten the time needed to bring an animal to market weight.

ANTIBIOTICS IN MILK

When infused into the udder of a cow, that part of the antibiotic that is not absorbed by the tissues is gradually eliminated in the milk over a period of about three days. When infused into one part of the cow's udder, penicillin, for example, often may be found in the milk of the other quarters.[9] Owing to the difference in construction in the udders of the cow and the goat, when penicillin is infused into one-half of the udder of the goat it does not diffuse to the other half.[2]

The use of feedstuffs containing antibiotic supplements may cause them to appear in the milk, especially if the dosage fed is more than 0.1 mg per pound of body weight.[14]

Traces of penicillin in the milk consumed by a person very sensitive to the drug may produce a severe allergic reaction. Some pathogenic bacteria tend to develop a tolerance for an antibiotic. If such an organism should be present in milk or other foodstuff and cause disease, treatment with the antibiotic concerned would be ineffective.

Of direct interest to the dairy industry is the effect that traces of an antibiotic may have on various milk-processing procedures.[1] By inhibiting the growth of bacteria, the presence of an antibiotic may increase the time needed for dye reduction tests and so tend to grade milk higher than its quality may warrant.[12] The growth of lactic acid organisms is inhibited in milk that contains traces of antibiotics and so this is of concern in the manufacture of cheese and cultured milk products (see Chapter 15). As little as 0.1 unit of penicillin per ml of milk is sufficient to inhibit completely the sensitive strains of the bacteria used in starter cultures.[13] A six-fold reduction in bacterial count was found after 48 hours at 45°F (7.2°C) in raw milk that contained 0.01 unit per ml of penicillin.29

One International Unit, or 1 USP Unit, is defined as the equivalent of 0.6 microgram of pure benzylpenicillin sodium; one milligram has the potency of 1667 units.

Antibiotics in oil-based ointments, used in the treatment of diseased udders, may be detected in the milk for a week after last treatment. Antibiotic preparations used to treat cows by infusion into the udder must contain not over 100,000 units of penicillin per dose. The product must be labeled to warn that milk from the treated cow must not be used for human consumption for at least 72 hours after the last administration. Milk from cows that have been given an intramuscular injection of an antibiotic must not be used for human consumption within 96 hours after the last treatment (U.S. Food and Drug Administration, November 1960). Raw Grade A milk that contains more than 0.05 unit per ml. as determined by the *Bacillus subtilis* disc assay method is considered to be adulterated. The presence of an antibiotic in milk indicates that it originated in a diseased cow.

Concern that even as little as 0.05 unit per ml. of penicillin may be of public health concern has led to the development of a test capable of detecting 0.002 unit. It is a disc assay procedure in which the color change of a tetrazolium dye indicates the presence of the antibiotic.[22]

The heat treatment given milk during processing has so little effect in reducing antibiotic activity that it does not warrant practical consideration.[1] Many chemicals can destroy antibiotic activity but their use in milk would constitute adulteration, or add a toxic substance. The enzyme *penicillase* has been proposed for use in milk to destroy traces of penicillin, but use of it also would be adulteration even if the cost were not prohibitive.

As yet there is no quick, reliable test that is simple enough to use in the routine testing of milk for the presence of an antibiotic. In one type of test an actively growing culture of *Streptococcus thermophilus,* which is very sensitive to penicillin, is mixed with the sample of milk to be tested and an indicator such as resazurin or TTC (triphenyltetrazolium chloride) is also added. The time needed to reduce the indicator is noted and is compared to that of a control milk similarly treated but known to be free of antibiotic.[21] The test is not specific, because an antibiotic, preservative, sanitizing agent, or even a high leucocyte count in the milk, may give a positive test. In a modification of this test, the rate of growth of the test organism, in the milk under test and in the control, is measured by the rate of increase in acidity or formation of curd.

In another test, the disc assay,[28] the growth of a test organism in an agar plate culture is observed. Usually *Bacillus subtilis* is used, being grown from a standardized spore suspension. Sterile discs of filter paper, one-fourth inch in diameter, are dipped into the milk to be tested and carefully placed on the surface of the agar plate. Control discs are dipped into antibiotic-free milk and into penicillin-in-milk standards that contain 0.05 unit and 0.10 unit penicillin per ml. These discs are obtainable commercially. The plates are incubated for two and one-half to three hours at 37°C. If an antibiotic is present in the milk under test, it diffuses into the agar and, depending upon the amount present, growth around the disc is inhibited or prevented. By comparing the amount of growth obtained with that around the control discs, an estimate of the antibiotic content of the milk is obtained.

Discs impregnated with penicillinase are used to determine whether the antibiotic that may be present is penicillin. If no zone of

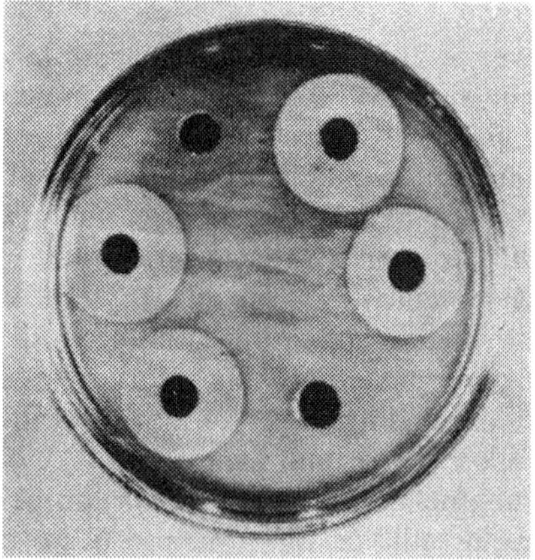

Fig. 8-1. Disc Assay for Antibiotics: White zones (no growth) indicate presence of penicillin in some samples of milk; diameter of zone measures amount of penicillin present.

inhibition appears around the penicillinase disc and one does surround the test sample disk, then penicillin is present. If a zone of inhibition surrounds both discs, an inhibitor other than penicillin probably is present. Other than the procedure with penicillin-penicillinase discs, no specific tests for other antibiotics in milk have been developed to date.

A modification of the disc assay makes use of a culture of *Sarcina lutea*, because this is more sensitive to penicillin than is *Bacillus subtilis*. Lower concentrations of penicillin may be detected, but an incubation period of 16 to 18 hours at 26°C is needed, rather than the three hours needed when *B. subtilis* is used. (Fig. 8:1)

PESTICIDE RESIDUES IN MILK

A number of chemicals are in use today in the production of agricultural commodities. They are essential to modern agriculture. The maintenance and improvement of the present nutritional status of the American people is contingent upon the continued production of an adequate food supply. Plant and animal pests rank among the foremost causes of the destruction, deterioration, and contamination of food. Hence, the absolute necessity for protecting growing crops and produce from serious attack by insects, plant diseases, and other pests is recognized as essential from the standpoint of both quantity and quality of the food produced. Since 1964, synthetic organic chemicals have been of prime importance in this area. In 1966, farmers in the U.S. spent $561 million for pesticides, mostly for use on corn and cotton. Pesticides for use on dairy cattle and dairy buildings cost $9.2 million. Some pesticidal chemicals, such as parathion, nicotine, and potassium cyanide, are very toxic to man. Others, such as pyrethrum, rotenone, methoxychlor, and TDE are relatively non-toxic.

Governmental regulations provide for the safe use of the various pesticides. Much emphasis is placed upon the *tolerance*, which is the amount of a pesticide that may remain in or on a foodstuff without causing any injury to the consumer. A specific tolerance is set for each pesticide and commodity. The Food and Drug Administration has declared a few pesticidal chemicals as safe and these do not require a tolerance. In setting a tolerance, many factors are considered, such as the amount of pesticide that may be present in or on the foodstuff, and the cumulative effect of the substance. A large safety factor is included to account for possible difference in susceptibility between test animals and the human consumer. In studies with DDT, the claim was made that about 200 times the

TABLE 8:1

Acceptable Daily Pesticide Intake from total diet. [a]

DDT-0.01 mg./kg. body weight per day.
Dieldrin-0.0001 mg./kg. body weight per day.
Lindane-0.0125 mg./kg. body weight per day.

a-Food and Agriculture Organization - World Health Organization

amount contained in the average person's diet can be consumer daily for a year with no ill effect.[10]

The maximum amount of pesticide residues that may have been derived from dairy products in the daily diet was reported to be as follows—DDT 0.004 mg., DDE 0.002 mg., TDE 0.002 mg., Dieldrin 0.002 mg. These amounts are well below the limits set forth by the World Health Organization, as shown in Table 8:1.

Much that was of concern to Rachael Carson in her book "Silent Spring" has become apparent following the extensive use of DDT. Contamination of the environment with persistent pesticides has become a hazard to fish, wildlife and possibly to human health. Stringent legislation, both Federal and State, as well as the opposition from groups with environmental interests, and the application of new techniques have all contributed to this decrease in pesticide usage. Rising costs have also had an influence in promoting the use of smaller amounts of pesticide application.

The use of chlorinated hydrocarbon pesticides has decreased markedly. The use of DDT in the United States ended in 1972, except for export sales and a limited use for public health application. The emphasis on licensing and restricted-use permits has favored a more careful application of pesticides, and decreased the possibility of their entering the food-chain. In 1971, one billion, 140 million pounds of synthetic organic pesticides were manufactured in the United States, valued at over $979 million.

Microbial and viral insect pathogens, such as B. thuringenisis and Heliothis, have found increased use. Other control measures involve the use of sterile insects, insect attractants (Pheremones) and hormones which interfere with insect growth.

The dairy farmer knows if his cows have been treated with an antibiotic drug, but may not know if the cow's feed is contaminated with a pesticide. As with the antibiotics, it is important that pesticides be used in accordance with the directions for their specific application. It is the misuse of the pesticide that created problems associated with their presence in milk. The pesticides based on chlorinated hydrocarbons have been given the most attention because they have been used in such large amounts and also because they are relatively easier to detect in milk than some other pesticides.

Chlorinated Hydrocarbon Pesticides

The pesticides based on chlorinated hydrocarbons were of special interest to the dairyman. The chlorinated hydrocarbon pesticides are fat-soluble. When sprayed on the skin of the cow, applied as a powder, or ingested through contaminated feed, most of these products are absorbed and stored in the fatty tissues of the body. Some of the pesticide is eliminated, some is metabolized by the body to some other form, and some is transferred to the body fat and the the milk fat.

Contamination of milk by inhalation of toxic vapors or by contact with dairy sprays is today of minor importance because in most States only certain non-toxic sprays are permitted for use in dairy buildings and on dairy animals. Of greater significance is the use of feed that had been treated with a pesticide or accidentally contaminated by drift from the treatment of nearby fields. The use of waste products, such as apple pomace, vegetable tops, and corn husks at one time were a serious source of contamination of milk. Although sources of contamination may ever be present, the combined efforts of the producer and the various governmental agencies are bringing about a situation in which pesticidal residues in milk are of decreasing significance.

Dairy farmers have been able to collect damages from the Federal government for milk which was destroyed because of contamination by chemicals over which the farmer had no control.

A report from the United States Department of Agriculture stated:

> When insecticides were fed to beef cattle and sheep as a contaminant of their feed, at dosages likely to occur as residues on forage crops, all except methoxychlor were stored in the fat. The order of their storage was as follows: aldrin, dieldrin, heptachlor epoxide, BHC, DDE, chlordane, lindane, endrin, heptachlor, toxaphene (listed in decreasing order).

Methoxychlor does not readily appear in the milk if present in the cow's feed, but it does quickly appear in the milk if the pesticide is applied as a spray or powder to the skin of the cow.

In a study made in Illinois, cows were fed known amounts of different pesticides for 16 weeks and the amount found in the milk was determined. DDT appeared when less than 1 ppm was in the ration, heptachlor when 4 ppm were present but methoxychlor did not appear until 640 ppm were in the ration. The writer has found that as little as 0.1 ppm of DDT in the feed will allow the pesticide to appear in the milk.

As yet, no procedure has been reported which will eliminate pesticides residues from contaminated dairy products. Pasteurization has no effect. Application of ultra-violet irradiation of milk was found to reduce methoxychlor by one-third but had no significant effect on other residues. Ordinary processing and storage also is without effect, although some loss was found to occur during drying. Prolonged, vigorous steam deodorization treatment of butteroil did reduce its content of some pesticide residues, but did so at the expense of its quality.[15]

Only fractional parts per million of certain pesticide residues are now permitted in milk. Two exceptions are a tolerance of 1.25 ppm of Methoxychlor or of Ronnel in milkfat. The tolerance for DDT is that not more than 0.05 ppm of the sum of DDT and its degradation products DDD and DDE may be present in milk or 1.25 ppm in manufactured dairy products. Although it is no longer used on or near dairy cattle, enough DDT is left in the soil, water, and air so that milk and other food products may become contaminated with a trace of DDT or its metabolites.

A total diet study in the United States (Pesticide Monitoring Journal, March 1972), of 12,989 samples of fluid milk, showed 592 contained pesticide residues. On a milk fat basis the average content of DDT was 0.03 ppm; DDE 0.07 ppm; Dieldrin 0.04 ppm. In manufactured milk products, such as butter, cheese, and ice cream, out of 6,231 samples 56.4% had an average residue content of DDT 0.03%, DDE 0.06%, and BHC (benzene hexachloride) 0.02 ppm. Imported products, of 1981 samples, 67.7% had pesticide residues, DDT 0.07 ppm, DDE 0.05 ppm, and BHC 0.776 ppm.

Pesticide residues in human milk are given in Chapter 9.

Organic Phosphate Insecticides

Some of the organic phosphate insecticides, such as parathion and TEPP, are very toxic when ingested, inhaled, or absorbed through the skin. Others, such as malathion, are much less toxic. They are not used to the same extent as the chlorinated hydrocarbon insecticides. Only by gross negligence in the use of them would they come into contact with milk or dairy products. Small amounts are rapidly metabolized by the cow should they be ingested. As a result they are not secreted in the milk and so do not present the problem associated with the use of the chlorinated hydrocarbon insecticides.

RADIOACTIVE MATERIALS IN MILK

Life on earth always has been exposed to natural radioactivity or ionizing radiation. The radioactive fallout from the testing of nuclear weapons is not now of major concern. Steadily developing programs for nuclear energy have produced radioactivity much over that which is of natural origin. The entire environment, air, water, and all foodstuffs have been affected. Affluents from nuclear power reactors and fuel reprocessing have become possible sources of radioactive contamination of food and water. Most of the ionizing radiation received by the populace today, other than that of natural origin, arises from the use of X-rays and radioactive materials used in medicine and industry. The maximum worldwide non-military use of nuclear energy that can reasonably be envisioned during the next several decades will produce an average radiation exposure much less than that already caused by weapons testing.

Under normal conditions, radiation within the body originates in radioisotope potassium-40—a normal constituent of potassium as it exists in nature—and in carbon-14, an isotope of carbon. An important natural source of external radiation is the cosmic rays from outer space.

An early belief was that, just as for toxic chemicals, there was a tolerance or dose of radiation that would not produce any deleterious effect. Many scientists now believe that even the smallest amount of exposure does some harm to living cells. Others hold that there is no evidence of deleterious effects from the quantities that occur naturally in the diet and in the human body. Depending upon the amount of exposure, ionizing radiation may destroy cells, impair their functions, accelerate the aging process, produce cancer and leukemia, and cause alteration or mutation in the genes of the reproductive cells. Such mutation, which generally is harmful, is hereditary. Besides the lack of specific knowledge of low-level radiation effects and of long-term experience with radioactivity in food and water, there is an additional element of confusion, that is, the lack of agreement of acceptable risk on which to base decisions in regard to radioactivity in food.

The *Pasteurized Milk Network* sponsored by the Office of Radiation Programs, Environmental Protection Agency, and the Office of Food Sanitation, Food and Drug Administration, Public Health Service, consists of 63 sampling stations throughout the United States, one in Puerto Rico, and one in the Canal Zone. Many State health departments also conduct local milk surveillance programs. Milk was chosen for this purpose because it is not only an important item of the diet, but is also available throughout the year, rather than seasonally. Results of the test are tabulated and published monthly.

With the constant expert surveillance given to milk and other foods available to the public, there is every assurance that they are safe to use. If for any reason a supply should become contaminated with a pesticide or through radioactive fallout, in such amounts as to create a hazard to health, the milk or other foodstuff would be removed from the market (See under iodine-131).

Relatively few of the many radionuclides that are formed as a result of nuclear fission become incorporated in milk. Most of the possible radio-contaminants are eliminated by the selective metabolism of the cow, which restricts gastro-intestinal uptake and secretion into the milk. The five fission-product radionuclides which commonly might occur in milk are strontium-89, strontium-90, iodine-131, cesium-137, and barium-140.

The recommended permissable levels on radiation are held conservatively low, in order to provide maximum safety for infants and children from the total of all sources of radiation in the air, water, and foodstuffs. In order to provide for control, three levels of radiation have been specified, each known as the R.P.G. or *Radiation Protective Guide* for a specific range of radioactivity. These guides or *safe standards* are defined as

> The radiation dose which should not be exceed without careful consideration of the reasons for doing so; every effort should be made to encourage the maintenance of radiation doses as far below this guide as practical.[17,26]

Each guide is based upon the total radiation from all sources over a period of one year. Higher levels may be present over a short period of time but the total average exposure and intake over one year is used to determine whether or not a situation is dangerous. The following table shows the ranges and actions to be taken, based on an average intake of one liter of milk per day.

Data on the radioactivity of foods give values that vary from less than one up to 17,000 pCi per kilogram of food. Generally values are lower for milk products, fruits, and vegetables than for cereals and nuts. The usual western diet probably contains from two to five pCi of alpha-particle activity per day. A 1962 United Nations Report gave the Radium-226 content of milk in the United States as 46 to 50 pCi per year.

TABLE 8:2
Graded Scales of Actions for Specific Radionuclides [25]

Range pf Daily Intake	Picocuries per day for Iodine-131	Cesium-137	Strontium-90	Strontium-89	*Action to be taken*
Range I	0.10,	0-137	0-20	0-200	Periodic confirmatory surveillance
Range II	10-100	360-3600	20-200	200-2,000	Quantitative surveillance and routine control
Range III	100-1,000	3600-36,000	200-2,000	2,000-20,000	Evaluation and application of additional control measures

*The unit of radioactivity is the **curie** (Ci), the radiation given off by one gram of radium-226. This is equal to 3.7 x1010 nuclear disintegriations a minute. The microcurie (mCi) is one-millionth of this value and the picocurie (pCi) is one millionth of the microcurie or 3.7 disintegrations per minute. The unit of measurement of ionizing radiation is air is called the roentgen (R); in any other medium than air, it is called a rad.

Potassium-40

In all natural materials, Potassium-40 is present as a constant percentage of 0.0118% of the total potassium. The amount present in any foodstuff can be estimated from the total potassium content. Potassium-40 contributes the largest natural radiation exposure by by way of food. Potassium-40 has a very long half-life.* It now probably exists at about one-half to one-fourth the levels that existed when simple life first appeared on earth.

Strontium-90 and Strontium-89

Strontium resembles calcium in its chemical properties, and like calcium, it is accumulated in the bony tissues when it enters the body. The bodily mechanism however, can discriminate somewhat between the two elements so that their relative concentration when deposited in bone is different from their relative concentration in the diet. Strontium-89 does not present so much hazard as does strontium-90 because the average energy absorbed in the disintegration of strontium-89 is one-half that of strontium-90, and its half-life is 52 days compared to 28 years for strontium-90.[3]

In man and animals, the deposition of strontium-90 in bone structure is related to the calcium content of the diet. The less calcium, the more strontium-90 is deposited. The cow further discriminates against strontium in favor of calcium, so that only about 1 or 2 percent of the strontium-90 ingested is excreted in the milk.[5] As milk is rich in calcium, the body assimilates calcium and discriminates against the strontium. As a result, only one-fourth to

*The half-life of a radioactive element is the length of time in which its radioactivity is reduced by one-half.

one-half of the strontium-90 per unit of calcium is available for deposition in the bones, because of the selective action of the alimentary canal and kidneys in favor of absorption of calcium. Should the diet be deficient in calcium, then relatively more strontium-90 is absorbed. Even though as much as 80% of an individual's intake of calcium may come from milk and only 10% from plant foods, the plants may contribute as much strontium-90 to the diet as does milk. Evidence indicates that in areas where the populace uses little or no milk, the strontium-90 in their bones is about as much as that in people in the United States, probably because of less calcium in their diet. It has been shown that most of the strontium-90 and cesium-137 present in plant tissues comes from rain by direct absorption, rather than from the soil through the roots.[16]

TABLE 8:3

Contribution of Various Foods to Strontium-90 of Diet in New York City (1968)[27]

Food	*pCi-90 Strontium per year*	*Percent of Calcium*	*Total Intake Strontium-90*
Dairy products	2080	58	38
Vegetables	1212	9	22
Fruits, fresh and canned	1192	3	22
Cereals and bakery products	588	20	11
Meat, poultry, eggs	178	8	3
Fish	5	2	-
Water	200	-	4

If an attempt is made to reduce radio-strontium intake by eliminating milk from the diet and increasing the vegetable portion to make up the deficit of calcium, the resulting intake of strontium-90 is greater than that from the diet containing the milk products. It will be noted that these amounts are much below those listed in Range I of the Radiation Protection Guide and do not approach a dangerous level.

In a new fission product, the amount of strontium-89 may be up to 180 times as high as the strontium-90. Since strontium-89 has a 51-day half-life (compared to the 28-year half-life of strontium-90) the amount of strontium-89 is soon depleted.

Iodine-131

Iodine-131 is a fission-product which appears at the time of nuclear testing and during the operation of power reactors. It diminishes relatively quickly because its half-life is about eight days. The ease with which radioiodine may be detected makes it a good indicator of its escape in nuclear power plants.

As does ordinary iodine, iodine-131 accumulates in the thyroid gland, but in high levels radioactive iodine is cancer-forming. Since an infant's thyroid gland weighs about two grams, an accumulation of iodine-131 may be of greater significance than if the same amount were present in the adult whose thyroid gland weighs about 20 grams. About 30% of the iodine-131 taken into the body is concentrated into the normal thryoid within 24 hours, but there may be wide variations among individuals. It is generally held that iodine-131 is not an important long-term fallout hazard, but is the most hazardous one during the first 60 days after a nuclear blast or reactor accident.

Atmospheric tests conducted in China have been responsible for most instances of elevated radioiodine levels in the food chain in the United States since 1963. Only levels of radioiodine over 30 pCi per liter are reported by the U.S. Public Health Service Pasteurized Milk Network, and on this basis there were about 12 occasions between 1963 and 1972 when this amount was exceeded. On three times the levels reached 200-300 pCi per liter. On two occasions more than 2000 pCi of radioiodine per liter of milk were detected in Great Britain.

Range III for iodine-131, which has an upper limit of 1000 picocuries to the liter has been exceeded in some areas of the United States. Following nuclear tests in Utah, in 1962, the levels of contamination in milk ran over 2500 picocuries to the liter. The contaminated milk was withdrawn from distribution and converted into butter and cheese, because in these products, the radioactivity would diminish to a negligible amount by the time they reached the consumer.

In October, 1957, an accident in the nuclear reaction at Windscale, Cumberland, England, caused a fallout of iodine-131 over an area of about 200 square miles. Cows pastured in the area produced milk which in some cases contained more than 100,000 picocuries of iodine-131 per liter, the maximum concentration permitted in a milk supply in England. The milk produced in this area was not permitted to be used for human consumption for six weeks after the accident.[7]

Other Radionuclides in Milk[23]

Some radionuclides are introduced into effluent water from nuclear reactors. Crops irrigated with water from the Columbia river have been found to contain radioactive zinc-65 and chromium-51. Zinc-65 is also a constituent of fallout. Milk produced on a farm that was irrigated with the same water contained 1.9 picocuries of zinc-65 and less than 0.23 of chromium-51. Alfalfa and pasture grass contained 8.9 and 3.9 picocuries of zinc-65, respectively. The body of the individual that operated this farm held about 0.2 microcuries of zinc-65, and no chromium-51 was detected. These values are much below the exposure limit recommended by the National Committee on Radiation Protection.[20]

Manganese-54, a gamma-ray emitter, has been observed in milk after the cows had consumed feed contaminated by fall-out with the radionuclide. The rate of secretion into milk is very slow, due to the poor absorption from the intestinal tract. Manganese-54 is not considered to present any hazard to the consumer.[30]

TESTS FOR AND REMOVAL OF RADIONUCLIDES

The procedures used to determine radionuclides in milk are highly specialized and make use of equipment not usually found in the analytical laboratory. Measurement of strontium-90 and iodine-131 is usually made but many other radionuclides may be present, such as iodine-132, -133, -135 and -139, strontium-89, cesium-137, and zinc-65.[4,24]

In the determination of strontium-90, interfering elements, such as calcium, potassium, and phosphate must be first be removed, not only because they cause chemical interference but also because some of them, like potassium-40,, which is normally present in milk in minute amount, are naturally radioactive. Strontium is isolated as the carbonate and is stored for about two weeks, during which time part of it changes to radioactive yttrium-90. The yttrium-90 is removed from the mixture, its radioactivity is measured, and by calculation it is expressed in terms of the strontium-90 originally present. Radionuclides such as iodine-131, cesium-137, barium-140, and the naturally occurring potassium-40, may be measured more directly.[18]

Even though such treatment is not practicable on a large scale, some experiments have been made to remove radioactive materials from milk that may have become contaminated. Most of these efforts have been directed towards treatment of the milk with an ion-exchange resin. About 90% of the calcium with its accompanying strontium-90 may be removed by this treatment.[3,6] Workers in the United States Department of Agriculture[19] have succeeded in removing 90-95% of the strontium-90, 85-90% of the barium-140, and 75% of the cesium-137 from contaminated milk. Such milk was first acidified with citric acid to ionize the radionuclides and then passed through an ion-exchange column. The product was then restored to the *p*H of normal milk (6.6) by being passed through another exchange column in the alkaline cycle.

A review for procedures for the removal of radionuclides in milk by acidifying with citric acid, ion-exchange and subsequent neutralization with sodium hydroxide was published in 1968.[11]

In another procedure, electrodialysis is proposed as a method to remove strontium-90 and cesium-137 from contaminated milk.[8] The device consists of a series of compartments, separated by cation-permeable membranes, which permit only positively charged ions to flow through. Milk is circulated through one series of compartments; through the other a solution of salts similar to those present in milk. A direct current is used to accelerate the transfer of cations into and out of the compartment containing the milk. In a demonstration of the device, about 90% of the strontium-90 was removed from milk.

According to U.S. Patent 3, 186,849 (1968),the claim is made that the addition of three ounces of calcium phosphate to one quart of contaminated milk will remove about ninety percent of any Strontium-90 present. The excess of calcium phosphate does not dissolve in the milk and is removed by centrifuging, carrying with it the Strontium-90.

In a series of experiments, workers in the U.S. Public Health Service and the Department of Agriculture found that on the basis of chemical analyses and feeding experiments, the ion exchange treatment of milk for removal of cation radionuclides does not seriously affect the nutritive qualities of bovine milk.[31]

REFERENCES

1. Albright, J. L., Tickey, S. L., Woods, G. T., Antibiotics in milk: a review, *J. Dairy Sci.*, 44:779 (1961).
2. Albright, J. L., Ormiston, E. E., Brodie, B.O., Witter, L. D. Observations on the infusion of penicillin in the mammary gland of the goat, *J. Dairy Sci.*, 44:2103 (1961).
3. *Annotated Bibliography of Strontium and Calcium Metabolism in Man and Animals*, Misc. Pub. 821; Agr. Res. Service, U.S. Dept. Agr., Washington (1961).
4. Cohn, S. H., Love, R. A., Gusmano, E. A. Zinc-65 in reactor workers, *Science*, 133:1362 (1961).
5. Cox, G. W., Morgan, A., and Taylor, R. S., Strontium-90 from fallout in the diet and milk of a dairy herd, *J. Dairy Res.*, 27:47, (1960).
6. DeMott, B. J. and Easterly, D. G., Removal of iodine-131 from milk, *J. Dairy Sci.*, 43:1148 (1960).
7. Dunster, H. J., Howells, H. and Templeton, W. L., District surveys following the Windscale incident, October, 1957. *Proc. Second Intern. Conf. on Peaceful Uses of Atomic Energy*, 18:296. United Nations, Geneva. (1958).

8. Electrodialysis rids milk of strontium-90, *Chem. Eng. News*. July 16, (1962).
9. Evans, D. A., and Stern, D. N. Observations on the incidence of penicillin transfer from treated to untreated quarters of cows' udders following infusion of penicillin for treatment of mastitis. *J. Dairy Sci.*, 43, 1886 (1960).
10. *Food, The Year Book of Agriculture*, 1959, p. 454; U.S. Dept. Agr., Washington, D.C.
11. Glasscock, R. F. et al, A pilot plant for the removal of cationic fission products from milk, *J. Dairy Res.*, Vol. 35; 257-286 (1968).
12. Hunter, G. J. E.—A note on the effect of penicillin on the reductase test for milk quality. *J. Dairy Res.*, 16:149, (1949).
13. Hunter, G. J. E.—The effect of penicillin on lactic streptococci, *J. Dairy Res.*, 16:39 (1949); *ibid.*, The effect of penicillin in milk on the manufacture of cheddar cheese, *ibid.*, 235 (1949).
14. Kennedy, H. E., Antibiotics in milk, *Milk Products J.*, July 1960.
15. Kroger, M., Effect of various physical treatments on certain organochlorine hydrocarbon insecticides found in milk fat. *J. Dairy Sci.*, 51:196 (1968).
16. Kulp, J. L., Schulert, A. R., Hodges, E. J., Anderson, E. C., Langham, W. H., Strontium-90 and cesium-137 in North American milk, *Science*, 133:1768 (1961).
17. Morgan, Karl Z., Permissible exposure to ionizing radiation, *Science*, 139:565 (1963).
18. Murphy, G. K. and Campbell, J. E.—A simplified method for the determination of iodine-131 in milk, *J. Dairy Sci.*, 43:1042 (1960).
19. Murphy, G. K., Masurovsky, E. B., Campbell, J. E., Edmondson, L. F., Method for removing cationic radionuclides from milk, *J. Dairy Sci.*, 44:2158 (1961).
20. *National Bureau of Standards Handbook* No. 69, (1959).
21. Neal, C. E. and Calbert, H. E., The use of 2,3,5-Triphenyl tetrazolium chloride as a test for antibiotic substances in milk. *J. Dairy Sci.*, 38:62 (1955).
22. Palmer, J. M. A. and Kosikowski, F. V., Simple ultrasensitive test for detecting penicillin in milk. *J. Dairy Sci.*, 50; 1390 (1967).
23. Perkins, R. W., et al, Zinc-65 and chromium-51 in foods and people, *Science*, 132:1895 (1960).
24. *Radioactive Fallout in Time of Emergency*, Agr. Res. Service, 22-55, U.S. Dept. Agr., Washington, D.C., (April 1960).
25. *Radiological Health Data*, U.S. Dept. Health, Education, Welfare, Public Health Service, Washington, D.C.
26. *Radiation Protection Standards*, Report No. 2, Federal Radiation Council, Washington, D.C. (Sept. 1961).
27. *Radionuclides in Foods*, National Research Council, Washington, D.C., 1973.
28. *Standard Methods for the Examination of Dairy Products*, 12th Ed.; Amer. Public Health Assoc., New York (1967).
29. Stolz, E. J. and Hankinson, D. J.—The effect of antibiotics on the standard plate count of normal raw milk, *J. Milk Food Technol.*, 16:157 (1953).
30. Wilson, D. W. and Ward, G. M. Transfer of fall-cut Manganese-54 from feed to milk, *J. Dairy Sci.*, 50:592 (1967).
31. Isaacks, R. E. Hazzard, D. G., Barth, J., Fooks, J. H. and Edmondson, L. F., Nutritional Evaluation of Milk Processed for Removal of Cationic Radionuclides. Chemical Analyses. Feeding Studies, *J. Agr. Food Chem.* 15:295-299, 300-304 (1967).
32. Knoll, W. and Jayaraman, On the Contamination of Human Milk with Chlorinated Hydrocarbons, *Nahrung*, 17(5) 599-615 (1973).

chapter 9

Grades and Classes of Milk

Milk generally is classed into grades according to the care with which it is produced. The bacterial content is used to establish the sanitary quality of the product, but other factors also are considered in the grading of milk and cream. More stringent control is placed upon *market milk*, that is, milk supplied to the consumer in its fluid state than upon *manufacturing milk* which is used in the preparation of evaporated milk, cream for making butter, milk powder, cheese, and similar uses. Most health departments have sanitary codes which cover the grading of milk and cream. The Milk Ordinance and Code of the United States Public Health Service[2] modified to fit local conditions, serves as a standard in many localities.

Dairies and farms engaged in the commercial production are visited by inspectors who observe the various details involved in milk production. The cows are watched for symptoms of disease and for assurance that, at the time of milking, their flanks, belly, tail, and udder are washed free of visible dirt and loose hairs. The udders and teats of the cows must be cleaned and treated with a sanitizing solution just prior to the time of milking. The milking barn or stable and the milk house in which the milk is cooled and stored usually must be built to conform to approved specifications which provide for proper sanitary conditions. All utensils and equipment that come in contact with milk during handling, storage, or transportation must be made of corrision-resistant, non-toxic metal such as stainless steel. Heat-resistant glass or approved plastic or rubber-like materials may be used. Between periods of use, the equipment and utensils must be cleaned and treated by an approved bactericidal process, such as steam, hot water, or chemical sterilization, and stored in the milk house.

Further safeguards are used in plants where milk is pasteurized. Close observation is kept over the type of equipment used to pasteurize the milk and the manner in which this equipment is employed. The time and temperature of pasteurization is watched, the accuracy of the thermometers is checked, and automatic recording charts are made to show the treatment given the milk during pasteurization. All processing usually is done by licensed operators. Few food industries have a sanitation program comparable to that of the milk industry.

Most states and the larger cities have established numerical standards for the bacterial count of various milk products. Cream usually is permitted to have a higher bacterial count than milk because, in its separation, bacteria are carried mechanically from the milk into the cream. As a rule, the number of bacteria permitted in cream is twice that allowed in the corresponding grade of milk. In some places an allowance is made for an increased count in milk and cream during the summer months, but this is not an approved practice. With proper care in the production and handling of the milk there should be no exceptional increase in the rate of bacterial growth during warm weather. A number of states have established regulations that milk be held at a low temperature from the time of processing until it is purchased by the consumer. In California this temperature is 7.2°C (45°F). The Federal standard is 10°C (50°F).[2]

Milk regulations originally were designed to prevent fraud and adulteration, but they now cover many factors that affect sanitation and public health. The courts have continually upheld the legality of milk ordinances to protect public health, especially if the ordinance confines itself to the requirements of public health and sanitation and does not interfere with or restrict trade. The marketing of milk and its products is regulated by Federal, State, and local laws and ordinances. Dairy products entering interstate commerce must have been produced and pasteurized under regulations equivalent to the United States Public Health Service standards.[2] The Grade "A" Pasteurized Milk Ordinance describes the steps needed to protect the milk supply. It gives the reasons for these procedures and how satisfactory compliance is determined.

The bacterial standards quoted in the following paragraphs are those suggested by the United States Public Health Service Milk Ordinance and Code. Many states and municipalities have more stringent requirements and the sale to the public of milk and cream below that of grade A is illegal. Milk to be pasteurized generally is permitted to have a higher bacterial content before pasteurization than the corresponding grade of raw milk.

GRADES OF MILK

Grade A Raw Milk

This is raw milk produced upon dairy farms that conform with all the items of sanitation required by the local and state authorities. The milk must be obtained from cows tested and found free of tuberculosis and brucellosis. The milk must be cooled, immediately after completion of milking to the legal temperature requirement and maintained at that temperature until delivery. Depending upon local regulations, the bacterial content of the milk at the time of delivery to the consumer may range from 15,000 to 50,000 per milliliter.

Grade A Raw Milk for Pasteurization.[2]

Grade A Raw Milk for Pasteurization is raw milk from producing dairies that meet the required sanitary regulations. The milk must be cooled to the legal requirement and maintained there at until processed. The bacterial content of the milk from an individual producer must not exceed 100,000 per milliliter prior to commingling with other producer milk. Prior to pasteurization, the bacterial limit of commingled milk must not exceed 300,000 per ml. The antibiotic content must be less than 0.05 units per ml.

Grade A Pasteurized Milk.[2]

This is a Grade A Raw Milk for Pasteurization, which has been pasteurized, cooled, and placed in the final container in a milk plant that conforms with the state and local sanitary requirements. The milk must pass a phosphatase test and at no time after pasteurization and before delivery shall the milk have a bacterial count exceeding 20,000 per ml, (15,000 in California), or a coliform count exceeding 10 per ml. At no time before pasteurization shall the bacterial count exceed 300,000 per ml, (50,000 in California).

In return for meeting the required standards of sanitaton on his farm, the producer of Grade A milk receives a higher price for the milk than do other milk producers.

Guaranteed Milk

Guaranteed Milk is not included in the grades of milk covered by the United States Public Health Service Milk Ordinance and Code, but it is produced in a number of localities. The conditions regulating production are similar to those required for Certified Milk, except that the local inspection authority has supervision. A typical regulation applying to raw milk is that it contain not more than 10,000 bacteria per ml. at the time of delivery to the consumer and guaranteed pasteurized milk not more than 3,000 bacteria per ml.

Other Grades of Milk

Prior to the adoption of the 1965 U.S. Public Health Pasteurized Milk Ordinance, a Grade B and a Grade C milk was recognized. Today, the quality of the milk available to the consumer is of such superior quality that only Grade A milk is recognized. Milk not wholly meeting the Grade A standard but otherwise acceptable may be used as manufacturing milk.

TABLE 9:1
Chemical, Bacteriological and Temperature Standards for Grade A Milk and Milk Products 2

Grade A raw milk for pasteurization.	Temperature...... Cooled to 50°F. or less and maintained thereat until processed. Bacterial limits.... Individual producer milk not to exceed 100,000 per ml. prior to commingling with other producer milk. Not exceeding 300,000 per ml. as commingled milk prior to pasteurization. Antibiotics........ Less than 0.05 unit/ml. by the *Bacillus Subtilis* method or equivalent.
Grade A pasteurized milk and milk products (except cultured products).	Temperature...... Cooled to 45°F. or less and maintained thereat. Bacterial limits..... Milk and milk products—20,000 per ml. Coliform limit...... Not exceeding 10 per ml. Phosphatase ...... Less than 1μg per ml., by Scharer Rapid Method (or equivalent by other means).
Grade A pasteurized cultured products.	Temperature...... Same as above. Coliform limit...... Do. Phosphatase....... Do. Bacterial limits..... Exempt.

Certified Milk[3]

The standards for Certified Milk were devised at a time when dairy sanitation had not yet attained the excellence that now prevails in many areas producing Grade A milk. The quality of Grade A milk produced today is so high that the demand for Certified Milk has decreased and it is produced in relatively few localities.

Certified Milk is the copyrighted name given to raw and pasteurized milk produced by dairies that operate according to the rules and regulations published in the *Methods and Standards for the Production of Certified Milk* issued by the American Association of

Medical Milk Commissions, Inc., Milwaukee, Wisconsin. These rules are intended to ensure the purity, uniformity, and adaptability of the milk for infants and growing children. The production of Certified milk involved the veterinary examination of cows and the sanitary inspection of the dairy farms and equipment, the medical inspection of the employees that handle the milk itself for quality and purity. This supervision is conducted under the auspices of a Medical Milk Commission appointed by the local Medical Society or other approved agency.

All cows whose milk is used for the production of Certified Milk must be free from tuberculosis, brucella infection, mastitis, or any signs of other disease. The milk from any cow must not be used for a period of 45 days before and five days after parturition. After the milk is drawn it is removed from the milking place, strained through a modern, single-service filter, and cooled below 50°F Before it leaves the farm where it is produced, the milk must be bottled, capped, and sealed with a sterile hood which covers the lip of the bottle. The milk may be transported to another Certified Milk farm for packaging.

Certified Milk may be pasteurized on the farm where it is produced or through freshly cleaned and sanitized equipment which may thereafter be used for other grades of milk.

All requirements covering the production, handling, and standards of Certified Milk, except those referring to milkfat, apply to other certified products such as Certified Cream, Certified Half and Half, Certified Fat-free milk, Certified Low Sodium Milk and Certified Acidophilus Milk.

The standards for Certified Milk require that it have a total bacterial count of not more than 10,000 colonies per ml. or a coliform colony count of not more than 10 per ml. The pasteurized product shall have a colony count of not more than 500 per ml. or a coliform colony count of one per ml. when delivered to the consumer.

The fat content is not standardized but the milk must contain an average of 3.5 % of fat with a tolerance for the average of plus or minus 0.2%, and the minimum of 3.3% for individual samples. The total solids content including milkfat shall not be less than 12%.

Although great care is taken to protect the quality of Certified Milk, it is not absolutely free from the possibility of transmitting pathogenic organisms. At least two epidemics are recorded that were traced to use of it. One of these was an outbreak of paratyphoid fever, the other an epidemic of septic sore throat. In 1968, raw Certified Milk produced in Southern California was found to contain organisms of Q-fever.

CLASSES OF MILK

Milk, as has been described, is given various grade names, according to the manner or purpose for which it is produced. A very involved economic classification is also applied to milk, in which milk production and sales competition is concerned. The cows give the farmer-producer a relatively large supply of milk in the spring months but much less in the winter. Feeding the cows costs less in the spring because green pasture is available, but in the winter more costly grain and hay must be fed. As a result, it costs more to produce each gallon of milk in winter than it does during the spring months. The consumer, however, uses about the same amount of fresh milk throughout each month of the year and makes little response to a change in its retail price, as he might for that of other processed dairy products.

The supply of fresh milk cannot readily be turned on and off to fit the supply to the demand. As surplus fresh milk cannot be stored for any prolonged period of time, it must be converted into less perishable products, such as evaporated milk, butter, cheese, and dried milk. These durable products come into direct competition with similar products that may have been produced elsewhere at lower cost. The market value of milk that must be converted into a manufactured dairy product is accordingly less than its value when produced for the consumer as fresh, fluid milk.

During the first half of the 1900's cooperatives tended to stabilize prices between producers and distributors. These arrangements often were disrupted by some farmers and dealers who traded in milk without regard to existing agreements. During the depression years of the 1930's many bargaining arrangements failed and dairy farmers turned to governmental agencies for help. The Federal Government and a number of States enacted legislation to stabilize milk markets and prices.[1,5] Each State can regulate milk prices within its borders but Federal authority applies to milk crossing State boundaries. Modern improvements in handling milk permit bulk milk shipments of over 1500 miles from the upper midwestern States to the Southeast and the Southwest.

In addition towards working for orderly marketing, the laws in a number of States have as their objective fair trade practices and a fair price return to the farmer.

A Federal milk marketing order regulates transactions between farmers and milk dealers (usually called handlers) within a specified geographic area. Unless regulated by State authority, prices are established through competition among handlers including cooperative associations. Sanitary regulations pertaining to the milk industry are administered by State and local health authorities.

In 1971, 143,400 producers delivered milk to 1,566 handlers regulated by 62 Federal milk marketing orders. These producers delivered about 65 billion pounds of milk to supply about 125 million consumers. In 1972, 18 states maintained control of fluid milk prices paid to producers. These states were Alabama, California, Hawaii, Louisiana, Maine, Massachusetts, Montana, Nevada, New Jersey, New York, North Carolina, North Dakota, Oregon, Pennsylvania South Carolina, Vermont, Virginia and Wyoming. These states price about 15 billion pounds of milk yearly, roughly a seventh of all fluid milk farmers sell to plants and dealers.

The Federal Milk Orders law requires that the minimum prices paid to producers be based on the "classified use" of the milk.[4,5] Milk used for fluid purposes is placed in Class I, the highest price class. Products usually included in Class I are whole milk, flavored whole milk, concentrated whole milk, skim milk, fortified skim milk, low-fat milk, buttermilk, milk drinks, cream, half and half, sour cream, yogurt and sometimes eggnog.

Milk supply and demand are the basis of establishing Class I prices. Two types of Class I pricing formulas are in use. One type is based on the average price paid to producers of manufacturing grade milk in Minnesota and Wisconsin, areas of heavy production. To this price is added a differential, which varies with the marketing conditions in the local area in order to insure an adequate supply of pure, wholesome milk.

A second pricing formula is based on general economic indicators, such as production cost, per capita disposable income in the area and to changes in the level of wholesale prices shown by the Bureau of Labor Statistics.

Milk used or diverted into any category other than Class I is usually placed into Class II. This includes manufactured products, such as cheeses, butter, evaporated milk, dry milk products, ice cream and frozen desserts and aerated cream. In some areas a Class III is used, in which milk is priced lower than that used for Class II. For example, cream and buttermilk might be put into Class II and other manufactured milk products are placed in Class III.

The classified price plans make allowance for the higher cost of producing and marketing milk for fluid use than it is to produce milk for use in manufactured products. The cost differential for Class I milk exists because of the expense needed to comply with the rigid sanitary standards for Grade A milk and also because milk in fluid form must be transported, usually over long distances, on a rapid daily schedule.

In some markets, milk orders provide marketwide pools, where class values are combined and averaged for all handlers in the market, and the uniform price reflects total usage by all handlers.[4] In marketwide pools, all producers in the market are paid the same uniform price per hundredweight. Prices are established for milk of 3.5 percent fat content, adjustments being made for milk which has a different fat test from that specified.

Federal milk orders apply only to the area concerned; they do not set resale prices, only the minimum price paid to producers by handlers that process the milk for various uses. Prices are adjusted upward or downward as local economic conditions may warrant. Excessive production of milk, lower feed costs or manufacturing economies may cause a downward adjustment in price. If these conditions are reversed, the price is adjusted accordingly. Federal orders are not a program on price support. Such programs are provided by Congressional action.

REFERENCES

1. Government's Role in Pricing Fluid Milk in the United States, Agr. Econ. Rep. No. 152, U.S. Dept. Agr., Washington, D.C., Dec. 1968.
2. Grade "A" Pasteurized Milk Ordinance, U.S. Public Health Service, Washington, D.C. 1967.
3. Methods and Standards for the Production of Certified Milk, The American Assoc. of Medical Milk Commissions, Inc., Milwaukee, Wisconsin.
4. Questions and Answers on Federal Milk Marketing Orders, U.S. Dept. of Agriculture, Consumer and Marketing Service, Washington, D.C. Oct. 1971.
5. The Federal Milk Marketing Order Program, Marketing Bulletin 27, U.S. Dept. Agr., Washington, D.C. April 1968.

chapter 10

Defects of Flavor and Other Defects

The Flavor of Milk

Physiologists have demonstrated that all reactions in taste are due to four sensations perceptible to the taste buds of the tongue. These are *sweet, sour, salty,* and *bitter*. Taste has been defined as the impression perceived in the mouth alone, and *flavor*, as the combination of taste and texture perceived in the mouth and of aroma or smell perceived through the inner nasal passages. The tip of the tongue is most sensitive to sweetness, the sides to saltiness or sourness, and the back is most responsive to a bitter flavor. There is little flavor response in the middle of the tongue. Although the tongue is sensitive to only the above four sensations. and their combinations, the nose of a trained person may distinguish among about 10,000 different odors. As it is difficult to separate these sensations, no clear distinction can be made between them in a discussion of the flavor of milk and most milk products. Taste and odor are evaluated by human judgment. As yet there is no instrumental method which will truly simulate human perception.

As only substances that are in solution affect the taste buds, those constituents of milk that are not in solution, principally the fat and protein, have little direct influence on flavor. The natural flavor of milk is scarcely discernible, yet it is pleasant and slightly sweet. It may be surmised that this flavor is due to the combination of sweetness, originating from the lactose and saltiness from the chlorides and perhaps the citrates and other mineral salts present. Normal milk contains no constituent that contributes a bitter flavor. Furthermore, a sour or acid flavor also is abnormal, as fresh milk does not contain enough acid to affect the taste buds. If lactic acid is formed by the activity of bacteria, milk tastes sour to most people when the acidity reaches 0.3%. An experienced person can detect the sour flavor at 0.2% acidity.

Milk rich in lactose usually is of better flavor than milk low in lactose; and one with a natural fat content between three and five per cent generally has a better flavor than milk with a higher fat content. This is in line with the observaton that as the fat content increases, the ash and lactose content decreases.

In addition to the factors just mentioned, the tactual property or sense of touch is an important factor in appraising flavor. This is shown by the more pleasing sensation imparted when drinking whole milk in contrast to skim milk. Milk of low fat content acquires an improved flavor when it is fortified with non-fat milk solids, and this is due in part to the added "body." In a similar way, milk with about 12% of total solids has less taste-appeal to many consumers than milk with 14 or 15% of total solids, such as may be obtained from Jersey cows. The sensation of smoothness in good ice cream, or of such defects as chalkiness, gumminess, and sandiness in ice cream and some other dairy products, also are manifestations of the tactual sensation.

Milkfat itself is an important contributor to flavor. In ice cream, cream and sweet butter the desired flavor is related to that of fresh milk. In processed dairy foods other flavors derived from milkfat are sought—such as cheese flavor, a butter-like flavor or a caramel-like flavor.

The normal flavor and off-flavors of the various milk products are discussed in the chapters dealing with the specific product. A comprehensive review of the literature dealing with flavor defects in milk was issued by the United States Department of Agriculture in 1953.[3]

Even though the flavor of milk may have little to do with its nutritive value, it is of much importance because it controls the relish with which milk is consumed. To this end, it is important that both the producer and the processor of milk supply a product of good flavor. Children have a keen sense of taste and are critical of any off-flavor in milk. The trend towards store purchase of milk rather than home delivery has increased the time between processing and consumption. This has made the use of good quality milk and proper sanitation a prerequisite to flavor control.

Defects of Flavor

Milk with an off-flavor may pass all quality tests pertaining to composition and sanitation but it will not be used so readily as milk with its normal, agreeable flavor. Many defects of flavor can be detected by the sense of smell alone, especially if the milk is warm. In general, these defects can be divided into four groups, as follows:

From Feed:

Off-flavors derived from the cow's feed are rather common. They often are more pronounced in the night's milk and in the cream than in the original milk. The flavor and odor imparted by alfalfa and by wild onions and other weeds is recognized easily. Workers in the College of Agriculture, University of California, showed that within a minute after a cow breathed the odor of garlic, it could be detected in her milk. Cows feeding on pastures containing peppergrass excrete a trace of indole in the milk, which may give it an off-flavor.

Flavor from feed may originate with some of the more common feedstuffs, such as ladino clover, corn silage, and grasses such as rye, fescue, and orchard grass. Cows that drink brackish water, or from tanks overrun with moss and algae, may give milk with an off-flavor that may be confused with a feed flavor. Some feed flavors that are volatile may be removed from the milk during processing, especially by vacuum treatment, as described in Chapter 11. To control feed flavor in the milk requires that the cows have no access to feeds for at least four hours before milking. An automatic feeding system in the milking-barn may have a helpful influence upon feed flavor since the cows are fed just before or during milking time. Concentrate feeds fed during milking normally do not have an adverse influence upon the flavor of milk.

From Biochemical Change:

A biological or chemical change that may occur after the milk has left the cow is an important cause of defective flavor. An off-flavor or taint that is present in the milk of a healthy cow at the time of milking probably is not of bacterial origin. Modern dairy processing and sanitary measures have practically eliminated a sour or acid flavor caused by the activity of lactic acid bacteria. Sour milk now is unusual and in its place milk with an *oxidized* flavor has become more common. This defect is of chemical, rather than bacterial origin.

Much evidence indicates that the off-flavor has its origin in an oxidative change in the phospholipids associated with the milkfat.[1] Oxidized flavor, also known as *cardboard, tallowy or oily* flavor, usually is found in milk of low bacterial content. During their growth in milk, bacteria have an antioxidative action, or they may produce a substance which protects the milkfat from undergoing the change which results in oxidized flavor. The reduction of the oxygen content of the milk by bacterial growth may also have a role in preventing an oxidized flavor.

Nonfat milk may develop an oxidized flavor because oxidation takes place on the surface of the fat globule membrane. The fat globules that remain in skim milk have about one-half of the surface area of the fat in 3.5% milk, as well as a higher percentage of polyunsaturated fatty acids which are most susceptible to oxidation.

The deaeration of milk and vacuum treatment, such as *Vacreation* (see Chapter 11) tend to prevent oxidized flavor. Homogenized milk also is resistant to this defect. Treatment with heat at 76.7°C (170°F) or higher liberates traces of sulfhydryl compounds from the milk proteins and these compounds are effective antioxidants in the preventing of oxidized flavor. The deliberate addition to milk of antioxidants or proteolytic enzymes to control oxidation is illegal in most places.

The presence of a trace of copper in the milk or contamination with iron is sufficient to hasten the development of oxidized flavor. As mere contact of the milk with equipment of copper or copper alloy such as brass and white metal is sufficient to favor development of this defect, copper has been eliminated from the construction of modern dairy equipment. Traces of chlorinated sanitizing solutions left in equipment may also trigger the oxidized flavor, perhaps because of their corrosive action on metallic surfaces, especially copper-bearing alloys, and subsequent solution of the metal in the milk. When cows are on dry feed their milk seems to be more susceptible to the development of oxidized flavor than when they are on pasture or green feed.

An off-flavor, very similar to an oxidized flavor, is caused by exposure of milk to direct sunlight. Homogenized milk is especially susceptible to this defect. Milk, in a clear glass bottle, exposed to a flourescent light or left standing in the sunlight for a few minutes may acquire this defect. The fluorescent light usually used in retail store dairy cases may cause a deterioration in the flavor of milk.

Oxidized flavor, especially if caused by contact with copper-bearing equipment, develops slowly over a period of one or more days, but the *sunlight* flavor may be apparent after an exposure to light for ten or fifteen minutes. The defect sometimes is called *burnt* or *cabbage* flavor. As milk in amber-colored glass bottles or in fiber containers receives some protection from light, it develops the defect much more slowly.

A *rancid flavor*, of bacterial origin, sometimes develops in raw milk. The causative organisms, known as *lipolytic bacteria* are active at a temperature below that favored by most other bacteria that may be present in milk. Lipolytic bacteria liberate a lipase which acts upon milk fat in the same manner as does the lipase found naturally in milk. Pasteurization will prevent the development of rancidity of bacterial origin since the causative organisms are destroyed easily at pasteurization temperature. If the rancid flavor is already present in the raw milk, pasteurization will not remove it. Another source of rancid flavor is described below. The effect of agitation in milk pipe lines upon the development of rancidity is described in Chapter 7; the effect of homogenization is described in Chapter 13.

From Condition of the Cow:

The physical condition of the cow has some influence upon the flavor of her milk. A *rancid flavor* sometimes is present in the milk secreted by an apparently healthy cow. The flavor is detectable in the milk at the time it leaves the udder or very shortly thereafter. Milk from individual quarters of the udder may show a difference in the intensity of rancid flavor. It is not of bacterial origin, but is caused by the action of lipase which liberates fatty acids from the milk fat. Rancid milk often is obtained from cows late in their lactation period and sometimes from cows with diseased udders. Green feed appears to favor a reduction in the incidence of rancidity in milk. During the winter months cows tend to produce milk that is resistant to rancidity. Induced and spontaneous rancidity is discussed in the section on lipase in Chapter 7.

A *salty* flavor may be present in the milk of a cow in the late stages of lactation. It is usually present in milk from cows with mastitis.

From Environment:

Milk readily absorbs odors to which it may be exposed. A "cowy" or "barny" flavor may be common in areas where cows are housed during the winter months. A cowy flavor is also present in milk from cows with *ketosis*, a condition in which there are traces of acetone in the milk and body fluids. Areas where milk is produced and stored must be free of strong odors. The household refrigerator is a common source of absorbed odor in milk. In general, however, absorption of odors is not so important a source of defects in flavor as may be commonly believed.[4] Absorbed odors are largely confined to the upper ten percent of the layer of milk in a container open to the air, provided the milk is not agitated.

Controlled experiments have shown that the direct absorption of odors during the normal production and handling of milk is relatively unimportant in its effect upon the quality of milk.[4]

TABLE 10:1

The following chart, issued by the New York State College of Agriculture, may serve as a useful guide on flavor control by the dairy farmer:

Off-Flavor	*Possible Causes*	*Prevention*
Oxidized cardboardy	Exposure to "white metal," worn tinned, or rusty surfaces on milk handling equipment	Use stainless steel, glass, plastic, or rubber on all milk contact surfaces
	Winter or dry lot feeding	Provide green feed
	Exposure to daylight or artificial light	Protect from artificial light and daylight
	Copper or iron in water supply	Water treatment may be necessary
Rancid bitter soapy	Late lactation (over 10 months) or low producing cows	Discard milk from low producing or late lactation cows
	Excessive agitation or foaming of raw milk	Keep fittings tight and air admission to a minimum Avoid risers and don't run milk pumps in starved condition
	High blend temperatures	Cool milk to at least 40 degrees F. and hold
Feed or Weed unnaturally sweet aromatic	Eating or inhaling odors of, strong feeds (grass or corn silage, green forage, wild onion, or other weeds) prior to milking	Feed after milking Ventilate barn Withhold objectionable feed or remove cows from pasture 2 to 4 hours prior to milking Store silage carts out of barn
	Sudden feed changes	Change feed gradually
Unclean barny cowy	Damp, poorly ventilated barns	Keep barns clean and well ventilated
	Dirty cows or barn	Clean cows
	Dirty milk-handling equipment	Clean and sanitize all milk-handling equipment
	Improper preparation and milking	Wash and dry cow's udder prior to milking; handle milker to avoid sucking up bedding
	Cows with Ketosis (Acetonemia)	Withhold milk, treat cows
Malty Or High Acid grapenut-like sour	Dirty milk-handling equipment	Clean milk-handling equipment after each use Sanitize milk-handling equipment prior to use
	Slow or insufficient cooling	Promptly cool milk to 40 degrees F. and hold
Other Off-Flavors		Use according to directions
Medicinal	Medications, insecticides	Use odorless medications
Disinfectant	Certain disinfecting or sanitizing agents	Avoid strong smelling disinfectants Use sanitizers properly
Salty	Mastitis, late lactation cows	Discard milk
Flat	Low total solids	Evaluate feeding program Thoroughly drain equipment before use

Cooked Flavor

A defect of flavor in milk that frequently results from modern procedures in processing is the "cooked" flavor. Slight as this change may be in milk of good quality, persons used to drinking raw milk can detect the change in flavor produced by pasteurization. Modern processing tends to use relatively high temperatures for pasteurization, but for only a very short time. Heat treatment, especially in the range above 170°F, involves changes in the milk proteins. Volatile sulfur compounds, such as hydrogen sulfide and certain mercaptans, have been identified in milk with cooked flavor. Only a minute amount of these sulfhydryl compounds need be present to produce a detectable flavor defect.[2]

Chalky Texture

Milk that is homogenized before it is pasteurized by an ultra-high heat process tends to have a chalky texture. If the milk is homogenized after a treatment with high heat this defect is not apparent.

DEFECTS OF MICROBIOLOGICAL ORIGIN

Ropy Milk

At times milk may acquire a viscous body and long threads of slime may be drawn from its surface, hence the name *ropy*. Except for the development of acidity, ropy milk shows no change in flavor. There is no evidence to show that consumption of it is harmful; in fact in Scandinavia, a fermented milk preparation actually is made with ropy milk, known as *taette* (tight) milk.

The ropiness is caused by the presence of gums and mucins formed by bacteria. It is closely associated with the formation of the capsule or gelatinous membrane that surrounds the bodies of some bacteria. The defect becomes apparent after milk has been held for several hours, during which time the bacteria grow. Ropy milk should not be confused with the *stringy* milk found in some cases of mastitis or udder disease, where the viscous condition is apparent as soon as the milk is drawn from the udder.

Ropy milk often is caused by the use of contaminated water into which milk cans are placed for cooling, or which is used to wash cans and other equipment. Material from wet or muddy fields which collects on the flanks and udders of the cows is another possible source of contamination. Ropy milk also has been traced to infection from dusty feed, infected bedding, and manure. Organisms of the coliform group and certain streptococci commonly are associated with the development of ropiness.

To control the defect it is necessary to identify and remove the source of contamination. Usually, the contamination is removed when the milk cans and equipment are properly sterilized. Before milking, the cow's udder should be wiped with a solution that contains about 200 parts per million of available chlorine. As the bacteria concerned are not spore-formers they are destroyed readily by pasteurization. Some of the organisms may survive and cause an outbreak in the pasteurized milk but this is infrequent.

Unless it has also undergone a lactic acid fermentation, ropy milk has no abnormal flavor and is safe to drink.

Sweet Curd

Sweet curd is a coagulation of milk which occurs without any immediate increase in its acidity. It is caused by bacterial action and often appears in the form of small, coagulated particles floating on the surface of the milk, but the entire mass of the milk may coagulate or assume a jell-like body. Later, acidity may develop and then decomposition occurs, especially if the milk is held in a warm place. In some localities, sweet curd is known as *broken, or bitty cream*.

Most milk may contain some bacteria that can form sweet curd, but conditions usually are not favorable to their growth. These organisms form an enzyme which can digest the phospholipid membrane around the fat globules and so permit them to coalesce and rise in the milk. The defect occurs most often when the milk is held at a temperature of 104° to 109°F. The enzyme is destroyed during pasteurization, but some of the organisms may survive and become active when suitable conditions prevail.

REFERENCES

1. Greenbank, G. R., The oxidized flavor in milk and dairy products—A Review, *J. Dairy Sci.*, 31:913 (1948).
2. Hutton, J. T. and Patton, S., The origin of sulfhydryl groups in milk protein and their contribution to cooked flavor, *J. Dairy Sci.*, 35:699 (1952).
3. Strobel, D.R., Bryan, W. G., Babcock, C. J., *Flavors of Milk: A Review of Literature* (mimeo.), U.S. Dept. Agr., Washington, D.C. (1953).
4. Trout, G. M. and McMillan, D. V., Absorption of odors by milk, *Mich. Agr. Expt. Sta., Tech. Bull.* 181 (1943).

chapter 11

Pasteurization Procedures

A striking description of conditions that existed some 200 years ago was given in 1771 by the Scottish novelist, T. G. Smollett, in his novel *The Expedition of Humphrey Clinker.* In it he described the cream put on strawberries as:

> the worst milk, thickened with the worst flour into a bad likeness of cream, but the milk itself should not pass unanalyzed—carried through the streets in open pails, exposed to foul rinsings discharged from doors and windows—dirt and trash chucked into it by rougish boys for the joke's sake, the spewing of infants, who have slabbered in the tin measure, which is thrown back in that condition for the benefit of the next customer; and finally, the vermin that drops from rags of the nasty drab that vends this precious mixture, under the respectable denomination of milk-maid.

During the latter half of the last century it was shown that certain diseases, such as typhoid fever, diphtheria, and tuberculosis, can be carried from the afflicted to the healthy, either by personal contact or indirectly by the use of infected water or milk. When proper care was taken in the disposal of sewage, and water systems were safeguarded, a great advance was made in the protection of public health, but practically no attention was given to the sanitary production of milk. In Great Britain it was commonly said that "Typhoid follows the milk-man".

During the late 1800's, the problem of providing clean milk, especially to impoverished children, attracted the attention of socially-minded pediatricians and philanthropists. Abraham Jacobi, a professor of diseases of children at New York Medical College, in 1860, established the first pediatric clinic in the United States. While in Europe he had observed the success of Pasteur's heat-treatment of milk, using equipment made in Germany.

The results that Jacobi obtained in the prevention of diseases in children by the use of boiled milk was noted by other pediatricians. Dr. Henry Koplik, in 1889, opened a dispensary for the distribution of heated milk for infants. Nathan Straus, a member of a family of wealthy merchants and philanthropists, was impressed by these accomplishments. In 1893, he established a number of depots in New York, other cities, and later in Europe, where clean, heat-treated milk was distributed to the poor. This world was continued until 1920.

Conditions improved gradually, but even in 1901, W. T. Sedgwick[22] wrote:

> Among all vehicles of infectious disease there is perhaps none more dangerous than milk. This fact is the more remarkable because milk has always been one of the most trusted of human foods. Clothed in a veil of white, associated with the innocence of infancy, of high repute for easy digestibility, believed to represent perfection as a natural dietary, popular and cheap, milk has always deservedly held a high place in public esteem.

In recent years tremendous forward strides have been made, as shown by the United States Public Health Service Reports, mentioned in Chapter 7. Notwithstanding the care taken in production, milk may at times contain disease-producing organisms. The raw milk is transported in cans and tanks over long distances and opportunities for contamination may arise between the milking barn and the bottling plant which may be hundreds of miles away. In order to safeguard public health it is of prime importance that the consumer gets milk that does not contain any harmful organisms.

FILTRATION AND CLARIFICATION

Even under ideal conditions of production, it is almost impossible to prevent some dust or sediment from entering the milk. Inspection services examine milk for the presence of sediment by comparison with an arbitrary set of standard sediment discs.[23] Milk produced under good conditions gives no *visible sediment;* whereas milk with a content of sediment that exceeds a certain maximum is rejected for human consumption. The presence of sediment indicates insanitary methods of production or handling. The absence of it may mean only that the milk had been filtered. The mere removal of the evidence of contamination does not make the product clean or safe. Sediment testing is described in Chapter 20.

Often it is assumed that some of the mechanical procedures, used to prepare milk for the market, tend to reduce its bacterial content.

For example, milk usually is strained or filtered through discs of cloth, cotton, or synthetic fiber, but this process does not remove microorganisms. The purpose of filtration is to remove visible particles and dirt that may have entered the milk. A new, dry filter is used for each lot of milk, because a used filter may harbor vast numbers of bacteria that would be washed into the milk if the old filter were re-used. Usually the milk is heated within a range of about 32.2°C (90°F) to 43.3°C (110°F).

Instead of filtration, most milk plants use a mechanical clarifier to remove foreign matter from milk. The clarifier operates on the principle of the centrifugal cream separator, which is described in Chapter 13. It is so designed that the cream is not removed from the milk, but dirt, body cells, leucocytes and some bacteria, the so-called *clarifier slime*, are caught in the bowl of the apparatus. Normally, somewhat less than one-half of the leucocytes present are removed during clarification.

Clarification is an efficient way to remove dirt from milk, but it does not materially reduce its bacterial content. It sometimes has the effect apparently of increasing the bacterial count because its mechanical action tends to break up clumps of bacteria and release the individual members throughout the milk. Clarification is done preferably while the milk is cold and before it is pasteurized or just after it leaves the regenerator section at a temperature about 57.2°C (135°F) in the HTST method.

In line with modern practice, there is increasing use of "cleaned-in-place-(CIP)" equipment, which eliminate manual cleaning operations. These have a capacity of 20,000 to 65,000 pounds of milk per hour. Raw milk is pumped to the clarifer at about eight pounds pressure. Clarification continues until the bowl is filled with sediment. At this point, a timing device opens ports in the bowl wall. Water at about twelve pounds pressure is fed into the chamber, forcefully ejecting the sediment. When the water flow is stopped, the discharge ports are closed automatically, and clarification is resumed.

Fluid milk plants often use a standardizing clarifier. This clarifies the milk and at the same time removes a portion of the cream, ranging from 0.1% to 2.0%, depending upon the setting of the machine. If standardization is not needed, the standardizing value can be closed and only clarification will result.

PASTEURIZATION OF MILK

Pasteurization is the process of heating a foodstuff, usually a liquid, for a definite time at a definite temperature and thereafter cooling it immediately. The procedure is based upon the work of the French scientist, Louis Pasteur, who, between the years 1860 and 1864, noted that when beer or wine is heated for a few moments at a temperature between 50° and 60°C (122° and 140°F) there is no subsequent abnormal fermentation and souring.

Long before this time heat had been used to preserve foodstuffs. The Italian priest-naturalist, Spallanzani, noted in 1765 that boiling preserves meat extracts; and in 1782, the Swedish chemist, Scheele, preserved vinegar by boiling it. In France, the founder of the modern canning industry, Nicolas Appert, in 1810, discovered that he could preserve milk and cream, among other foodstuffs, by the application of heat to the foodstuff in a closed container. He sterilized milk in sealed, glass containers with such success that the product was used by the French navy.

Heat treatment was applied to milk by unscrupulous distributors before 1906, but only as a means of preserving the milk. In 1906, New York City began an inspection service and prohibited this practice. In 1908 the first compulsory pasteurization law was passed in Chicago, for all except milk from tuberculin-tested cows.

As applied to milk and its products, *pasteurization* means the exposure of all the product to a heat treatment which will destroy all pathogenic organisms and nearly all other bacteria, and yet not alter the flavor or composition of the product. All yeasts and molds generally are killed by pasteurization. Pasteurization is not the same as sterilization because it usually destroys only 95 to 99% of the bacteria present, whereas in sterilization the heat treatment is sufficient to destroy completely all living organisms present.

The U.S. Food and Drug Administration (The Federal Register, October 10, 1973) published new standards for milk, lowfat milk, skim milk, half-and half and cream products. (Table 11.1)

TABLE 11:1

Temperature		*Time*
C°	F°	
62.8*	145*	30 minutes
71.7*	161*	15 seconds
88.4	191	0.1 second
95.6	204	0.05 second
100.0	212	0.01 second

*If the fat content of the dairy ingredient is 10% or more or if it contains added sweeteners, the specified temperature shall be increased by 2.8°C (5°F).

Pasteurization is defined to mean that every particle of a dairy product shall have been heated in properly operated equipment to one of the temperatures specified in the table below and held continuously at or above that temperature for the specified time (or other time-temperature relationship equivalent thereto in microbial destruction).

Pasteurization done at the higher temperaures, above 71.7°C (161°F) sometimes is called *flash pasteurization*, but is better known as *high-temperature short-time pasteurization*, commonly abbreviated to *HTST* pasteurization.

After pasteurization, the product must be cooled quickly to 7.2°C (45°F) or less, sufficiently low to retard the growth of surviving organisms. It is also essential that the product be protected from contamination.

Heat is the important factor in pasteurization. It will be noted that by raising the temperature from 62.8°C (145°F) to 71.7°C (161°F),-an increase of only 8.9°C (16°F)-the milk processor is able to reduce the time of pasteurization from 30 minutes to 15 seconds. Little is gained in bactericidal efficiency if the temperature is raised over 87.8°C (190°F); it has been found that about as many bacteria are killed at 190°F as at a higher temperature.

In pasteurization, especially by the HTST and aseptic procedures, the relationship of the time and temperature needed to obtain the desired bactericidal efficiency is based upon a logarithmic function of these relationships. This relationship was established in the canning industry and was based upon a sterilizing temperature of 121°C (250°F). The time needed to destroy a given number of organisms or to establish a sterile condition at 250°F. is known as the F_0 value. The slope of the curve representing this relationship is referred to as the Z-value. This is the temperature range of the curve representing the thermal death time over a range of one log cycle-in the present example 18°F. The Z value may vary from 10 to 20, depending upon the type of microorganism under examination. The Z value needed to destroy enzymes may be much higher.

For example, if a product is held at 250°F to affect sterilization, using a Z factor of 18, and the temperature is increased by 18° to 268°, the time required for sterilization would be reduced by a factor of 10 from 10 minutes to one minute. If the temperature were increased by 36° to 284° the time needed would be reduced to 0.1 minute or six seconds to accomplish the same degree of bacterial destruction.

For many years it was considered that pasteurization at 60°C (140°F), for 30 minutes, or 71.1°C (160°F) for 15 seconds, was sufficient, because under these conditions *Mycobacterium tuberculosis* is destroyed. As this time-and-temperature relationship was found to be inadequate for the destruction of the organism causing Q fever, the United States Public Health Service recommends pasteurization for thirty minutes at 62.8°C (145°F), or at 71.7°C (161°F) for 15 seconds by the H.T.S.T. method.[14]

The procedure followed in pasteurizing milk may appear simple to one not acquainted with the dairy industry, but complicated engineering problems had to be solved before efficient and economical equipment became available. These problems included such factors as the design of the equipment, the kind of metal to use, and the means by which the milk was to be heated and then rapidly cooled.

An example of the kind of problems that had to be conquered was the formation of foam. As foam on top of a vat of heated milk has a lower temperature than that of the bulk of the milk, bacteria in the foam may not be destroyed. To ensure proper pasteurization, mechanical features had to be designed to prevent the formation of foam or to heat it and the air space above the vat higher than the temperature of pasteurization by means of live steam or hot air. Likewise, it is important that there are no leaky valves, protruding pipes, *dead-ends*, or other mechanical features which prevent the constant flow of milk during the pasteurization process. An important feature of the continuous-flow type of pasteurizers, is the flow-diversion valve which diverts the flow of heated milk back to the supply of raw milk, should the temperature of the heated milk fall below that required for pasteurization.

METHODS OF PASTEURIZATION

Batch-Holding Processes

As the name implies, in a batch-holding process the entire lot of milk is heated to a definite temperature for a given time, usually 62.8° to 65.6°C (145° to 150°F) for 30 minutes. The higher the temperature used, the shorter the holding period that can be employed. A holding period of 20 minutes sometimes is permitted if a temperature of 68.6°C (155°F) is used. A temperature over 65.6°C (150°F) is not favored because this is about the maximum temperature that will not adversely affect the flavor of the milk or lessen the volume of the cream layer, usually called the *cream line*, in unhomogenized milk. If the milk is heated too high, a *cooked* flavor is imparted and the creaming ability is impaired. As the consumer used to judge the quality of bottled milk by the depth of its cream layer, the milk dealer was careful not to treat the milk in a manner that would injure its cream line. The widespread use of homogenization and of paper containers has eliminated concern over the cream line. Actually no serious reduction in the volume of the cream is caused by temperatures up to 72.2°C (162°F), when applied for 15 seconds; but at 79.4°C (175°F) the milk acquires a distinct, cooked flavor and its creaming ability is practically destroyed.

In a batch-holding process, the entire operation of heating, holding and cooling the product may be done in the same unit. Plants that process less than about 2500 quarts per day can operate economically with the batch-holder process. With larger volumes of milk, the *HTST* method is more efficient.

Spray Vat:

In the *spray vat* system, water at 65.6°C (150° F) is forced through perforated pipes placed between the outer and inner walls of a double-walled vat. The milk is agitated by slowly moving paddles or blades suspended from and moved by a mechanism on the top of the vat. In a *coil vat* pasteurizer the milk is heated and at the same time gently agitated by a revolving coil through which hot water passes. In some installations the heating medium is low-pressure steam but this does not permit as easy control of the heating as does the use of hot water. The coil usually is rotated at about 125 rpm. The time and special attention needed to clean the coil has decreased the popularity of the coil vat.

Continuous-flow:

In the *continuous-flow system* the milk is impelled or pumped through a series of pipes or tanks while it is being heated. The length and capacity of the milk containers or holders and pipes is such that exactly 30 minutes are needed for the complete procedure of heating and holding. A regenerative heating system generally is used. In this process, the cold incoming milk flows over the tubes that contain the hot milk leaving the pasteurizer. Thus the raw milk is partially heated before it enters the pasteurizing unit. The outgoing pasteurized product is cooled before it enters the equipment where it is brought down to the final temperature 4.4°C (40°F). The pasteurized milk that flows through pipes is held at a pressure slightly greater than that of the incoming raw milk so that, in case of a leak in the system, the raw milk will not enter the pasteurized supply, but on the contrary the pasteurized milk is forced to the side of the raw milk flow. This simple device prevents accidental contamination of the pasteurized milk with the raw product, should the equipment become defective. The flow-diversion valve is an important part of any continuous flow system.

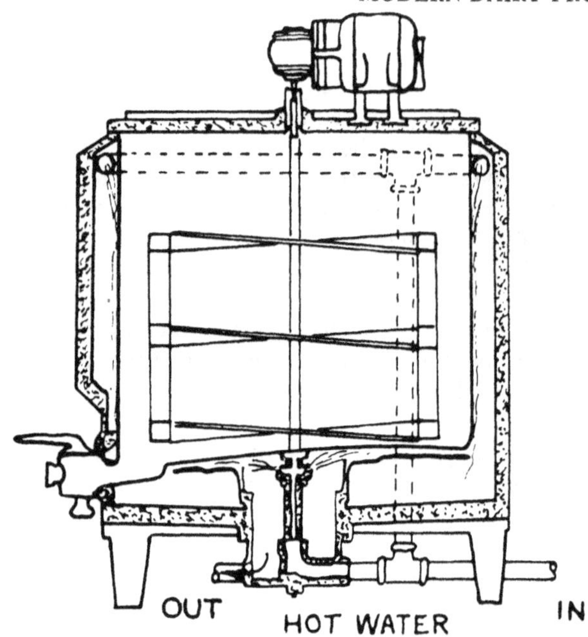

Fig. 11-1. Sectional View of Simple Batch-Type Pasteurizer (Showing spray-heating system and agitator)

High-Temperature Short-Time Pasteurization

High-temperature short-time pasteurization (H.T.S.T.) consists of heating milk rapidly to a temperature of not less than 71.7°C (161°F) for not less than 15 seconds. Usually the milk is held at a temperature between 71.7°C (161°F) and 74.4°C (166°F) for 15 to 16 seconds. The principle of the continuous-flow pasteurizer is an essential feature of high-temperature short-time pasteurization. In one method, the milk flows rapidly over the surface of flat, rectangular stainless steel plates. These plates are criss-crossed with grooves, knobs, or ripples, which direct the milk in a turbulent flow completely over the plates, so that all particles of the milk are in motion and evenly heated. On the other side of these plates, hot water flows in a direction opposite to that of the milk.

In another modification of equipment for H.T.S.T. pasteurization the cold, raw milk is heated in tubes that pass through a steam-heated cylinder. In both cases, the temperature of the milk is closely regulated by a sensitive thermometer connected to an electrical control mechanism. During the holding period, the milk is not allowed to drop a fraction of a degree below the temperature of pasteurization. Should the temperature drop below this point, a flow-diversion valve is acted upon instantly and all of the milk in the pasteurization unit at that time is emptied into the raw milk supply. The time needed for pasteurization is controlled by the length and capacity of the tubes as well as the rate of speed at which the milk is forced through the equipment. Usually one to three minutes are needed for pasteurization by this method, counting from the time the milk is preheated until it is cooled. The holding time is measured preferably with an electrical conductivity device, with which the time taken for a saline solution to pass through the holding tube is determined. Although satisfactory for determining the holding time for milk, a correction must be made for other products such as cream, ice cream mix, and condensed skim milk which have a higher viscosity than milk.[6]

A dye solution may be used instead of a salt solution and the presence of the dye at the discharge end is measured by an optical method. Less often, cold pasteurized milk may be passed through the unit, and the time needed for the flow diversion valve to operate is noted.

Ultra-High Temperature Process:

A development in H.T.S.T. pasteurization is the Ultra-High Temperature process.[20] It is used principally for processing milk but has found much application also in the processing of ice cream mix, cream, and cream toppings.

There is no legal definition of ultra-high temperature pasteurization throughout the United States, but in England it means heating the product to not less than 270°F for not less than one second.

The product is heated to 285°F for 15 seconds and then to 300°F for one-half second. In one procedure, the product is under a pressure up to 100 pounds per square inch, in order to produce high turbulence and prevent any burning onto the plates as it passes through the pasteurizer. After this practically instantaneous heating, the product may be further heat-treated by being diverted to a holding tube for about 25 seconds at 180°F. Products treated in this manner are essentially sterile at this stage. If a sterile product is to be made, it is first treated in a special homogenizer and then bottled or canned by an aseptic filling procedure, as described in this chapter.

Other than some loss of ascorbic acid, some folic acid, and of vitamin B_{12}, and about 20% of the thiamin content, there is little loss of the vitamin content. UHT treatment results in up to 70% denaturation of the serum proteins, especially lactoglobulin. The denaturization of the soluble proteins results in a whiter color of the milk. A homogenizing effect of the heat treatment, also reduces the size of the fat globules.[27]

Infra-Red Ray Treatment

A French process is reported to use infra-red rays to pasteurize or sterilize milk products. The product is passed in a continuous flow over horizontal stainless steel plates placed under infra-red tubes. The temperature is raised quickly to 85°C (185°F) or other desired temperature and held in a holder for the desired time.

Vacuum Pasteurization

A method of pasteurization by means of direct heating with steam in a vacuum chamber was first developed in New Zealand. The equipment is known as the *Vacreator*. In *Vacreation*, the product within a chamber under a low vacuum, about 5 inches, is flash-pasteurized at 90°-96.2°C (194°-205°F) by direct contact with steam. The product then passes to another chamber where the vacuum is somewhat higher, reaching to about 20 inches. In this chamber it undergoes distillation, wherein excess steam and gases are removed. Finally it enters a third chamber where there is a high vacuum—about 28 inches. Here extraneous flavors and odors that may be present in the milk or cream are eliminated, together with some water vapor. The product leaves this chamber at a temperature of 32.2°C-43.3C (90°C-110°F) and is then passed over a surface cooler. It takes about one minute to process milk from its initial temperature of 1.7°C (35°F) until it is pasteurized, homogenized and again cooled to 1.7°C (35°F).

Vacuum pasteurization is an efficient method to remove feed and weed flavors from cream for buttermaking. It reduces the bacterial count by over 99% and destroys all molds and yeasts that may be present.[24]

A number of vacuum-pasteurization systems have been developed since the introduction of the Vacreator. In a typical system, the raw product passes through a plate heat-exchanger into a vacuum chamber at a temperature of 73.8° to 76.7°C (165° to 170°F). In the vacuum chamber,the product is formed into a spray and it falls to the bottom over a series of baffles. At this time, air, odors, and flavors from feed are removed from the product. A double vacuum chamber may be used if much off-flavor or feed odor is to be removed.

From the vacuum chamber the product is pumped to the regenerative section of a heat-exchanger, wherein it is heated between 76.7°C (170°F) and 91.1°C (196°F). If treatment at higher heat is desired, the product is passed to a heat-treatment unit where it is heated to 104°C (220°F) or higher as needed. Where a holding time is needed, a holding tube is used. The product next passes a flow-diversion valve, at which point, if it is at the desired temperature, it passes back through the cooling section of the regenerative heat-exchanger and on to the final cooler. In another system, the product is first heated to the desired temperature and then passed to a vacuum chamber, wherein off-flavors and odors are removed. Vacuum pasteurizers for ultra-high heat are available, which will process up to 80,000 pounds of milk an hour.

Vacuum pasteurization tends to raise the freezing point of milk about 0.005°C., indicating some condensation of water vapor or removal of carbon dioxide. To meet regulatory requirements the operations must be so controlled that there is neither dilution nor enrichment of the pasteurized product. This requires that the steam condensed during heating is balanced by the vaporization of an equal amount of water from the sterile product.

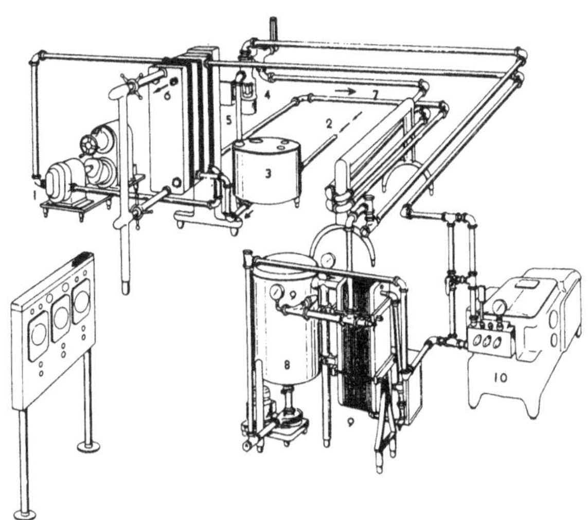

Fig. 11-2. Flow Diagram for High-Temperature, Short-Time Vacuum Pasteurizer (Aro-Vac) (Courtesy Cherry-Burrell Corporation)

1. Milk-timing pump
2. Raw Milk line
3. Ballast tank
4. Diversion valve
5. Regenerator unit
6. Pasteurized milk exit
7. Holding tube
8. Vacuum chamber
9. Water condenser
10. Homogenizer

Fig. 11-3. Milk Processing Room. From left to right - ballast tank, timing pump, heat-exchanger, holding tube, vacuum chamber, homogenizer with pressure gauge. Courtesy Crepaco, Inc. Chicago, Ill.

A flow diagram for a typical HTST, vacuum-pasteurizing installation is shown in Fig. 11.2. An actual milk-processing room is shown in Fig. 11.3.

Cooling Milk After Pasteurization

In order to prevent the growth of surviving bacteria in pasteurized milk, it is essential that the product be cooled rapidly after the heating period. Market milk and cream ordinarily are not cooled in vats agitated by paddles or coils, because agitation of warm milk favors separation or churning of the fat. As a rule, a counter-current flow of milk and cooling medium is used to cool the product. Market milk plants generally use surface-type coolers, plate coolers, or double-tube cooling equipment, of the following types:

Surface Coolers:

Surface coolers consist of a series of horizontal tubes, usually built in two or three sections. The hot milk flows in a thin layer down over the surface of the tubes. Through the tubes of each section, the cooling medium is circulated. Cold water flowing through the upper section is used to cool the milk to about the temperature of the water. Refrigerated brine, water or another cooling medium is used in the lower section in order to cool the milk to the desired temperature. As the cooling medium enters the bottom of each section of the cooler, the coldest milk comes in contact with the coldest section of the cooler.

Cabinet Coolers:

Cabinet coolers consist of a series of surface coolers placed close together, thus permitting a large area of cooling surface to be installed in a small floor space.

Plate-type Coolers:

Plate-type coolers operate in much the same manner as the plate-type milk heaters used to pasteurize milk. Instead of hot water, however, cold water or refrigerated brine is circulated between the plates in order to cool the milk. Fig. 11.4.

Unless maintained in a slightly alkaline and air-free condition, brine solutions tend to corrode surface milk coolers. Propylene glycol often is used as a cooling medium, but becomes too viscous for proper flow when used for cooling near -17.8°C (0°F) or lower.

Sterilized Milk

The commercial production of sterilized whole milk and of cream is a development that has increased in importance since it was introduced about 1940. Before this time, chocolate-flavored milk was available in cans. It was processed by the Heat-Cool-Fill (H.C.F.) system, in which the cans and covers were sterilized with steam and then passed through rotary valves, into a closed chamber, where they were filled with the cold previously sterilized, product. The

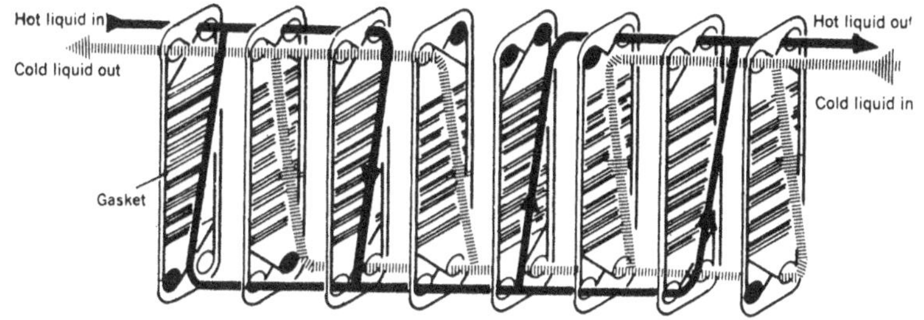

Fig. 11-4. Flow Diagram of a Plate Heat-Exchanger
Courtesy A.P.V. Company, Inc. Buffalo, N.Y.

filling-and-sealing chamber was sterilized before use with steam and while in use, was maintained with steam or a mixture of steam and sterile gas at a pressure of two to three pounds per square inch.

Shortly after 1940, sterilized milk, light cream, and whipping cream became available in glass containers. In the *Avoset* process, the product is sterilized by direct injection with steam followed by evaporative cooling. The glass containers, with their closures, are sterilized in a continuous hot-air sterilizer which discharges them into the filling room. The air for this room is sterilized before it enters and contamination during the filling operation is avoided by the use of germicidal ultraviolet lamps.

Sterilized milk has found its greatest use in places where refrigeration is not generally available. It is used in Europe, especially in Italy, aboard ships, and by the armed forces abroad.[2] Sterile whole milk in paper containers was introduced about 1959[2] in Canada, and about a year later in California.[8]

Direct injection with steam is the preferred method of heating the milk since at the high temperature used indirect methods employing tubes or plate heat-exchangers favors the deposition or "burn-on" of milk solids.

In one process, milk is heated to 74.4°C (165°F) in a plate heat exchanger and then live steam is injected to raise the temperature of the milk to about 150°C (300°F). After 2.5-3 seconds at this temperature in a holding tube, the sterilized milk is flash-cooled in a vacuum chamber to between 76.6°C (170°F) and 80°C (176°F) to remove condensation water. The low pressure vapor is used to preheat the incoming milk. The milk is then pumped under sterile conditions through a two-stage homogenizer, aseptically sealed by steam (see Chapter 12). Homogenization at from 5000-8000 pounds pressure tends to minimize development of a chalky flavor in sterilized milk.

The relatively severe heat treatment during sterilization causes some loss in the thiamine and vitamin B_2 content of the milk. In the canning industry the loss of thiamine is used as a measure of flavor destruction in the sterile product.

The liberation of sulfhydral compounds during sterilization produces an off-flavor which usually persists in the milk for several days. During this time an improvement in flavor occurs through oxidation of the sulhydral compounds. A method to introduce sterile oxygen into the milk to assure acceptable flavor at the time of packaging was patented in England in 1964. The addition to sterilized milk of 100-150 ppm of dioctyl sodium sulfosuccinate, a wetting agent, is reported to improve and maintain the flavor of sterilized milk during storage.

The bacterial content of sterilized milk is so reduced or destroyed that the product may be stored at room temperature, as is done with evaporated milk. When the sterilized product is placed aseptically into hermetically sealed containers, further heating in the container is avoided, resulting in improved flavor.

The addition of about 0.6 lb. of sodium hexametaphosphate per 100 pounds of milk solids present is an effective means for the prevention of gel formation or stratification of milk solids during storage.

ASEPTIC PROCESSING

A recent development in the processing of dairy products is packaging or canning under aseptic conditions (see Fig. 11:5). The procedure, conceived by W. M. Martin[13,17] and now known as the Dole Aseptic Canning System, is used in the commercial packing of many heat-sensitive foodstuffs, such as sauces, soups, fruits, and vegetable purees, baby foods, and milk products. Aseptic canning of milk products utilizes high-temperature short-time sterilization, following any of the conventional procedures, such as steam injection and subsequent vacuum treatment to remove the condensate by a heater of either tubular-type or plate type.

Before starting operations, the filler and the closing machine are sterilized with superheated steam at atmospheric pressure. In 40 to 60 seconds the empty cans are carried on a conveyer through a tunnel where they are sterilized with superheated steam at a temperature up to 288°C (550°F). The cans, themselves, reach a temperature up to 425°F, a little below the melting point of tin. The covers for the cans are sterilized in a similar way, in a separate, in-line unit.

Cold, sterile milk flowing from the heat-exchanger is filled at a temperature of about 12.8°C (55°F) into the cans as they pass under the filling nozzle. Cans of different sizes are filled at the rate of 20-25 gallons per minute. Before filling, the outside of the cans is cooled by a stream of cold, sterile water. An atmosphere of superheated steam or hot inert gas is maintained in the filling section and the related parts of the unit in order to maintain sterility and prevent the entry of airborne contamination. Even though the cans enter the filling section at about 12.8°C (55°F), the temperature in this area is so high that the contents of the closed containers may reach 21.1°C (70°F). As they leave, the filled cans are washed and dried. Usually the filled cans are stored for ten to thirty days in order to allow detection of any defective containers. Until recently the cost of the equipment had

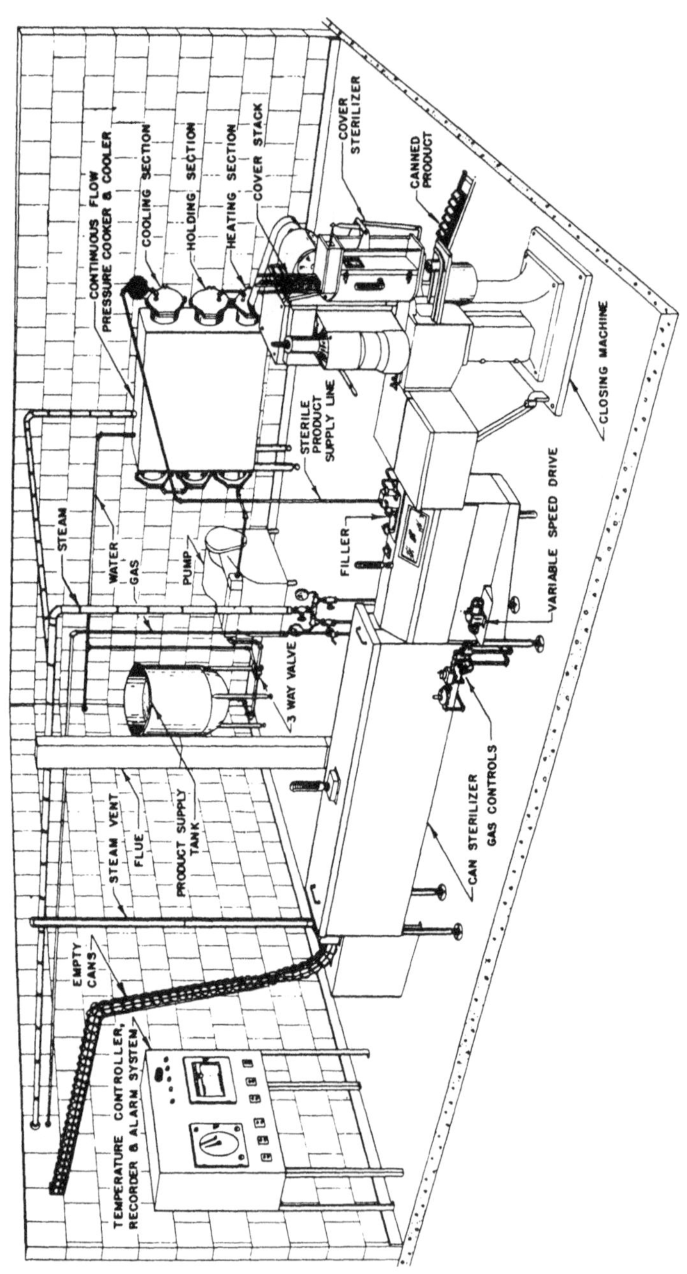

Fig. 11-5. Diagram of Martin Aseptic Canning Process (Courtesy of James Dole Engineering Corporation)

restricted the use of aseptically canned milk to areas where fresh milk is not available, such as for areas overseas, military use, and aboard ships.

Aseptic canning is used for whole milk, concentrated milk, chocolate milk, coffee cream, whipping cream, ice cream mix, and formulated infant foods. The product may be held for many months without refrigeration. As the product has not been subjected to severe treatment by heat, the color, flavor, and nutritional factors undergo relatively little change. At least one brand of evaporated milk is canned by the aseptic system, thus eliminating treatment with high heat for sterilization in the can. By a modified technique, whipping cream and similar products are pressure-canned as aerosols. They are filled under germicidal ultraviolet lights in a sterile room that is fitted with air cleaners.

The Tetra Pak Container

The Tetra Pak is a milk container developed in Sweden in 1952. The machine on which the Tetra Pak is made forms, fills, and seals the finished carton. The original, tetrahedral-shaped carton is made from a roll of polyethylene-coated paper. In making the carton, such as the small, individual cream-serving carton, the first operation is to punch a hole in the paper and apply a heat-sealed tab to form a pull-tab opening, for emptying or for the insertion of a straw to drink the contents. The paper then passes through a sterilizing bath, containing either a solution of hydrogen peroxide or a solution containing 400 ppm of available chlorine. The excess of sanitizing solution is removed by a squeegee. As the paper is formed into a tube and heat-sealed longitudinally at a temperature of about 400°F, it is sterilized and any remaining sanitizing solution is decomposed.

The milk or other product to be filled is pasteurized at ultra-high temperatures and enters the tube from above, through a stainless steel pipe. The tube is then sealed transversely by two sets of jaws at right angles to each other which heat-seal the package under about two tons of pressure (7 in Figure 11.7). The container, literally, is formed around the column of milk. The level of milk is kept above the end of the supply pipe to eliminate foaming. The containers, having been formed and filled in a continuous chain, are sheared apart into individual units and packed automatically into the shipping carton.

The Tetra Pak Brik equipment makes completely rectangular cartons, with no protuding flaps or gables. A continuous strip of paper laminated with a vapor-barrier of aluminum foil and plastic is formed into an impervious, air-type container with a pouring-spout which folds down the side to give protection to the contents and

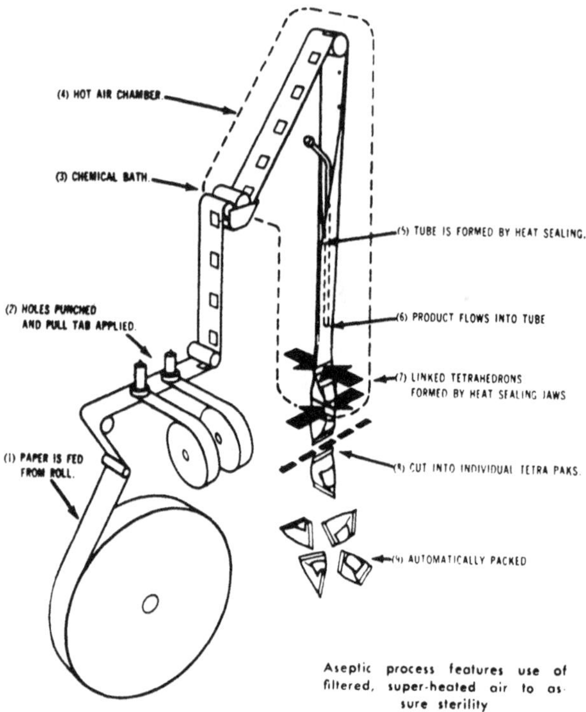

Fig. 11-6. Principles of Tetra Pak Process (Courtesy of Milliken Tetra Pak, Clemson Industries, Inc.)

pouring lip. The carton is filled completely, without any air space, and is given its final shape after it is filled. The rectangular shaped cartons are gradually replacing those that are pyramid shaped.

RADIATION

The processing of foodstuffs by radiation, as a substitute for pasteurization, has received much attention. Only two sources of ionizing radiation need be considered for this purpose:

1) An electron beam from a machine such as the Van de Graaf generator, or from a linear accelerator:
2) Gamma rays from a radioisotope, such as cobalt-60.

It appears that the higher the form of life, the more susceptible it is to injury from radiation. Bacteria, for example, can withstand over 1000 times the ionizing radiation that would be lethal to man. About 5% of the ionizing energy needed to sterilize a product usually is

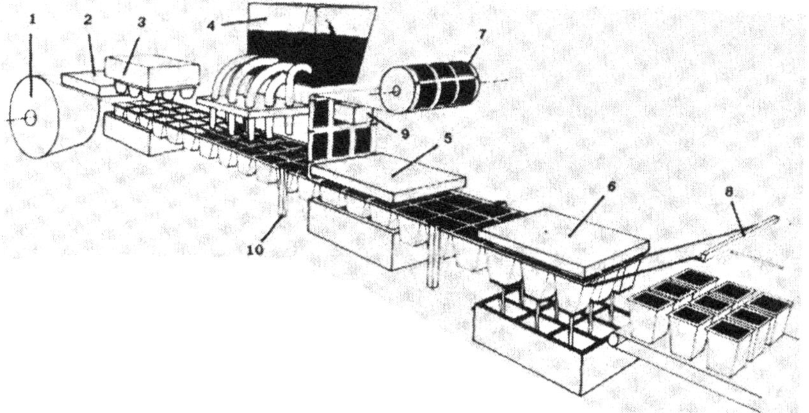

Fig. 11-7. Principles of the Benz & Hilgers Formseal packaging machine. It is capable of forming, filling and sealing numerous styles of containers for liquid or solid materials. (Courtesy of Len E. Ivarson Inc., Milwaukee, Wisconsin)

1. Web roll of thermo-forming plastic material.
2. Heating unit
3 Forming press
4 Filler
5 Sealing press
6 Trimming unit
7 Aluminium roll for lids.
8 Scrap trimming
9 Foto electric cell
10 Web clamps

sufficient to "pasteurize" it, by reducing the bacterial population 90 to 99%.

The unit used to measure ionizing radiations is the *rep*, an abbrevition for *roentgen-equivalent(s)-physical.* *

Between one-half and three-quarters of a million rep are needed to effect a reduction in the bacterial content of milk equivalent to that from pasteurization, but at least two million rep are needed for complete sterilization. As little as 7000 rep will produce an off-flavor in milk, and 20,000 will do so in skim milk. At these energy levels little or no radioactivity is induced in milk.[11] Very little heat is

*The *rep* represents an absorption of energy of slightly less than 100 ergs per gram of material of unit density. For practical purposes it is about equivalent to the roentgen unit, i.e., the amount of X-or gamma radiation which, when passing through pure air under standard conditions, produces one electrostatic unit each of positive and negative ions per cubic centimeter. One roentgen liberates 83 ergs per gram of energy in air.

produced during radiation of a food. A raw food is still raw after being treated.

The potential toxicity of irradiated milk products has been studied in tests on the feeding of animals.[18] In most tests, no gross toxic manifestations were found, but in one investigation[11] rats that were fed cooked irradiated milk developed heart leisons. Change in flavor is an important obstacle to the use of radiation by the dairy industry. Most of the off-flavors produced by the irradiation of milk are volatile and in large part may be removed by vacuum distillation. With some products, a color change may occur; thus, a colored dairy product, such as yellow cheddar cheese, is bleached by radiation. Changes in texture also may occur; irradiated milk jells in storage. As enzymes resist destruction by radiation, enzymic activity may produce undesirable changes in the milk.

Radiation, like cooking, is not considered to alter materially the nutritional value of the major constituents of a foodstuff, but it does act upon the vitamins. Of the water-soluble vitamins, thiamine is most sensitive, and niacin most resistant, to destruction. Gamma radiation of fresh milk results in marked reduction in its content of vitamin B_6, riboflavin, ascorbic acid, and thiamine, but little or no change in other B-complex vitamins. Vitamins A, E, and K are also sensitive to ionizing radiation.[18] The extent of the loss of vitamins is related to the dose of radiation. The United States Food and Drug Administration holds that irradiated foods are unknown quantities and each product must be tested individually.[10] Radiation of dairy products is not permitted.

Ultraviolet Light

Equipment to pasteurize milk by exposure to *ultraviolet* light was developed in Germany during World War II. The milk was pumped through spiral units made from quartz tubing, about 125 feet in length. The tubing was housed in aluminum cylinders, about seven feet long and two feet in diameter. The cylinder and its spiral was flooded with light from a series of eight ultraviolet light tubes. Each unit had a capacity of about 100 gallons of milk an hour.

For most effective pasteurization of milk by ultraviolet light, the principal wavelength is about 2537 Angstrom units.

The United States Department of Agriculture found that irradiation with ultraviolet light imparted an unpleasant flavor and odor to milk. Bacterial spores normally present in milk were not killed. The application of heat sufficient to inactivate lipase in the milk intensified the off-flavor. It was concluded that milk sterilized by ultraviolet radiation would not be acceptable to the consumer.[3]

Ultraviolet light is used in commercial units which can sterilize up to 7,000 gallons of water an hour for the manufacture of some dairy products and for washing dairy equipment. The U.S. Public Health Service has issued specifications for the equipment used to disinfect water. Among the requirements, radiation at 2537 Angstrom units must be used at a minimum dosage of 16,000 microwatt-seconds per square centimeter through the water disinfectant chamber.

It takes about ten times as much ultraviolet radiation to kill gram-positive bacteria than the gram-negative organisms. Ultraviolet lamps are sometimes used in dairy plants to inhibit atmospheric contamination during the preparation of cultures, cultured products, and during the packaging of butter and cheese.

TREATMENT WITH HYDROGEN PEROXIDE

Much of the milk produced in tropical and undeveloped areas often is of very poor quality, especially from the bacteriological viewpoint. Such milk may hardly be fit for processing by the time it arrives at a dairy plant. When delivered to the consumer directly, the milk usually is boiled before it is used—a treatment that serves to destroy any harmful organisms that may be present and also improves the keeping quality of the milk. The use of a preservative, such as formaldehyde or boric acid, constitutes an adulteration that could become a health hazard.

Much investigation has shown that, under certain conditions, a potable grade of hydrogen peroxide may be used as an acceptable preservative.[21] Such hydrogen peroxide is manufactured by an electrolytic process and is free of poisonous, heavy metals, such as barium, which may be present in some grades. Hydrogen peroxide is unique as a preservative in that it may be removed from milk before it is consumed.

The use of hydrogen peroxide to preserve milk was introduced in Germany about 1900. By 1903, Budde received a German patent for Buddized milk in which hydrogen peroxide was added to milk. After 15 minutes the milk was heated to 50-55°C to promote destruction of the added hydrogen peroxide by the catalase present in the milk. Impurities in the hydrogen peroxide used and in the catalase preparation sometimes added, led to undesirable defects of flavor and limited the use of the milk. Today, improved procedures permit the preparation of an acceptable product. During the war years, hydrogen peroxide was used in Germany to treat milk, instead of pasteurization by heat, thus saving fuel.

In Italy, treatment with hydrogen peroxide is officially recognized as a substitute for pasteurization. The Food and Agricultural Organization of the United Nations[19] recommends that under some circumstances, hydrogen peroxide may be used to improve the keeping quality of milk, but should not be substituted for pasteurization, because, in the amount used (0.01 to 0.08% of the weight of the milk), some pathogenic organisms are not destroyed. Morris[16] found that a concentration of 0.3% hydrogen peroxide for 20 minutes was needed to destroy *Salmonella typhosa.*

A solid source of hydrogen peroxide has found limited use. One tablet made with carbamamide peroxide (64% urea and 36% hydrogen peroxide) will furnish 0.06% hydrogen peroxide, enough to treat 500 ml. of milk.[25]

The hydrogen peroxide should be added to the milk within an hour or as soon as possible after milking; because within the first hour there is usually little bacterial growth, and the catalase activity of the milk is lowest (see Chapter 7). The catalase normally present in raw milk will decompose added hydrogen peroxide in 12 to 20 hours and pasteurization will speed the destruction. Usually between one and two millliliters of 33% hydrogen peroxide, diluted with water, to the liter of milk is sufficient to inhibit souring for one day under adverse conditions. When the milk is pasteurized or boiled, the peroxide is destroyed. Milk that contains hydrogen peroxide should be allowed to come in contact with only stainless steel, because such milk corrodes copper and tin.

Although the use of fluid milk treated with hydrogen peroxide is not permitted in the United States, milk, cream, and ice cream mix treated by the Winger Process (U.S. Patents, 2,596,753; 2,622,983, Dec. 1922) is made for export use. In this process, heated milk is treated with hydrogen peroxide for 30 minutes, cooled to 43.3°C (110°F), and a catalase preparation is added. After 15 minutes, the treated milk is pasteurized and poured into cans, which are sealed and sterilized as in the manufacture of evaporated milk. Such milk has remained usable after two years in storage without refrigeration.

There is no evidence to indicate any hazard to health in the use of milk treated with hydrogen peroxide and catalase, even though its nutritive value is slightly reduced. It has been shown[7] that there is no significant effect on vitamin A, E, carotene, or the various B vitamins—except vitamin B_{12}—in the milk. The loss in nutritive value is about 6%, and it appears to be due to changes in the milk proteins, especially a loss of methionine, which is unstable in the presence of hydrogen peroxide.

The use of milk treated with hydrogen peroxide is permitted in the

United States for the manufacture of cheese; this is discussed in Chapter 19.

OTHER METHODS OF PASTEURIZATION

Electropure

It is possible to pasteurize milk in equipment heated by electricity. In the Electropure system, a thin film of milk was forced upwards between the walls of a narrow, rectangular chamber, the opposite walls of which consist of two flat carbon electrodes. These are separated from each other by plate glass partitions. As the milk flows past the electrodes it completes the electrical circuit between them. Heat is liberated directly into the milk, which is thus pasteurized as efficiently as with methods where heat is applied indirectly. There is no evidence that the pasteurization is due to any specific electrical effect other than the heat produced. This method is no longer used commercially.

In-the-Bottle

The system of pasteurizing milk in the bottle is not used in the United States, but it has been used to some extent in England. The milk bottles or metal containers first are heated with hot water, then with superheated steam to sterilize them, and finally they are filled with milk that previously has been heated to 63.9°C (147°F). The filled and capped bottles are then held in a vat of water heated to 63.9°C (147°F) for 30 minutes or the time required for pasteurization. In a modified procedure, the containers are hermetically sealed and heated to 120°C (248°F) for at least five minutes. By this procedure a sterilized product may be obtained. The bottles are cooled by means of cold water. At no time does the water level in the heating or cooling procedure reach to the cap of the bottle. The *in-the-bottle* system of pasteurization is supposed to prevent possibility of contamination because the milk is always in the bottle until it is opened by the consumer. The difficulty of heating and cooling milk through the glass bottle adds to the cost of pasteurization, but more important is the fact that possibly not all the milk will be heated to the proper temperature. The procedure was not adaptable to large scale operation.

Uperisation

Uperisation, a Swiss process, and Palarization, a Danish procedure, are names applied to H.T.S.T. processes for sterilizing milk. The trade-marked name *Uperisation* is derived from

Ultra-Pasteurization. The milk to be processed is pumped through two pre-heaters in which its temperature is raised to 77.4°C (173°F). It is then passed to a high-pressure unit where its temperature is raised to 150°C (302°F) by direct injection of steam. The heated milk is cooled quickly to 80°C (176°F) as it is passed into a vacuum chamber where the excess water derived from the injected steam as well as foreign odors are removed from the milk.

To avoid concentration or dilution of the product, the amount of live steam injected is balanced by the amount of vapor removed while in the vacuum chamber. This is done by electronically comparing the temperatures of the product during processing. The equipment is so calibrated that the specific gravity is controlled to within 0.04% of the required value. The accuracy attained is claimed to be greater than that obtained by a comparison of freezing point values before and after processing.

Bactofugation

Bactofugation is the name applied to a method for removing bacteria from milk by a certrifugal process, developed by Professor Simonart of Belgium and the Alfa-Laval Company in Sweden. The

Fig. 11-8. Bactofuge. Courtesy of The De Laval Separator Company Poughkeepsie, N.Y.

process does not constitute pasteurization in-as-much as heat treatment is not involved, Fig. 11:8.

When milk is centrifuged in a clarifier, the bacterial count is lowered during the first few minutes of operation and then becomes constant. In the Bactofuge process, the centrifuge is operated at about 20,000 rpm and is modified so that a small amount of skim milk, about 2.5%—6% by volume, is discharged continuously[26]. When operated around 76.6°C (170°F) about 90% of the bacteria present in the milk are removed. The centrifuged milk is next passed through a second Bactofuge and about 90% of the remaining 10% of bacteria is removed, making a total of 99% removal. The 2.5% discharge from this operation is returned to the raw feed, keeping the loss of milk to about 2.5%. The discarded milk is used for animal feed. Sediment and bacteria concentrated in the milk is gathered in the bowl casing, from which it is removed through a special sludge groove. The bowl is double jacketed so that cooling water may be circulated during operation. This prevents the sludge from baking onto the inner surfaces of the bowl.

The Bactofuge process may have application in areas where the bacterial count in the raw milk reaches 20 million or more. Pasteurization of such milk may still leave a large number of viable

Uperisation plant capable of processing up to 1300 gallons per our. (Courtesy A.P.V. Company, Inc., Buffalo, N.Y.)

Fig. 11-9.

organisms in the milk and these would lower its keeping quality. Removal of 99% of the bacteria present before the milk is pasteurized has a beneficial effect on the product.[9,12]

Bactofugation has found its major use in the preparation of milk for cheese making, especially cheddar cheese. Swiss cheese made from Bactofuged milk may develop defects. The method may also be useful to remove yeasts and molds from milk to be used in the manufacture of cottage cheese.

REFERENCES

1. Burton, H., Progress in the aseptic filling of milk, *Dairy Industries*, 8:30. November 1961.
2. *Canadian Dairy and Ice Cream J.*, 38:32 (1959).
3. Curran, H. R. and Tamsona, A., Some observations on the ultraviolet irradiation of milk, *J. Dairy Sci.*, 53:410 (1960).
4. Dahlberg, A. C., The relationship of the growth of all bacteria and coliform bacteria in pasteurized milk held at refrigeration temperatures, J. Dairy Sci., 29:651 (1946).
5. De Mott, B. J., Influence of vacuum pasteurization on the freezing point of milk, *Milk and Food Tech.*, 30:230 (1967).
6. Dickerson, R. W., Scalzo, A. M., Read, R. B. and Parker, R. W., Residence Time of Milk Products in Holding Tubes of High-Temperature Short-Time Pasteurizers, *J. Dairy Sci.*, 51:1731 (1968).
7. Gregory, Margaret E., et al., The effect of hydrogen peroxide on the nutritive value of milk, *J. Dairy Res.*, 28:177 (1961).
8. Havighorst, C. R., Now, Sterile packing "Tetra" way, *Food Eng.*, 32:48 (1960).
9. Houran, G. A., *Bactofugation*; Convention Proceedings, Milk Industry Foundation, Washington, D.C. (October 1962).
10. Jamison, A, Irradiated Food-FDA Blocks AEC, Army requests for approval, *Science* 161:146 (1967).
11. Kraybill, H. F. and Brunton, D. C., Commercialization technology and economics in radiation processing, *J. Agr. Food Chem.*, 8:349 (1960).
12. Mann, E. J., Bactofugation of milk, *Dairy Ind.*, 33:178 (1968).
13. Martin, W. M., U.S. Patent 2 549 216.
14. *Milk Ordinance and Code, Public Health Bull.* 229, U.S. Public Health Service, Washington, D.C. (1965).
15. Monson, H. Heart lesions in mice induced by feeding irradiated foods, *Federation Proc.*, December 1960.
16. Morris, A. J., A comparative study of the treatment of milk with hydrogen peroxide and pasteurization, Report No. 9, p. 1230; *Utah State College*, Logan, Utah (1948).
17. Pflug, I. J., Hall, C. W., Trout, G. M., Aseptic canning of dairy products, *Dairy Engineering*, London, November, 1959.
18. Read, M. S., Current aspects of the wholesomeness of irradiated food, *J. Agr. Food Chem.*, 8:342 (1960).
19. *Report on the Use of Hydrogen Peroxide and Other Preservatives in Milk;* Food and Agriculture Organization, United Nations; 57/8655 (1957).
20. Roberts, W. M., Trends in ultra-high temperature pasteurization, *J. Dairy Sci.* 44:559 (1961).
21. Rosell, J. M., Hydrogen peroxide-catalase method for treatment of milk, *Can. Dairy and Ice Cream J.*, 50: Aug. 1961.

22. Sedgwick, W. T., *Principles of Sanitary Science and the Public Health;* Macmillan, New York (1902).
23. *Standard Methods for the Examination of Dairy Products,* 13th Ed.; American Public Health Assoc., New York (1971).
24. Wilster, G. H., Vacuum pasteurization of cream for butter, *Oregon Agr. Expt. Sta. Bull.* 368 (1940).
25. World Health Organization, Mimeo. Series 48, Geneva (1962).
26. Simonart, P., Bactofugation, *Netherlands Milk Dairy J.* 7:81 (1962).
27. Burton, H., Ultra-High Temperature Processed Milk. *Dairy Sci. Abstr.* 31:287-297 (1969).

chapter 12

Fluid Milk Products

HOMOGENIZED MILK

According to the United States Public Health Service Milk Code, homogenized milk is "milk which has been treated to insure breakup of the fat globules to such an extent that, after 48 hours of quiescent storage at 45°F., no visible cream separation occurs on the milk, and the fat percentage of the top 100 milliliters of milk in a quart, or of proportionate volumes in containers of other sizes, does not differ by more than 10 percent from the fat percentage of the remaining milk as determined after thorough mixing. The word "milk" shall be interpreted to include homogenized milk."[10]

The homogenizer, invented in France about 1902, was patented in the United States in 1904. The use of homogenized milk did not become popular until 1932, although it was being sold in Canada in 1927. In 1939 the United States Public Health Service Milk ordinance defined homogenized milk, but the requirement of only a 5% fat differential was too stringent for commercial use. The 10% differential explained in the preceeding paragraph was adopted in 1947.

Milk is homogenized by pumping it under high pressure through the very small opening between a valve and its seat, or between the narrow spaces of a series of discs pressed against one another by means of a heavy spiral. The openings through which the milk is forced may be adjusted within 0.001 inch. Two-stage homogenizers have two separate valves through which the milk is pumped. The pressure on the second stage usually is less than that used on the first stage. (Figs. 11:3, 12:1).

Milk, in which the fat globules average two microns or more in size, enters the chamber of the homogenizer at a rate of about 30 feet a second. As it is forced between the seat of the valve and the plug the flow of milk undergoes a rapid increase in velocity to about 300 feet a

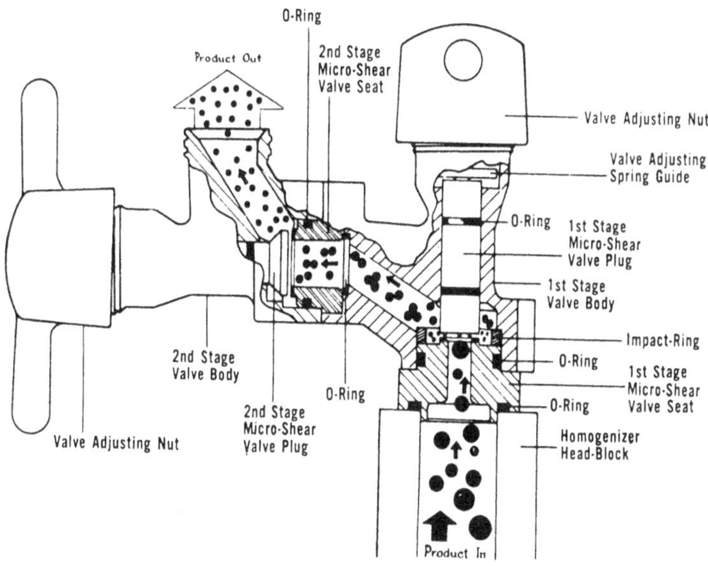

Fig. 12-1. Diagram of Two-stage Homogenizer Valve Assembly (Stainless Steel Micro-Shear Multi-Flo Homogenizer; Courtesy of CP Division St. Regis)

second at a pressure of about 2,000 pounds per square inch. At the same time, the product shifts 90° in direction during which the fat globules are reduced in size and evenly distributed throughout the product.

Various theories have been proposed to explain the reduction in size of the fat globules during homogenization, among them a shearing effect, explosion, and cavitation. The cavitation theory[16] holds that at the point where the valve clears there is a very sharp drop in pressure and a vapor space forms. This vapor cavity is collapsed as it enters the high-pressure area of the homogenizer, resulting in a shattering of the fat globules. The collapse of the cavitation bubbles, under pressure, against the homogenizer valve results in the valve system vibrating in a natural harmonic frequency of about 21,000 cycles per second. This is not unlike the vibration of an ultrasonic device. The greater the pressure used, the smaller will be the size of the fat globules. The pressure generally is 2,000 to

3,000 pounds per square inch, but 1,500 pounds often is sufficient to prevent the formation of cream on milk.

During homogenization, the temperature of the milk is raised about 1°F for each 350 pounds of pressure used; thus the homogenizer can add 5 or 6°F to the heat treatment of the milk.

Homogenizers used in the manufacture of sterile milk products have the plungers enclosed in a chamber under constant steam pressure. No part of the homogenizer that comes into contact with the milk is exposed to the atmosphere. The product in the chamber is held at a higher pressure than that of the steam, thus making it impossible for the steam to enter the product.

The ability of the globules to combine or coalesce is practically destroyed owing to the presence of a film of protein which forms around them. No appreciable layer of cream is formed on homogenized milk because of the small size of the fat globules and their ability to rise to the surface. Although homogenization will prevent the formation of large individuals fat globules, it does not prevent the small globules formed in the process from clumping together or adhering to one another. This clumping action increases the viscosity of the product, especially homogenized cream and ice cream. It is also employed in the manufacture of cream cheese and to incorporate vitamin concentrates into fortified milk. Evaporated milk is homogenized in order to delay the rising or separating of fat during storage.

Clumping is promoted by homogenization at a high pressure and relatively low temperature as well as by a high fat content. A second homogenization at a low pressure (500-1000 psi) reduces the viscosity by disrupting the clumps of fat first formed.

The greatly increased number of fat globules and their much greater surface area makes homgenized milk very susceptible to the action of lipase, with consequent development of rancidity (see Chapter 7). To prevent this, homogenized milk must be made from pasteurized milk or the milk must be pasteurized promptly after it is homogenized. A considerable saving in homogenizer capacity and time may be obtained when milk is first separated and only the cream is homogenized. The homogenized cream is added to the skim milk, thereby producing the equivalent of homogenized milk.

Homogenization alters the physical condition of the protein in milk and makes it more easily coagulated by heat or acid. This change, together with that produced by the subdivision of the fat globules, gives homogenized milk the characteristics of a soft-curd milk. The lowering of curd tension and the change in protein stability cause

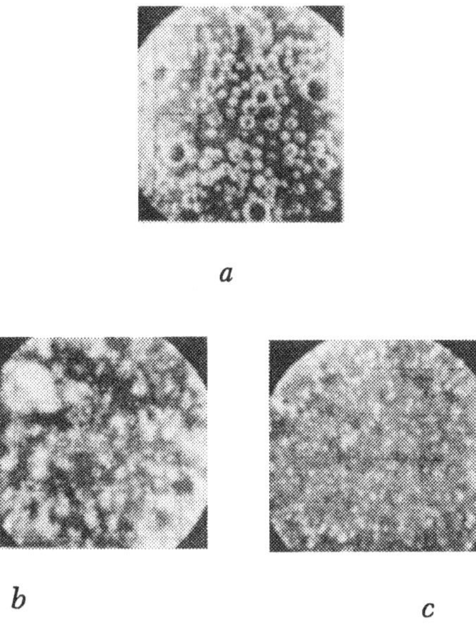

Fig. 12-2. Fat Globules in Milk (X 1000) *a*) Normal milk *b*) Poorly homogenized milk *c*) Properly homogenized milk. Each space between small vertical lines represents 2 microns; note clusters and large globules in poorly homogenized milk.

home-made custards to take several minutes longer to cook than if made with non-homogenized milk. Soups, gravies, and other cooked dishes may curdle more readily when made with homogenized milk. As homogenized milk is more opaque than the untreated product, it gives a more desirable color when used in coffee.

Homogenization makes milk less susceptible to copper-induced oxidized flavor, but more susceptible to the off-flavor induced by sunlight.

Sediment in Homogenized Milk

It is not unusual to find a layer of sediment on the bottom of a bottle of homogenized milk. This sediment consists of a mixture of the leucocytes and other cells normally present and any foreign material which may have been in milk before homogenization. Ordinarily these substances,especially the leucocytes, remain suspended in a milk or rise with the cream, but as homogenized milk does not form a cream layer, the cells and other material sink to the bottom as a grayish sediment and become noticeable, especially if the milk is packaged in glass. Clarification is the best method to remove sediment, especially if done on the cold, incoming milk at a temperature of about 4.4°C (40°F). Filtration is not efficient for this purpose.[26]

SOFT CURD MILK

The observation that milk from different cows may vary according to the kind of curd it will form has been made the subject of considerable research. It long has been known that heated milk will produce a soft coagulum with rennet whereas milk that is high in acid will form a firm curd. A somewhat similar reaction occurs with such milks during digestion in the stomach.

A number of ways have been devised to measure the hardness or tension of the curd formed by milk. In one commonly used method, the milk is coagulated by the addition of a mixture of dilute acid and pepsin which simulates the gastric juice. Before it is coagulated, a multi-bladed knife is placed in the milk and after coagulation has taken place, the tension, measured in grams, is determined by the force needed to pull the knife through the curd. This force is known as the *curd tension* of the milk and is a measure of its firmness.[5]

Fig. 12-3. Comparison of Curds (Courtesy of Evaporated Milk Association)

The curd tension of most cow's milk ranges from fifty to ninety grams. The tension of soft-curd milk is thirty grams or less, milk with a curd tension over sixty grams is classed as *hard-curd* milk. Cows that give natural soft-curd milk are relatively few in number. As a rule, Holstein and Ayrshire cows give soft-curd milk more commonly than do Guernsey and Jersey cows.

The principal chemical difference between natural soft-curd and hard-curd milks has been found to lie in their content of casein. Hard-curd milk contains more casein and if this difference is equalized by the addition of water, the curd tension of the milk is reduced. As natural soft-curd milk is relatively low in casein, its nutritive value is less than that of ordinary milk. A number of authorities believe that if soft-curd milk is desired for feeding purposes, it should be obtained by processing milk of normal composition. Among the simplest procedures for this purpose are heating or boiling the milk, diluting with water, adding dilute acid, citrate, gelatin or a cereal. Infants with weak digestive systems probably are benefited by the use of milk formulas that result in the formation of a soft, loose curd in the stomach. Such a coagulum presents a large surface to the digestive juice, in contrast to the hard and relatively impenetrable curd formed by hard-curd milk.

The ability of children to digest milk varies with their age and physical condition. Altering the character of the curd in milk for infant feeding is a practice of long standing. Cow's milk, even if boiled, is more difficult to digest than breast milk, but acidified milk, buttermilk, and diluted evaporated milk are practically as easily digested as breast milk. These milks have a curd tension of zero. Pasteurization reduces the curd tension of cow's milk about twenty percent and if the milk also is homogenized, the reduction amounts to about sixty percent.

Sonic vibration sometimes is used to produce a soft-curd milk. The process consists of passing a thin film of milk over a stainless steel diaphragm, about two feet in diameter, which is vibrated electromagnetically at a frequency of 360 vibrations a second. The mechanical agitation changes the physical character of the milk so that the fat globules are reduced in size and the curd tension is lowered.

Other methods that have been used to produce soft curd milk are the partial digestion of the milk proteins by the action of enzymes or the removal of some of the calcium from the milk by an exchange system similar to that used to soften water. A detailed study of the digestive characteristics of various types of milk compared with human milk showed that the measurement of the curd tension is of little value in predicting the digestibility and curd character of a milk.[5,27]

BUTTERMILK

Genuine buttermilk is the liquid that remains after the fat is removed from milk or cream by the process of churning butter. If the butter is made from sweet cream or milk, the buttermilk does not differ materially from ordinary skim milk. If the milk or cream churned were sour or fermented, lactic acid is present and somewhat less of milk sugar than is found in buttermilk from sweet cream.

Except for a much lower content of fat and vitamin A, genuine buttermilk has about the same composition and nutritive value as milk. When milk or cream is churned a great part of the phospholipids associated with the fat globules are retained by the buttermilk, thereby adding to its value as a phosphorus-bearing food. Buttermilk is easily coagulated in the stomach and the small curd formed facilitates its digestion. The composition of buttermilk is given in Table 12:1.

TABLE 12:1
Average Composition of Buttermilk

	From Sweet Cream	*From Sour Cream*
Water	90.83%	91.30%
Protein	3.45	3.40
Fat	0.55	0.65
Lactose	4.40	3.40
Ash	0.73	0.65
Lactic Acid	0.04	0.60

A great amount of buttermilk is dried for use in bakery goods, pancake flour, and animal feed. It has found some use as an ingredient of ice cream, especially if butter is used as the major source of fat. The high phospholipid content of the buttermilk appears to improve the whipping property of the ice cream mix.

Churned Buttermilk

Churned buttermilk, the product obtained by churning milk or cream, contains particles of butter, retained after the bulk of the butter is removed. Particles of butter may be added to cultured buttermilk in a small amount of churned cream. In another procedure, melted butter may be added to the cultured buttermilk by spraying it on the surface of the milk while it is under gentle agitation. This modification is known as *flake buttermilk*. More than the usual amount of color is added to the melted butter in order to give it a deep yellow color and so be easily seen in the buttermilk.

Most buttermilk sold today is cultured buttermilk and usually contains no butter granules. Cultured buttermilk is described in Chapter 14.

SKIM MILK—NONFAT MILK

Skim milk is that portion of milk that remains after the cream has been removed, in whole or in part. The average composition of skim milk, as coming from a cream separator, is given in Table 12:2.

TABLE 12:2
Composition of Non-Fat Milk

	Plain Skim Milk	*Condensed Skim Milk*
Water	90.42%	70.32
Protein	3.68	11.83
Fat	0.10	0.37
Lactose	5.00	15.10
Ash	0.80	2.38

Until recently, skim milk had been considered an unworthy foodstuff for human beings, although large amounts of it are used for animal feed. Skim milk contains all of the nutriments of milk with the exception of the fat and the vitamins associated with the fat. When these deficiencies are supplied from other sources, skim milk is a wholesome food for human beings. It is used in the manufacture of some varieties of cheese, especially cottage cheese. Casein and milk sugar are prepared from skim milk and vast amounts of it are manufactured into dry non-fat milk. Much skim milk is condensed for use by bakers, confectioners and for the preparation of ice cream.

Condensed skim milk is made by evaporation of skim milk to about one-third of its original volume. Its composition is approximately as given in Table 12:2.

Skim milk is used by persons that desire a low caloric value in their diet. Various State standards require a maximum fat content of 0.1 to 0.5% and a solids-not-fat content of not less than 8.0 to 9.0%. Sometimes the skim milk is fortified with vitamins A and D and a product with other added vitamins and minerals is also prepared. Canned, evaporated skim milk is available.

A concentrated sour skim milk is used for animal and poultry feed. It is acidified by the addition of a culture of bacteria that form lactic acid in amounts up to six per cent. Concentrated and dry skim milk are discussed in subsequent chapters.

WHEY

Whey is the product that remains after the removal of most of the casein and fat from milk. It is a by-product in the manufacture of cheese and casein. The greater part of the albumin, the lactose, and the mineral matter of the milk remains in whey. Cheese whey is a good source of lactose. About 22 billion pounds of whey are produced in the United States annually. The potential production of cottage cheese whey is around 450 million pounds. About 8.5 to 9 pounds of whey is obtained per pound of cheese made.

Cottage cheese whey has been used in the production of a protein-rich food supplement for use in soups and high-protein bread. Whey is sterilized and inoculated with a yeast (*Saccharomyces fragilis*), which utilizes the lactose in the whey. The yeast cells and precipitated whey protein are concentrated to about 15% total solids and then roller dried. Alcohol is produced during the fermentation and is converted to vinegar. Any dried product not suitable for food use is used as animal feed.

TABLE 12:3
Average Composition of Whey Products
(Percentage)

Product	*Water*	*Protein*	*Fat*	*Lactose*	*Lactic Acid*	*Ash*
Cheese Whey (Rennet) (*Sweet*)	93.2	0.8	0.6	4.7	0.2	0.5
Cottage Cheese Whey	93.6	0.8	0.2	4.3	0.6	0.5
Condensed Whey	55.5	8.0	1.5	28.0	1.5	5.5
Dried Whey	4.0	12.5	1.0	72.0	2.5	8.0
Cottage Cheese Whey (dried)	3.5	12.4	0.3	67.0	7.8	8.5

Whey at one time was supposed to have medicinal properties. In Europe, "Whey Houses" were established to treat various ailments. In Switzerland especially, at the present time, a popular soft drink called "Rivella" is made from an infusion of herbs and deproteinized whey. In Poland and Russia, carbonated, fermented whey beverages are used, such as "whey champagne", "whey kvass" and beer-like "Bodrost".

The procedure for concentrating whey by electrodialysis or by reverse osmosis is discussed under dry whey in Chapter 17.

In Europe, much liquid whey is fed to dairy cows. Large quantities of whey are condensed or dried for use in formulas for infant feeding and as animal feed (see Chapter 17). Its high content of both lactose and riboflavin make it a valuable component of poultry feed. Some kinds of cheese are made from whey (see Chapter 19). Dried whey is used as an ingredient in certain types of cheese spreads and in some canned soups. A modified dry whey, with reduced lactose content is described in Chapter 17.

SPECIAL MILKS

Chocolate Milk

Chocolate milk or chocolate flavored milk is an important dairy product and is used by many who otherwise would not drink milk. When it contains less than the legal amount of milk fat, the product is given a fanciful name, such as *chocolate dairy drink*. Chocolate milk has wide use in school milk programs. Various commercial preparations are available for the manufacture of chocolate milk. These contain cocoa or chocolate, other flavor materials such as vanillin, ammoniated glycyrrhizin, cinnamon, and the stabilizer. Sugar may be present or added during processing.

In general, chocolate milk contains about one percent of cocoa and from five to seven percent of sugar. Usually a stabilizer, such as vegetable gum, starch, or sodium alginate, is added to prevent the cocoa from settling out.

Some authorities on the nutrition of children have criticized the use of chocolate milk on account of its content of cocoa and sugar. It has been shown that the indiscriminate use of chocolate-flavored foods may lower the retention of dietary calcium and phosphorus, especially if the diet is already low in calcium. If the diet supplies at least 0.6 gram of calcium, the consumption of one ounce of cocoa per day has no effect upon calcium availability, according to a study made by the University of Illinois in 1965.

Frozen Milk

Milk may be frozen and so preserved for a considerable time. In the past, a difficulty in the use of frozen milk has been the destabilization of the protein during storage, so that upon thawing, particles of curd separate from the milk. Some fat may also separate. Pasteurized, homogenized milk for home use, may usually be frozen in a deep-freezer and held for several months without any appreciable defect appearing after it slowly thaws overnight in the household refrigerator. The thawed milk often is very susceptible to the development of an oxidized flavor.

A commercial method for preparing frozen milk was developed in 1956 by the National Institute for Research in Dairying, of Shinfield, Reading, England. The product, called "Frosonic" milk, is prepared by treating pasteurized milk for five minutes to sonic vibration at one megacycle a second with an ultrasonic generator, and then freezing the product within two hours. This treatment stabilized the proteins in the milk so that, when thawed, the resulting milk has essentially the flavor and texture of the original product. The frozen product may be held below -23.3°C (-10°F) for over a year without showing deterioration.

Metal containers and gable-topped paper containers, which provide space for the expansion of the frozen product, are used. The container must not hold over one gallon of the product, because of difficulty in the quick freezing and thawing of larger amounts. The milk is frozen in an ice cream freezing tunnel, but gallon-sized containers of metal may be frozen in brine tanks at -28.8°C (-20°F).

Milk concentrated to 3:1 ratio may be frozen and held at a temperature of -26°C (15°F) or lower for about six months without serious deterioration. The milk should not contain more than 40% of total solids.

Multi-Vitamin Mineral Milk

The Council of Foods and Nutrition of the American Medical Association is opposed to the indiscriminate fortifying of general-purpose foods. It is held that, except for the addition of 400 units of vitamin D per quart of the general milk supply, further fortifying treatment is not in the interest of public health. In 1971 the production of Multi-Vitamin-Mineral Milk was illegal in four states and in nine others there was no provision for its preparation.[7]

In order to meet a demand from some consumers, a multivitamin mineralized milk is available in some areas. The milk usually is fortified so that it contains the minimum daily adult requirement as established by the United States Food and Drug Administration for certain vitamins and minerals. The levels generally used are shown in Table 12:5.

These ingredients usually are added in the form of concentrates obtained from commercial sources. No satisfactory procedure is available for fortifying milk with vitamin C. The ascorbic acid (vitamin C), both what is normally present in milk and what might be

TABLE 12:4
Factors in Multi-Vitamin Milk and in Normal Milk[31]

Factor	*Amount in Quart*	*Average in Natural Milk, per Quart*
Vitamin A	4000 I.U.	500-1000 I.U.* (winter)
Vitamin D	400 I.U.	5-15 I.U.
Thiamine	1 mg	0.26-0.4 mg
Riboflavin	2 mg	1.5 mg
Niacin	10 mg	0.2-1.2 mg
Iron	10 mg	0.6-2.26 mg
Iodine	0.1 mg	0.015-0.07 mg

* 2000-3000, pasture.

added, is in most part lost during the processing and storage of milk. Iron is added in the form of ferric ammonium citrate, and iodine as potassium iodide.

Low-Sodium Milk

A daily intake of one gram or less of sodium is sometimes prescribed as part of the medical treatment in diseases accompanied by high blood pressure or edema. Such a severe reduction in sodium calls for the absence of added salt in the diet as well as the use of specially prepared foods low in sodium. The United States Food and Drug Administration requires the label on a *low-salt* or *low-sodium* foodstuff to state the number of milligrams of sodium in 100 grams of product.

Low-sodium milk is available in a number of cities. It usually is prepared in a dairy franchised by the owners of the patented process, and it is supplied to the consumer through a local distributor. The sodium content should not exceed 5 mg. per 100 ml. of milk compared to 56 mg. in normal milk.

In one process the milk is prepared by an ion-exchange procedure using a phenolsulfonic acid type of resin which contains both potassium and calcium in such concentration that the minimum of calcium is removed from the milk but the sodium is substituted by potassium.[4] The milk obtained by this process contains from 3 to 10 mg of sodium in 100 ml and is normal in all other respects except for a higher potassium content.

In another procedure[3] the milk is passed through an ion-exchange resin in the potassium form and then through a calcium-hydrogen resin to replace part of the potassium that was lost and to restore the calcium lost in the processing. The milk obtained by this method may contain less than 3 mg of sodium in 100 ml.

Care must be taken not to pass so much milk through the resin that its sodium-exchange capacity is exceeded, else the product may contain too much sodium. A measure of the sodium content can be made rapidly with a flame photometer, using a solution obtained by dissolving the ash from about 10 g of the treated milk.

Immune Milk

Working in collaboration with B. Campbell, Dr. W. E. Petersen, while with the University of Minnesota Dairy Department, announced the commercial preparation of *Immune Milk* in October, 1959.[24] The statement was made that the mammary gland will respond with the production of antibody when it is infused with an antigen. When enough of the milk produced is consumed at one time, the antibody will be absorbed from the gastrointestinal tract into the blood.

The milk was produced by infusing the udder tissue of the cow, via the teat canal, with a preparation of polyvalent antigens. The infusion, given at weekly intervals before the cow freshens, was made from phenol-killed pathogenic organisms. Another type of milk was made for the treatment of allergies and was obtained by infusing a preparation made from different pollens, yeasts, molds, and house dust.

It was claimed that the antibody is absorbed by ingesting at least one pint of Immune Milk at one time. This large amount serves to dilute the digestive enzymes and promotes the passage of the milk from the stomach into the small intestine from which the antibody is absorbed.

A number of cases of rheumatoid arthritis were reported to have been relieved by use of Immune Milk, but cases diagnosed as osteoarthritis did not respond. Immune milk, in 1963, was available in very few areas, usually only with a physician's prescription. Where it was sold, the milk was offered as a frozen product. Two portions were consumed daily, one pint at a time.

In general, the medical profession has not been favorable to the concept of Immune Milk. Both the American Medical Association and the Minnesota Medical Association disapproved of the product. In some states, it cannot be produced and distributed as a dairy product. It was considered to be a biological medicinal product and as such, it would require a Federal license for manufacture and interstate distribution. According to available information, this product is no longer prepared.

TABLE 12:5
U.S. Federal Standards for some Fluid Milk Products

Product	*Pasteurization* [1]	*Ultra-Pasteurization* [2]	*Homogenization* [3]	*Fat % (min./range)*	*MSNF min.*	*Fat % Labeling Mandatory*	*Vit. D*	*Vit. A*	*Nutrition Labeling*	*Protein Fortified*	*MSNF/Milk Derivatives*	*Emulsifiers/Stabilizers*	*Flavorings* [7]	*Sweeteners*
Milk	M	Opt.	Opt.	3.25	8.25	No	Opt.	Opt.	Opt.[4]	NP	NP	NP	Opt.	NP
Lowfat Milk	M	Opt.	Opt.	0.5-2.0	8.25	Yes	Opt.	M	M	Opt.	Opt.	Opt.[6]	Opt.	NP
Skim Milk (Nonfat Milk)[5]	M	Opt.	Opt.	<0.5	8.25	No	Opt.	M	M	Opt.	Opt.	Opt.[6]	Opt.	NP
Half-and-half	M	Opt.	Opt.	10.5-18.0	NP	No	NP	NP	Opt.	NP	NP	Opt.	Opt.	Opt.
Light Cream (Coffee Cream of Table Cream)[5]	M	Opt.	Opt.	18-30	NP	No	NP	NP	Opt.	NP	NP	Opt.	Opt.	Opt.
Light Whipping Cream (Whipping Cream)[5]	M	Opt.	Opt.	30-36	NP	No	NP	NP	Opt.	NP	NP	Opt.	Opt.	Opt.
Heavy Cream (Heavy Whipping Cream)[5]	M	Opt.	Opt.	36	NP	No	NP	NP	Opt.	NP	NP	Opt.	Opt.	Opt.

Legend — Opt. — use of process/ingredient is optional
M — use of process/ingredient is mandatory
NP — no provisions specified in standards

Footnotes:

1 — For all products. Pasteurization is mandatory but declaration of term PASTEURIZED is optional
2 — For all products "ultra-pasteurization" is an optional process — declaration of term ultra-pasteurized is mandatory on principal display panel, if applicable
3 — For all products Homogenization and label declaration (Homogenized) are both optional
4 — No vitamins are added
5 — Names(s) in [brackets] are acceptable alternate product names
6 — Limited to 2% by weight of the added milk derivatives
7 — Flavoring ingredients may contain coloring, nutritive sweetners, emulsifiers and stabilizers

HUMAN MILK

Human milk differs considerably from cow's milk in its composition and properties.[15] Especially notable are its low casein content, higher lactalbumin and lactose content. Differences in the fat also exist. As shown in Table 2:1, Chapter 2, human milk contains none or very little of the lower fatty acids that are characteristic of cow's milk. As far as the fatty acids are concerned, human milkfat resembles a margarine fat blend rather than milk fat.[13] The average composition of human milk is given in Table 12:6 but the milk of an individual woman may differ much from the average value.

TABLE 12:6
Average Percentage Composition of Human Milk
(Based on published data)

Water	87.6	Lactalbumin	0.22
Fat	3.8	Lactoferrin[22]	0.28
Lactose	7.0	Cholesterol	0.34
Casein	0.44	Serum albumin	0.07
Globulin	0.05	Ash	0.2

Composition of Ash of Human Milk

Calcium	16%
Magnesium	2
Sodium	7
Potassium	26
Phosphorus	7
Chlorine	20.5

The vitamin content is given in Table 6:2, Chapter 6. In common with the apes and monkeys, the human body cannot synthesize vitamin C. The vitamin C content of the mother's milk varies with the vitamin C content of her diet. An adequate supply of vitamin C is needed by the infant and must be furnished by supplementing the infant's diet.

Cow's milk contains about 2.5 to 3 times as much protein as does human milk. The cystine contents are about the same because human milk contains about twice the alpha-lactalbumin and serum albumin as does cow's milk. An iron-binding red protein, known as *lactoferrin*, is present in human milk. It has also been found in cow's milk as well as in some human body secretions.

The practice of diluting cow's milk with water for infant feeding is based on the fact that the protein content of cow's milk is about three times that of human milk.

The carbohydrate of human milk is *lactose*, but traces of other compounds which contain galactose, glucose, and fructose, have been identified [2,17]. These *oligosaccharides* and the *bifidus factor* are of interest in that they promote the growth of *Lactobacillus bifidus*, which is the predominating organism in the intestinal tract of breast-fed infants, in contrast to the mixed flora that are present in the tract of infants fed on formulas based on cow's milk.[12] Human milk has about forty times as much bifidus factor activity as does cow's milk. Infants nursed on human milk rarely suffer from constipation.

By use of various tests and electron microscopy, virologists have detected "virus-like" particles in human milk. No correlation has been found between the presence of these particles in the milk and a family history of breast cancer.[32]

In some developing countries, especially in Africa, urban mothers consider bottle and formula feeding to be superior to breast feeding. Malnutrition of the infants is becoming apparent.[33]

According to the Committee on Nutrition of the American Academy of Pediatrics (Ped. 48:483, 1971) about twice as many breast-fed infants suffer from hemorrhagic disease than do those fed cow's milk formulas. Nearly all newborn infants are given a parenternal dose of 0.5 to 1.0 mg of vitamin K, which is necessary for prothrombin formation needed for normal blood clotting.

Infants assimilate their mother's milk more readily than that from any other source. Whenever possible it is a distinct advantage to the child to be breast-fed, as this gives the child a better chance of survival through its first year of life than does artificial feeding.[18] Human milk at times may be overrated in its nutritive value as it may vary greatly in quality and fall short of the infant's requirements.

Healthy infants are able to assimilate more phosphorus and calcium from breast milk than from cow's milk. Although its vitamin D content is low, the high content of lactose and oligosaccharide in milk may be a factor in the low incidence of rickets among breast-fed infants. Lactose in the diet tends to increase the acidity of the intestinal tract and so creates a condition which favors the assimilation of calcium.

The ratio of calcium to phosphorus is 2.2 times higher in human milk than in cow's milk. Infants that are fed cow's milk during their first two weeks of life may develop a tetany due to a low calcium content of their blood serum. This is caused by the inability of the infant's kidneys to excrete the excess of phosphorus absorbed from the cow's milk, so the phosphorus accumulates in the serum, combines with calcium, and thus lowers the calcium content of the serum.

Young infants absorb much less of the fatty acids and calcium from modified cow's milk than from human milk even when fatty acid compositions are similar.[27] Infants fed human milk and those fed cow's milk differ in their body composition and pattern of weight gained. Breast-fed infants gain weight more rapidly during the first four or five months of life, then are surpassed by those fed artifically. Eczema, tetany, and respiratory infections are more frequent in artifically fed infants.[1]

Human milk is relatively rich in amylase, the enzyme that digests starch. The absence from human milk of xanthine oxidase is used as a test to distinguish it from cow's milk.[25]

At various times the question concerning the effect of smoking by the mother upon the suckling infant has been discussed. One investigation showed that the amount of milk secreted apparently is not affected by smoking. Nicotine was found in the milk of all mothers that smoked, whether it was only one or more than twenty cigarettes per day. In no case however, was there any evidence that the nurslings were affected by the quantity of nicotine they ingested. As a conclusion, it was stated that although the presence of nicotine in the milk may not cause demonstrable effects in the infant, it cannot do the child any good.[23]

In a January 1973 report, the U.S. Public Health Service stated that in addition to significant amounts of nicotine, smoking may pass on carcenogenic nitrosamines to the developing fetus. Unlike certain hazards, such as rubella, which do damage during early pregnancy, cigarette smoking is most dangerous in the last six months, when the central nervous system of the fetus is developing.

In some hospitals, milk is obtained from nursing mothers and stored for use by infants that for one reason or another cannot be nursed by their mothers. Usually the milk is stored in a refrigerator but in some cases it is put into small cans and then sterilized for future use. In a few instances the milk has been dried.

Although the results do not apply to breast-fed infants, experiment has indicated that human milk may have a protective effect against pathogenic organisms.[11] Mice were injected with sterilized human milk and then given a sublethal dose of virulent *Staphylococcus aureus*. Controls were given cow's milk. It was noted that human milk gave significant protection whereas cow's milk did not. The findings were interpreted to denote that human milk imparts an immunological reaction, possibly by encouraging specific production of antibodies.

The wide-spread use of pesticides has resulted in the presence of pesticide residues in human milk. A report of the contamination of human milk in the Griefswald area in Germany indicated that in 164 samples of human milk, the average chlorinated hydrocarbon content was 0.46 ppm. Cow's milk in the same area had 0.0038 ppm, indicating an accumulation of DDT residues in the mothers during lactation.[32]

A survey of 67 samples of human milk in Victoria, Australia was done in 1970. All contained traces of DDT, with an average of 0.14 ppm. In Colorado, in 1973, 40 samples showed DDT, 7 to 109 ppb, DDE, 19 to 386 ppb, and DDD, a trace to 38 ppm.

Prepared Infant Formula Products

The use of infant formula foods is common in the United States (Fig. 12:4). By the age of four months, 80% of the total daily caloric intake comes from formula or market milk. This may well lead to an imbalance in nutrition, if these products do not supply all needed nutrients. Iron deficiency is prevalent; especially among non-white and low-income groups. The use of soy protein products may require an increased requirement for zinc, unless fortified with zinc. Experimental animals fed diets with soy protein isolates have an increased need for zinc.[8]

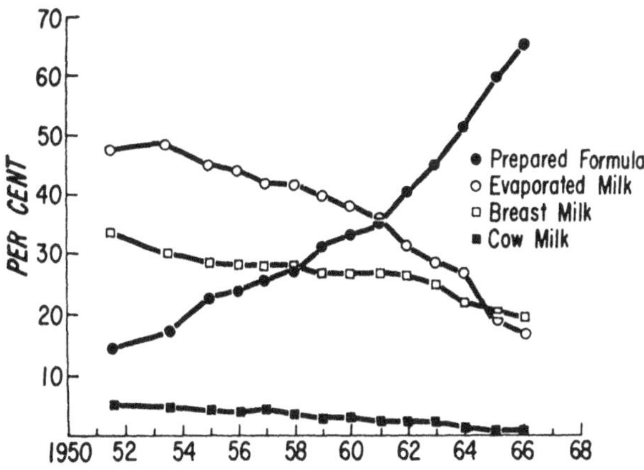

Fig. 12-4. Change since 1951 in method of infant feeding at time of discharge from the hospital[8]

GOAT'S MILK

Although there are more than four million goats in the United States, kept for mohair, meat, or as pets, only about one million are used as dairy animals. The production of goat's milk is not so important an industry here as it is in southeastern Europe and other parts of the world. Many of the short-haired goats produce but little milk but the better type of Swiss goats, especially the Saanen and Toggenburg breeds, give four to six quarts a day. This milk approaches in its composition that of Holstein cows. The Nubian goat produces milk of high fat content, similar to that of the Jersey cow.

Goat's milk is almost always white in color, as it contains little carotene, the yellow pigment that is converted in the body into Vitamin A. Although lacking in carotene, the milk is as good a source of vitamin A as is cow's milk because the deficiency in carotene is compensated by the presence of the vitamin itself.

The small size, about 2 microns in diameter, of the fat globules in goat's milk is one of its chief characteristics. This makes it impractical to obtain cream by allowing the milk to stand, because the fat rises slowly and in small amount. Goat's milk may be separated without difficulty in a cream separator. The average chemical composition of goat's milk, which is very similar to that of cow's milk, is shown in Table 12:7.[8,28]

TABLE 12:7
Average Composition of Goat's Milk

	Whole	*Evaporated*
Water	87.1	74.8
Fat	4.1	7.2
Protein	3.3	8.0
Lactose	4.7	8.4
Ash	0.8	1.6

The composition of the milk of an individual goat varies more than that of the cow, probably because of greater variation in diet.

The fact that goat's milk-fat contains about twice as much as each of capric, caprylic, and caproic acids as does cow's milk has been claimed to be a reason for the characteristic odor and flavor sometimes associated with goat's milk.

Goats are milked away from the vicinity of the billy goat in order that his odor may not be absorbed by the milk. The characteristic odor of the billy goat may be eliminated by the removal of the odor-producing glands on his head.[19]

The vitamin content of goat's milk is essentially the same as that found in cow's milk, as shown in Chapter 6. The folic acid content is lower. The vitamin B_{12} content (0.2 microgram per litre) is less than in cow's milk. This is reported to be a reason for an anemia ("goat's milk anemia") present at times in infants fed almost entirely on goat's milk.

Although it generally is used as fresh milk, cheese, especially in Europe, is made from goat's milk. Neufchatel and Roquefort-type cheeses often are made wholly or in part from it. Sheep milk cheese often has up to five percent of goat's milk added. These cheeses have a different flavor from that made from cow's milk, primarily on account of differences in the fatty acid content of the milk fats. The fatty acid content of the milk fat of various mammals is given in Table 2:1.

Goat's milk may be utilized for practically every purpose for which cow's milk is used. It is often recommended for infants and invalids because the milk forms smaller and more flocculent curds in the stomach and therefore may be more easily tolerated than cow's milk. One investigation has indicated that goat's milk is more slowly digested than cow's milk. The curd tension of goat's milk is about 30 to 40 grams, it is therefore about 30% softer than Holstein milk, and about 50% softer than Jersey milk.[9]

Compared to cow's milk, the protein of goat's milk is relatively insoluble in an alkaline,alcoholic solution. A test based on this reaction to distinguish goat's milk from cow's milk is described in Chapter 20.

Canned, evaporated goat's milk is made commercially in the United States. Its average composition is shown in Table 12:7.

TONED MILK[14,30]

After World War II, the cost of milk in India, for reasons beyond the control of the producers, went out of reach of the populace. The government, after considering several measures determined the best remedy to be the addition of water and imported non-fat dry milk to the locally produced milk. The resulting mixture, called "toned milk", results in a product that is cheaper than locally produced milk.

Toned milk is defined as a combination of fresh milk, nonfat dry milk, and water, in such quantities as are required to obtain the percentages of fat and nonfat milk solids desired in the finished product. The final product may vary from a fat content of 3% to 1%

and from 8.5 to 10.5% solids-not-fat, usually to supply a total of 11.5% total milk solids.

After its preparation, the product is filtered, pasteurized, sometimes homogenized, and then bottled. Holding it under refrigeration for twelve hours tends to improve the product.

The production of toned milk is intended to provide a market for all milk produced locally, in non-dairying countries, such as India and some places in the Middle East, and South America.

REFERENCES

1. Bakwin, H. and Bakwin, R., Clinical Management of Behavior Disorders in Children, W. B. Saunders Co., Philadelphia, 1960.
2. Bell, D. J., Oligosaccharides of milk; their relation to the bifidus factor, *Ann. Repts. on Prog. Chem.* (Chem. Soc., London) 52:333-5 (1955).
3. Carnation Co., *U.S. Patent* 2 793 953; May 28, 1957.
4. Chaney, A. L., *U.S. Pat.* 2 707 152; April 26, 1955.
5. Doan, F. J., Soft curd milk. A critical review of the literature, *J. Dairy Sci.*, 21:739 (1938).
6. Doan, F. J. and Dizikes, J. L., Digestion characteristics of various types of milk compared with human milk, *Penn. Agr. Expt. Sta. Bull.* 428 (1942).
7. Federal and State Standards for the Composition of Milk Products, U.S. Dept. Agr., Agr. Handbook 51, Washington, D. C. 1968.
8. Filer, L. J., Jr., Enrichment of special dietary food products, *Agr. Food Chem.*, 16., No. 2: 184-189 (1968).
9. Gamble, J. A., Ellis, N. R., Bede, A. K., Composition and properties of goat's milk as compared with cow's milk. U.S. Dept. Agri. Tech. Bull. 671 (1939).
10. Grade "A" Pasteurized Milk Ordinance U.S. Public Health Service, Washington, D. C., 1967.
11. Gyorgy, P., Dhanamitta, S. and Steers, E., Protective effects of human milk in experimental staphylococcus infection, *Science*, 137:338 (1962).
12. Gyorgy, P., Kyhn, R., Rose, C. S., Zilliken, F. Bifidus factor occurrence in milk from different species and other natural products, *Arch. Biochem. Biophys.*, 48:202 (1954).
13. Hilditch, T. P., *The Chemical Constitution of Natural Fats;* John Wiley & Sons, New York, 1954.
14. Khurody, D. N. Preparation and Marketing of Toned and Double Toned Milk, 17th International Dairy Congress, B-67, Munich, 1966.
15. Kon, S. K. and Cowie, A. T., *Milk, The Mammary Gland and Its Secretion*, Academic Press, New York, 1961.
16. Loo, C. C., Slatter, W. L., Powell, R. W., A study of the cavitation effect in the homogenization of dairy products, *J. Dairy Sci.*, 33:692 (1950).
17. Malpress, F. H. and Hytten, F. E., *Nature*, (London), 180:1201 (1957).
18. McCollum, E. V., Orent-Keiles, E., Day, H. G., *The Newer Knowledge of Nutrition;* The Macmillan Company, New York, 1939.
19. MacKenzie, D., Goat Husbandry, 2nd Ed., Faber, London, 1967.
20. *Milk Dealer*, 49:158 (June, 1960).
21. Mueller, W. S. and Cooney, The effect of cocoa upon the utilization of the calcium and phosphorus of milk, *J. Dairy Sci.*, 26:951 (1943).
22. Nagasawa, T. et al, Amounts of Lactoferrin in Human Colostrum and Milk, *J. Dairy Sci.*, 55:1651 (1972). Ibid, 56:177 (1973).

23. Perlman, H. H., Dannenberg, A. M. and Sokoloff, N., The excretion of nicotine in breast milk and urine from cigaret smoking, *J. Am. Med. Assoc.*, 120:1003 (1942).
24. Petersen, W. E., Immunity from milk, *Milk Dealer*, 36:53, (April 1963).
25. Rodkey, F. L. and Ball, E. G., Test for distinguishing human from cow's milk, *J. Lab. Clin. Med.*, 31:354 (1946).
26. Trout, G. M., Homogenized milk and public health, *J. Milk Technology*, 6:214 (1943).
27. Turner , A. W., Digestibility of milk as affected by various types of treatment, *Food Research.*, 10:52 (1945).
28. Whittle, E. G., Goat's milk and human milk, *Analyst*, 68:247 (1943).
29. Widdowan, E. M., Absorption and excretion of fat, nitrogen and minerals from "filled" milk by babies one week old. *Lancet*, 2, 1099 (1965).
30. Toned Milk, the Why and How of it, Dairy Society International, Pub. No. 32, Washington, D. C. 1967.
31. Coulter, S. T. and Thomas, E. L., Enrichment and fortification of dairy products and margarine, *Agr. and Food Chem.* 16, No. 2:158-162 (1968).
32. Vaidya, A. Molecular Biology of Human Milk, *Science*, Vol. 180, 776 (1973). Also see Page 180 (1973).
33. Wade, N., Bottle-Feeding, Adverse Effects of a Western Technology, *Science*, Vol. 184:45-48 (1974).

chapter 13

Cream

Cream is that portion of milk, rich in milk fat, that rises to the top of milk while it is standing, or is separated from it by means of a centrifugal separator. When intended for direct consumption, rather than in the manufacture of butter or ice cream, it is called *market cream* and it must meet the standards for composition and bacterial content of the area in which it is sold.

Grades of Cream

The Federal standards for cream of different fat contents are as follows: light, coffee, or table cream has 18% to not over 30% fat; light whipping cream, 30% to 36% fat; and heavy whipping or pastry cream has a minimum of 36% milkfat. The approximate composition of cream of different fat content is given in Table 13:1. *Sour Cream* is discussed in Chapter 14.

TABLE 13:1
Approximate Composition of Cream

	Light Cream	*Whipping Cream*
Fat	19.00%	36.00%
Protein	2.94	2.20
Lactose	4.05	3.15
Ash	0.60	0.46
Water	73.41	58.19

Gravity-separated cream is cream that has risen naturally to the surface of milk. The fat content varies from 10 to 28%, with an average of 20 to 22%.

By means of a special type of cream separator it is possible to prepare cream of 65 to 85% fat content. This product, known as *plastic cream*, has a heavy body, and when the fat content is 80% or more, its composition is about that of unsalted butter. It is sometimes used in the manufacture of ice cream.

Half and Half is a mixture of milk and cream, usually homogenized, which generally contains about 11.5% milk fat; but this may vary from 10.0% to 12%, depending upon local standards.

The *cream line* or depth of the layer in unhomogenized, bottled milk is of considerable commercial importance because the consumer often considers a deep layer of cream as indicative of rich milk. As a result, the attempt is made to produce as deep a cream line as possible for given fat content of the milk. Agitation of milk while it is cold, or processing and pasteurization at a high temperature, causes a lowering of the cream line. Milk that is marketed in glass bottles, therefore, is not heated above 145°F during pasteurization, or above 163°F in high-temperature short-time pasteurization. As milk distributed in cartons does not show the cream line, its distributors may at times heat the milk somewhat higher during processing.

The thick layer of cream often found on the top of the bottle of milk or cream is called the *cream plug*. It is caused by the rise of large fat globules and usually contains from 60 to 80% of fat. This defect occurs more often in milk pasteurized by the HTST method and then homogenized at a temperature lower than that used for pasteurization.

When cream is added to coffee or some other hot drinks, the cream sometimes coagulates with the formation of small, feathery flakes which float on the surface of the liquid. This is known as *feathering* and it happens rather often if homogenized or very thick cream is used. If the coffee is made with hard water or the cream is slightly sour, the tendency for the cream to feather is increased.[3] Sodium citrate, where lawful, has been used to stabilize cream against feathering. It is added in the proportion of two to five ounces for 1000 pounds of cream.

Properties of Cream

The consumer often judges the quality of cream by its appearance. If it is heavy or thick, it is considered to be of high fat content and good quality; if thin, the reverse is assumed to be true. In reality, different lots of cream of the same fat content may vary greatly in their viscosity, but, in general, an increase in fat content also increases the viscosity of the cream. Creameries usually control the processing of cream in order to obtain a product of considerable viscosity. If the cream is held a day or so at a low temperature—about 40°F—it undergoes an increase in viscosity. Sometimes the cream is homogenized, as this increases its viscosity considerably. Pasteurized cream does not thicken with age as much as raw cream does, and homogenized cream undergoes practically no increase in viscosity upon aging.

The *viscosity* of cream may be increased by controlled treatment with heat. This method, which is used to some extent, consists of heating the cream to about 87°F for twenty minutes and then cooling it slowly, over a period of twenty minutes to two hours, to a temperature of 70°F and then rapidly to 40°F.[1] The addition of dry milk solids or other materials to cream in order to thicken it is an adulteration.

Whipping Cream

Unlike most milk products, the processing of whipping cream usually is completed by the consumer. Inasmuch as this procedure does not always lead to a uniform product, the use of pressurized cream, cream toppings and imitation products has become widespread. These products often do not have the flavor and desirable properties of true whipped cream.

The whipping of cream is essentially the first step in the churning of butter, the churning process being stopped before the emulsion is broken and the fat globules separate. At this stage, a semi-stable foam is formed, with an increase in volume (overrun) and little tendency for serum separation ("bleeding" or leakiness). Cream that has been separated from cold milk yields a better whipping cream than milk separated at temperatures around 90°F or higher.

Whipping cream should contain not less than 30% fat. There is no advantage in a fat content over 38% as the stiffness of the whipped cream is not increased if the fat content is over this amount. Whipped cream thickens because its fat globules, stabilized by a film of milk protein, form a continuous structure or bridgework which maintains a stable foam when air is forced into the cream.

For best results, both the bowl and the cream to be whipped should be a temperature below 45°F. Only small amounts should be whipped at one time; successive whippings are better if large amounts are to be prepared. Cream that has been aged whips better than fresh cream, and the addition of dry skim milk or calcium saccharate also improves its whipping ability. Calcium saccharate, a compound of lime and sugar, known as *Viscogen* and other trade names, was formerly used as an adulterant in cream. Sugar should not be added until the cream is whipped because it inhibits coagulation of the milk proteins and so reduces the stability and stiffness of the cream.

The type of beater used is an important factor in preparing whipped cream. One in which flat, perforated blades revolve at the bottom of the bowl is more efficient than the more usual type of egg beater, especially if only a small amount of cream is to be whipped.

A kind of whipped cream known as *aerated cream* is prepared by placing cream in a container with carbon dioxide or nitrous oxide gas under high pressure. The Food and Drug Administration has approved the use of a fluorocarbon, octafluorocyclobutane or Freon C-130, as a propellant for use in foods. A mixture of 15% of this gas and 85% of nitrous oxide is available, because at present manufacturers of whipped cream find the fluorocarbon gas alone too costly, even though it yields a whipped cream of good stability and stiffness. The amount of whipped cream obtained from the pressurized container may be six to eight times that of the original volume of the cream.[2] The product is used by fountains and restaurants but in some localities is not popular because the fluffy product is not similar to normally whipped cream.

Sterilized Cream

Owing to its high fat content, attempts to prepare a canned, sterilized cream have, until recently, been unsuccessful. By avoiding the procedures used to can evaporated milk and employing new techniques, it has been possible to prepare a sterilized cream that can be canned or bottled, and which will maintain its flavor and body for at least two years.

Cream, of 18% or 30% fat content is used. Its acidity must be low, preferably not over 0.14 to 0.15%, expressed as lactic acid. In order to prevent coagulation during sterilization, an edible stabilizer, such as from 0.10 to 0.3% of sodium alginate, a gelatinous gum obtained from seaweed, is added. Sterilization is accomplished by flash pasteurization at 280-300°F, or by adding steam under high pressure directly to the cream, which is heated in this matter to 260° to 280°F for about four minutes. The cream is then homogenized at not over 2000 pounds pressure, followed by a second homogenization at about 500 pounds pressure. By carefully controlling the pressures, the cream of the desired viscosity is obtained.

After homogenization, the cream is cooled and held in sterile tanks from which the bottles are filled. The cream is run into bottles or cans that have been sterilized with steam at 275°F or hot air at about 400°F. Bottles are filled and capped in a room equipped with ultra-violet lamps to protect the product against air-borne contamination. The air in the bottling room is washed and filtered before it enters the room. It will be noted that the product is sterilized before being placed in sterile containers, in contrast to the method of making canned, evaporated milk, which is sterilized after being placed in the can.

More details on aseptic canning (of milk) are given in Chapter 11.

Manufacturing Cream

Much cream, especially that not obtained from Grade A milk, is classified as *manufacturing cream*. It is used in making butter, ice cream, some varieties of cheese, aerated cream, and similar products. Although the quality of much manufacturing cream is very good, the bulk of it does not meet the standards required of cream for home use. Manufacturing cream, which is often separated at the farm, usually is shipped in cans, and these are often delivered by the producer to a centrally located cream station. Here it is weighed, graded, tested for fat content, and shipped by truck or rail to the manufacturing plant. It may be several days old at the time of arrival.

The grading of manufacturing cream is a quick and simple procedure, but it calls for considerable skill on the part of the grader. A small amount of cream of poor quality mixed with some of good quality may degrade an entire churning of butter. Much manufacturing cream has an acidity between 0.2 and 0.5% expressed as lactic acid, which may place it in second grade; higher acidity would place it in the category of third place.

In addition to acidity, the grader must consider the odor and flavor of the cream. By these properties he can separate cans of cream which will yield butter of different grades or score. Various methods used in testing cream and other dairy products are described in Chapter 20.

A skilled grader can judge the conditions under which the cows were fed and the sanitary condition under which the cream was produced and held before delivery to the dairy.

Cream that is decomposed or with an odor or flavor that cannot be readily removed during processing is rejected. Such cream may be degraded by adding to it a harmless food-dye, such as brillant blue, which is not decolorized by enzymic action in the cream. This coloring prevents the use of the cream in food for man but permits use of it as animal feed.

THE CREAM SEPARATOR

The cream separator is essentially a bowl that can be rotated at a speed of 3000 to 20,000 revolutions a minute (Fig. 13:1). The bowl consists of a series of conical discs, separated from one another by about 0.02 inch by projections upon their surface.

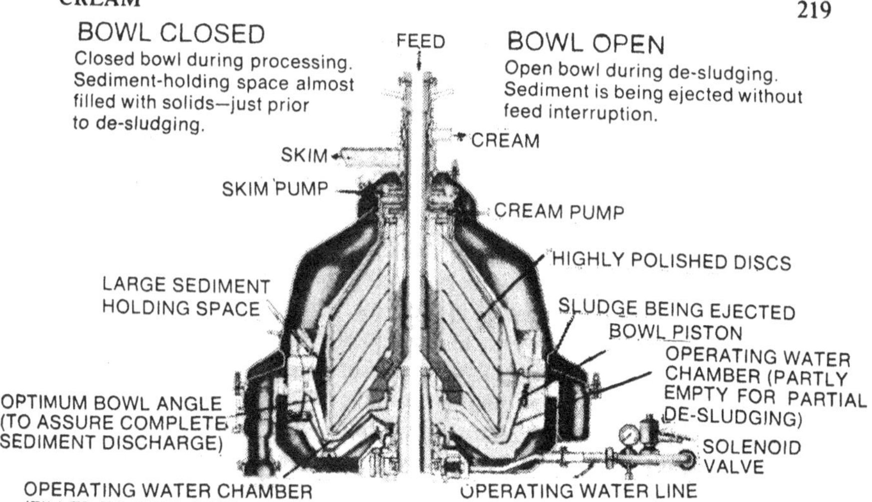

Fig. 13-1. Courtesy Centrico, Inc. Westfalia Separator.

Milk enters the separator in one type of machine by gravity; in another type, the separator is sealed in an air-tight casing and the milk is pumped to it by a feed pump. Milk enters the machine through holes placed near the center of the discs. As the bowl revolves, the cream, being lighter than the milk or skim milk portion, is driven by centrifugal force towards the center of the bowl, while the skim milk is directed outwards. As the space between the discs is small in relation to the length over which the cream must pass, there is ample time for the cream to rise or separate before it reaches the center of the bowl from which it flows out. An efficient separator will remove all but about 1 to 1.5% of the fat in milk so that the skim milk will contain as little as 0.02 to 0.05% fat.

Fat globules smaller than two microns in diameter are not separated efficiently. The separator may be adjusted so that the ratio of the volume of skim milk to cream is constant, thus permitting separation of cream of a desired fat content from milk of a given fat content.

During separation, solid matter, known as *separator slime*, which consists of any dirt that may have been in the milk and some of the cells and bacteria, collects along the inside walls of the separator drum. Fig. 13:1 shows a CIP separator, that is, one which may be *cleaned-in-place* without being dissembled. The CIP separator has a bowl which automatically ejects accumulated sediment. During processing the bowl is in a closed position until the sediment space is filled. Then ports in the bowl wall open and the sediment is ejected in such a short time that process flow is not interrupted. When processing is completed, hot caustic and then acid solution is recycled through the machine which is then rinsed with clean water.

The clarifier has many of the features of the cream separator. Equipment is available which combine the features of a separator, clarifier, and standardizer. Separators usually have a standardizing or clarifing capacity about twice that of their separating ability. The clarifier has one spout through which the milk exits. The separator has two spouts, one for the cream and the other for the skim milk. The separator may be adjusted to yield cream of different fat contents.

Usually the separator, which may have a capacity up to 50,000 pounds of milk an hour, is operated with milk around 90°F, but separators for cold milk are in use which give good separation at 50°F. The cold milk separator has discs about 0.05 inch apart, with the holes near the edge. Hot milk is less viscous and easier to separate efficiently than cold milk. Separators used in the manufacture of dry non-fat milk often handle milk at 110 to 125°F, producing a skim milk with very low fat content.

Milk that is separated around 50°F gives skim milk that has much less tendency to foam than if the milk were separated at 120°F. The cream obtained at 50°F is of better viscosity than that separated at the higher temperature.

The Bactofuge, a special type of separator for reducing the bacterial content of milk is described in Chapter 11.

REFERENCES

1. Burgwald, L. H. Increasing the viscosity of cream, *Milk Dealer*, 29, 6:52-54 (1940).
2. Getz, C. H., Smith, G. F. and Tracy, P. H., Instant whipping of cream by aeration. *J. Dairy Sci.*, 19:490 (1936).
3. Tracy, P. H. and Corbett, W. J., Coffee as a factor in the feathering of cream. *J. Dairy Sci.*, 19:490 (1936).

chapter 14

Cultures—Fermented Milk Products

Fermented milks were used by the people of Eastern Europe and Asia Minor long before the discovery of bacteria. Undoubtedly they observed that milk usually spoiled after it was drawn, but if it were allowed to sour in a controlled manner the resulting product had not only a pleasing flavor but also improved keeping quality. It was not until about 1840 that scientists began to believe that the souring of milk and similar fermentations were caused by the activity of microorganisms; following the work of Pasteur, about 1857, notable progress was made.

The basis of a fermented milk is lactic acid fermentation but special types of bacteria and yeasts produce the typical characteristics of the fermented milks used in various parts of the world. There are two principal kinds of fermented milk: 1) the *acid* type, such as cultured buttermilk; and 2) the *kefir type*, in which a combined acid and gassy fermentation takes place. In the latter case a mild alcoholic fermentation also occurs.

It has been demonstrated that an acid milk is somewhat more easily digested than ordinary milk. It makes little difference whether the acidity is the result of bacterial activity or an acid is added artificially, as in the preparation of some infant foods. Fermented milks have been used for their possible therapeutic value in stomach and intestinal disorders, a use based upon the assumption that the acid-fermenting bacteria and the lactose in the milk are able to create conditions in the intestinal tract which are unfavorable to the growth of putrefactive bacteria. Putrefactive organisms favor the decomposition of protein material and the formation of gas and an alkaline reaction in the large intestine which condition may be associated with the formation of substances which cause a so-called *auto-intoxication*. Only certain strains of the *lactobacilli* can implant themselves in the intestinal tract.

It often has been assumed that pathogenic organisms cannot survive in contaminated fermented milks on account of the high acid content of the latter. The United States Public Health Service has reported on an outbreak of gastroenteritis caused by *Staphylococcus aureus* in buttermilk placed in unclean containers, and experimental work has demonstrated that various pathogenic organisms can survive for several weeks in fermented milks of high acidity.[10]

CULTURES

About 1860, creameries in Denmark and Holland began to add buttermilk to cream in the attempt to hasten souring and to control the flavor of the butter to be made from the cream. If the proper organisms happened to be present in the buttermilk, the desired effect was obtained. On the other hand, if gas-forming or proteolytic bacteria were present, the butter would be defective or spoil quickly. Dairy scientists sought a solution to this problem and by 1890 developed the use of *cultures*, or *starters*, which consist of actively growing strains of selected bacteria, usually in pasteurized skim milk. They can produce desirable flavor, aroma, and acidity in milk or cream and in the butter or cheese made from them.

When it is desired to produce lactic acid, the culture consists predominantly of *Streptococcus cremoris* or *Streptococcus lactis*. The lactic acid organisms present in a culture are chosen for their ability to produce acid under various conditions. The organisms mentioned are active at room temperature. At the higher temperatures needed in the manufacture of cheese, other strains are used. When high yields of lactic acid are desired, organisms such as *Lactobacillus bulgaricus* are used. When a combination of acidity, flavor, and aroma is desired, the cultures contain *Leuconostoc citrovorum* (*Streptococcus citrovorum*) or *Leuconostoc dextranicum* (*Streptococcus paracitrovorum*).[8]

These organisms not only produce lactic acid but also act upon some of the constituents of the milk, especially citric acid, to form a small amount of *diacetyl* as well as some acetylmethyl carbinol, acetic acid, alcohol and carbon dioxide. Diacetyl is a yellowish, oily liquid, traces of which are found in roasted coffee, honey, tobacco smoke, and other aromatic substances. It has a very pronounced odor, and as little as 0.0002% of it is sufficient to give butter its characteristic aroma. The production of diacetyl is reported to be increased at least 100% if the milk to be cultured is first given a hydrogen peroxide-catalase treatment.[5]

Ordinarily, a culture does not contain enough actively growing bacteria to ripen a vat of milk quickly, so it is used to inoculate about a quart of whole milk or skim milk, previously sterilized or heated to a high temperature in order to destroy bacteria that it might contain The inoculated milk, known as the *mother starter* is held overnight in a warm place or incubator in order to encourage active growth. The ripened mother starter often is transferred to a larger quantity of pasteurized or sterilized milk, which then is ripened to make a *second starter*. Sometimes the process is carried on for a third or even a fourth starter. The continued transfer through several lots of milk increases the activity and rate of growth of the original culture. Freeze-dried cultures are in common use. These often are low in activity and may require a number of transfers to obtain proper results. Commercial cultures often are tested for the presence of bacteriophage (See below).

The taste and odor alone of a culture are not reliable tests of its quality because its acidity may mask the true flavor and aroma. A starter should have a clean, acid flavor and a firm, smooth body. The coagulated milk should leave the side of the container cleanly when slightly tilted. Very little or no whey should be present on top of the culture. Chemical tests for titratable acidity and for diacetyl are sometimes desirable.[2] A bitter taste or a lumpy, gassy, or ropy body, indicates the presence of undesirable contamination.

Starters fulfill a number of functions in the preparation of fermented milks and in the manufacture of some varieties of cheese. The acidity they produce helps rennet to coagulate the milk in cheesemaking. Cultures that contain *Streptococcus thermophilus* and other special types of bacteria are used in the manufacture of Swiss and other kinds of cheese. *Propionbacterium shermanii* is important in the manufacture of Swiss Cheese because it uses the lactic acid produced by other organisms to form propionic acid, acetic acid and the carbon dioxide gas which give the cheese its characteristic flavor and forms the "eyes".

CULTURED MILK PRODUCTS

Sour Cream

The Grade "A" Pasteurized Milk Ordinance of the U.S. Public Health Service defines sour cream or cultured sour cream as the fluid or semifluid cream resulting from the souring, by lactic acid producing bacteria or similar culture, of pasteurized cream, which contains not less than 0.20% acidity expressed as lactic acid. The acidity may also be obtained by the addition to pasteurized cream of a food-grade acid, such as lactic acid or phosphoric acid.

Sour cream is a favorite milk product among the people of Central Europe and the Slavic nations. In this country it is used as a salad dressing, a spread for bread, for "dips," and for mixing with cheese and other foodstuffs. It is a cultured-milk product, made by ripening pasteurized cream of about 18 to 20% fat content with a lactic culture in the manner used to make cultured buttermilk. The cream is pasteurized at 180°F for 30 minutes, cooled to 155°, and homogenized at 2000 lbs pressure. After cooling to 70°F, from 1 to 3% of starter is added and thoroughly mixed with the cream.

Where such additions are permitted, from 0.2 ml to 0.5 ml of rennet extract or a stabilizer such as gelatin or sodium alginate is added to 10 gallons of cream in order to give the finished product a heavy body and smooth appearance. The finished product should have about 0.6%-0.7% acidity, within 12-14 hours, with a clean, acid flavor. As a heavy body is desired, agitation must be avoided as much as possible. The inoculated cream sometimes is packaged in consumer-sized containers and allowed to ripen in the container, thus avoiding undue handling.

The precise flavor of sour cream used in making "dips" is not as important as is a firm body in the product. A consumer does not use the dip as is, but in combination with some other food. A fat content of 16% in the dip usually is satisfactory, with a total milk solid content of 24%. Distinctive flavoring such as fruit, cheese, or spice may be added.

In some places the name "Sour Cream" is considered a hindrance and a fanciful name is used, such as *Salad Cream, Hampshire Cream,* or *Cream Dressing*. A sour cream dressing has about 16% milk fat and up to 0.6% of stabilizer.

A spray-dried sour cream, for use in food products in which the flavor is desired, is made by drying a mixture of sour cream, vegetable gum stabilizer, and an antioxidant.

Cultured Buttermilk

Owing to variations in the churning process during the manufacture of butter, and differences in the quality of the cream churned during various parts of the year, a uniformly high quality of genuine buttermilk is not always obtainable. To supply the demand for the beverage, a large amount of cultured buttermilk, sometimes called *artificial buttermilk*, is made by the fermentation of milk or skim milk with lactic acid bacteria. Usually skim milk is used, but a small amount of cream may be added to improve its flavor. Practically all buttermilk sold to the consumer today is cultured buttermilk.

The pasteurized milk used for the preparation of cultured

buttermilk is inoculated with up to 1% of starter, as described previously, and held at a temperature around 70°F until the desired acidity is reached. This usually is between 0.7% and 0.9%, expressed as lactic acid. Often a little common salt is added to improve the flavor. When legal regulations do not prevent the use of it, a small amount of gelatin or a similar substance may be added to give the buttermilk a smooth body, prevent the separation of whey, and so improve the appearance of the product. When the buttermilk does not contain a stabilizer, such a gelatin, the fine casein particles gradually settle as the product stands, leaving a layer of more or less clear whey on the buttermilk. The presence of gas bubbles, a sharp acid odor or an excessive amount of whey indicates that the buttermilk is contaminated with undesirable organisms.

Synthetic culture flavor concentrates often are used to prepare commercial buttermilk. The most important flavor constituents are reported to be lactic acid, diacetyl, acetaldehyde, dimethyl sulfide, and acetic acid. Consumers show little difference in their preference between naturally cultured and artificially flavored buttermilk.[4]

Acidophilus Milk

Acidophilus milk is milk that has been inoculated with a pure culture of *Lactobacillus acidophilus* and has been allowed to ferment under conditions that favor the growth and development of large numbers of the organism. A product that contains more than 200 million organisms per milliliter is obtainable commercially in a few localities.

Care must be taken not to contaminate the culture or the fermenting milk because *Lactobacillus acidophilus* does not multiply rapidly in milk. As the acid forms rather slowly, it may not be able to overcome contamination with other bacteria. It is necessary to sterilize the milk before inoculation in order to remove the possibility of contamination from this source. About 2% of acidophilus starter is used. The best flavor is obtained when 0.8 to 1.0% of acid is developed. Candy-like preparations of *Lactobacillus acidophilus* have been prepared, but unless they are fresh, they may contain very few viable organisms.

Acidophilus milk is recognized by the medical profession and by dietitians as a practical way to introduce large numbers of the organism into the intestinal tract. It has been found to be of some use in the treatment of certain intestinal disorders, such as constipation and colitis, but authorities differ in their opinion of its actual value. The therapeutic action of the milk is believed to be due in part to the activity of the bacteria that reach the lower intestinal tract, and not entirely to the acidity of the milk.[7]

The type of bacteria that inhabit the intestinal tract is influenced by the implantation of *L. acidophilus* which in the presence of lactose brings about an acid condition in the lower intestine by reacting with the unfermented lactose. This increase in acidity favors the destruction of bacteria of the putrefactive type, which often cannot tolerate an acid environment. One pint to one quart of acidophilus milk per day usually is taken. It is often recommended that about two ounces of lactose also be taken, especially if a laxative action is desired.

There has been a considerable decrease during recent years in the manufacture and use of acidophilus milk and related products with yogurt replacing it in popularity.

Bulgaricus Milk

Lactobacillus bulgaricus is used in the preparation of a fermented milk of high acid content. The method of manufacture is similar to that for acidophilus milk. Bulgaricus milk has been used in the treatment of intestinal disorders but as the organisms do not implant themselves in the intestinal tract acidophilus milk is favored.

Kefir

Kefir is cultured milk which has been subjected to both an acid and alcoholic fermentation. The reaction is induced by means of kefir grains which are gelatinous, popcorn-like particles consisting of the curded milk containing the fermenting organisms obtained from the fermented milk itself. Kefir grains are obtainable commercially from supply houses that maintain a stock of biological supplies and bacterial cultures. The fermenting organisms probably are varieties of *Streptococcus lactis*, *Lactobacillus bulgaricus*, and lactose-fermenting yeasts. Kefir grains are maintained in an active form by transferring them every few days to a fresh supply of pasteurized skim milk, held at about 70°F. The grains are carried to the surface of the milk by the carbon dioxide formed during the fermentation. The grains may be preserved by removing them from the milk, washing with water, and drying them at about 29.4°C (85°F).

In the countries of southwestern Asia, kefir is made from the milk of mares, goats, or cows, depending upon the region and on what kind of milk may be available. About four ounces of kefir grains will ferment a gallon of milk in about two days. The acidity that develops is about 1% as lactic acid, and about 1% of alcohol may be present. At this point the kefir grains are strained out of the milk and placed in a new lot of milk. The fermented milk is held at 10°C (50°F) or lower, and gently agitated before being transferred to containers which are then stored at 4.4°C (40°F).

Kumiss

Kumiss is the typical fermented milk drink of Russia and western Asia. Genuine kumiss is made from unheated mare's milk, which imparts a characteristic flavor to the product. The high lactose content of mare's milk also favors an alcoholic fermentation by lactose-fermenting yeasts. Various other lactic-acid-forming organisms, such as *S. lactis* and *L. bulgaricus*, also are present.

The yeasts ferment a part of the lactose to alcohol and carbon dioxide. As the fermentation takes place in an open vessel, most of the gas escapes. The alcohol content varies with the degreė of fermentation and may reach 3%. In some Siberian and other Asiatic places a brandy-like alcoholic drink, called "araka" is distilled from kumiss. It may contain up to 8% alcohol.

Sometimes cow's milk is used, which when up to 5% of sugar is added, may yield a kumiss with about 1% of alcohol.

Yogurt

Yogurt (also *yoghurt*) is the Turkish name for a fermented milk of the lactic acid type. It is known by different names according to the place where it is made; for example, in Armenia it is known as *Matzoon*; *Leben*, in Egypt; *Naja* in Bulgaria; *Gioddu* in Italy; *Tiaourti* in Greece and *Dadhi* in India. Unlike kefir and kumiss, yogurt contains little or no alcohol and it is comparatively high in acid. In Greece, yogurt often is made from ewe's milk.

The starter often is prepared by growing separate cultures of *Streptococcus thermophilus* and *Lactobacillus bulgaricus* and mixing them just before adding them to the milk that is to be fermented. In Europe, a starter containing *Lactobacillus bifidus* sometimes is used. When the organisms are present in about equal number, the yogurt develops a desirable flavor and mild acidity. Some producers obtain a satisfactory culture by growing a mixture of the organisms. The milk to be fermented is forewarmed to about 140°F; 3 to 5 % of low-heat dry non-fat milk solids are added, and the mixture is homogenized at 1800-2000 pounds pressure. The homogenized milk is pasteurized at 180°-190°F for about 30 seconds. It is then cooled to 110°F and inoculated with 2 to 3% of mixed culture. The inoculated milk is transferred to sterile containers which are placed in a water-bath or incubator held at 110°F. An active starter will coagulate the milk in about three hours, by which time the titratable acidity should be

TABLE 14:1
Proximate Composition and Mineral Content of Cultured Dairy Foods Compared with Milk [a]
(amount of nutrient per 100 gm edible portion)

	Major Constituents				*Macrominerals*					
	Food Energy kcal	*Protein gm*	*Fat gm*	*Total Carbo-hydrate, gm*	*Calcium mg*	*Iron mg*	*Magnesium mg*	*Phosphorus mg*	*Potassium mg*	*Sodium mg*
Buttermilk (skim milk)	36	3.6	0.1	5.1	121	Trace	14	95	140	130
Bulgarian Buttermilk	62	3.2	3.5	4.5	114	0.10	13	88	154	96
Sour Cream	211	3.0	20.4	4.3	102	0.04	9	77	56	40
Yogurt (partially skimmed milk)	50	3.4	1.7	5.2	120	Trace	—	94	143	51
Yogurt (whole milk)	62	3.0	3.4	4.9	111	Trace	—	87	132	47
II. MILK										
Skim, Fluid (vit. A & D enriched)	36	3.6	0.1	5.1	121	0.03	14	95	145	52
Whole, Fluid (vit. D enriched)	65	3.5	3.5	4.9	118	0.06	13	93	144	50

Dashes indicate lack of reliable data for a nutrient believed to be present in measureable amount.

a - Dairy Council Digest 43:4 (1972).

0.85-0.95%, or a *p*H value of 4.4 to 4.5. At this time, the product is cooled quickly to stop fermentation. Cooling is done by a cold-air blast or by transfer to a cold room held at about 35°F. The product has a storage life of about two weeks.

A number of yogurt-type products are made. In some regions, the product is agitated to render it fluid, but in the United States a firm, custard-like consistency is preferred. The latter type is obtained by adding condensed skim milk or dry milk solids to the milk before it is homogenized and fermented. Sometimes fruit flavor or chocolate is added, with or without 3 to 5% added sugar. In Swiss-type yogurt fruit is folded into the yogurt. A number of different fruits and flavors are used.

Yogurt often can be used instead of sour cream in most recipes. In Eastern Mediterranean countries it is often used instead of milk for cooking and baking purposes. A popular Near-East dish, known as tarhana in Turkey and Kishk in Lebanon is made by mixing yogurt with flour and then drying the mixture.

A type of soft cheese, called "Lebanie" is made by draining the whey from yogurt and holding the curd overnight at room temperature.

The composition of plain yogurt is about 2 to 3.5% fat, 4% protein, 5% lactose. The calorie content is about 54 per 100 grams.

BACTERIOPHAGE

At times a culture may not develop acidity, or it may do so very slowly. Such inhibition may be due to the presence in the milk of traces of an antibiotic, or a germicide, such as hypochlorite or

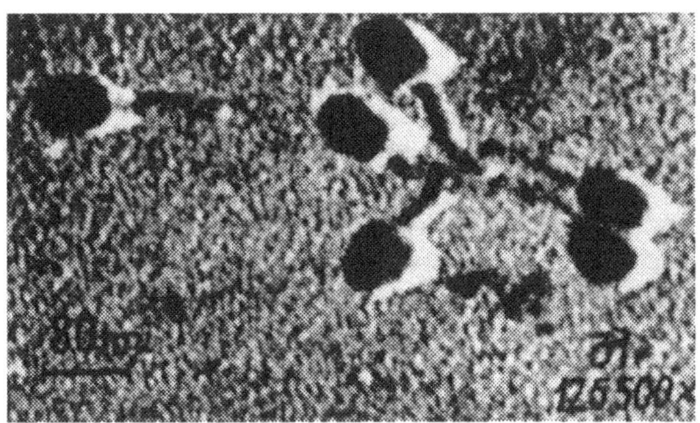

Fig. 14-1. View of bacteriophages in electron microscope. Magnification 125,000 X (Courtesy Dairy Research Institute, Olsztyn, Poland)

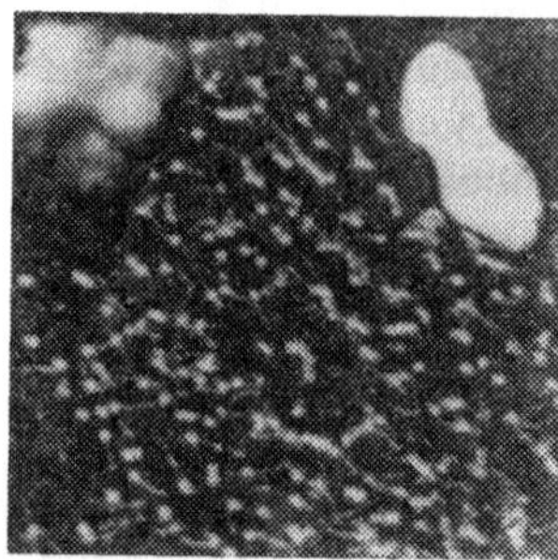

Fig. 14-2. Appearance of Streptoccocus Lactis and Bacteriophage as Shown by Electron Microscope (X 15,000): Upper right, normal cells; upper left, cells undergoing attack; center, completely lysed cell and liberated bacteriophage.

quaternary ammonium compound; or, perhaps most often, the presence of bacteriophage. (Fig. 14:1 and 14:2)

A bacteriophage is a virus that attacks bacteria. Examination under the electron microscope shows these organisms have a cylindrical or a pentagonal head and a tail-like appendage. They attach themselves by the tail to the bacterial cell, after which the contents of the head enter the bacterial cell through the hollow tail. Bacteriophage reproduce only after invading actively growing bacterial cells. The bacteriophage, often referred to as a "phage," generally attacks only a specific strain or species of baceria, and mixed bacterial cultures therefore show varying resistance to them. During a latent period of less than one hour after the bacterial cell is attacked, the bacteriophage has reproduced itself many times within the host cell. The bacterial cell is then disrupted or *lysed* and up to 500 newly formed bacteriophages may emerge, to again begin a cycle of infection. The very small size of the bacteriophage is evident from the number that may emerge from a single bacterial cell. An invasion by bacteriophage is so virulent that lactic acid organisms may be destroyed before any significant fermentation has occurred in the preparation of a dairy product.

Contamination with bacteriophage has not caused much difficulty in the United States, but it has in New Zealand, where the cheese industry uses single-strain cultures in starter. Slow development of acidity, or none, indicates contamination by phage. As a result of it, operations that depend upon the production of acid, such as cheese-making, come to a halt with subsequent loss of product time.

Bacteriophages probably are dust-borne and become active when they find a suitable host. In a cheese factory, conditions may be suitable for the lactic acid bacteriophage to reproduce often and in time to infect the equipment, air, and surroundings.[9] Stringent precautions must be followed. The rotation of bacterial cultures or the use of mixed cultures is sometimes advisable, so that if one strain is infected the other one may carry on the fermentation.

Some strains of lactic bacteriophage are resistant to pasteurization temperatures. Some are killed at 162°F in ten minutes or at 195-203°F in less than one minute.[11]

Sanitizing with a 500-ppm hypochlorite solution will destroy the organism on contaminated surfaces.

Certain divalent ions, especially calcium, are needed to support bacteriophage activity. The addition of sodium orthophosphate to the starter milk was found to be an efficient inhibitor of bacteriophage activity against the lactic streptococci. This procedure reduces the ionic calcium content of the culture and so inhibits phage development and permits normal bacterial growth. The procedure followed is as follows:[3]

To milk that has been forewarmed to 130°F, add sufficient of a 50% solution of sodium-potassium orthophophate buffer (*p*H 6.5-6.6) to yield 1.7% of phosphate salt by weight, and steam the treated milk for one-half hour. While it is still hot, add enough 50% solution of tetra-potassium pyrophosphate to yield 0.3% by weight of this salt. Cool and inoculate the phosphated milk in the usual manner, and incubate it for 18-20 hours at 72-77°F.

REFERENCES

1. Habaj, B. et al., The Influence of some properties of bacteriophages on the difficulties of controlling fermentation troubles, XVII Inter. Dairy Congress, Sec. D/2, 483, Munich, 1966.
2. Hammer, B. W. and Babel, F. J., Bacteriology of butter cultures; a review, *J. Dairy Sci.*, 26:83-168 (1943).
3. Hargrove, R. E., McDonough, F. E., Tittler, R. P., Phosphate heat treatment of milk to prevent bacteriophage proliferation in lactic cultures, *J. Dairy Sci.*, 44:1799 (1961).
4. Lindsay, R. C., Day, E. A. and Sather, A., Preparation and evaluation of butter culture flavor concentrates, *J. Dairy Sci.*, 50:25 (1967).
5. Pack, M. Y. et al., Hydrogen peroxide-catalase milk treatment and diacetyl in lactic starter cultures, *J. Dairy Sci.*, 51:511 (1968).
6. Parmelee, C. E., Carr, P. H., Nelson, F. E., Electron microscope studies of bacteriophage active against streptcoccus lactis, *J. Bacteriol.*, 57:391 (1949).
7. Rettger, L. F., Levy, M. N., Weinstein, L. Weiss, J. E., *Lactobacillus Acidophilus and its Therapeutic Application*; Yale University Press, New Haven (1935).
8. Symposium on Lactic Starter Cultures; Collins, E. B., Sandine, W. E., Elliker, P. R., Allen, L. K., Brown, W. C., Marth, E. H., Speck, M. L., Babel, F. J. and Foster, E. M., *J. Dairy Sci.*, 45:1262-1294 (1962).

9. Whitehead, H. R., Bacteriophage in cheese manufacture, *Bacteriol. Rev.*, 17:109-124 (1953).
10. Wilson, F. L. and Turner, F. W., Acid-milk hazard, *Food Research*, 10:122 (1945).
11. Zottola, E. A. and Marth, E. H., Thermal inactivation of bacteriophages active against lactic streptococci, *J. Dairy Sci.* 49:1338 (1966).

chapter 15

Concentrated Milk Products

FRESH CONCENTRATED MILK

Fresh concentrated milk is milk in which the volume is reduced three to one by evaporation, so that one quart of the concentrate, diluted with two quarts of water will produce three quarts of fluid milk. A higher concentration is not used because of possible thickening during subsequent heat treatment, and also because the concentration of lactose is essentially that of a saturated solution.

Because of its reduced weight and volume, the product can be placed in containers, shipped, and stored at less cost than whole milk. As it is a perishable product, it must be kept refrigerated. Concentrated milk finds some use by consumers as a substitute for cream, especially in coffee. It is usually sold at a lower cost per quart on the reconstituted basis, than fluid milk, partly to compensate for the inconvenience of reconstituting it in the home.

To prepare fresh concentrated milk, Grade A milk is clarified and standardized to about 3.5% fat content. It is pasteurized around 180°F for 16 seconds, then homogenized at 2500 pounds of pressure and cooled to about 125°F. The milk is then evaporated to slightly over a 3:1 concentration, preferably in a low-temperature evaporator wherein evaporation is carried on around 100°F. This part of the processing often is done at a central plant and the product is then shipped by tanker to distributing plants where it is pasteurized at 180°F for 20-25 seconds, cooled to 40°F, and bottled for distribution to the consumer.

The fat content of the finished product is 9.9 to 10.5%, but if the fat content is too high it is adjusted by the addition of pasteurized skim milk or water. The milk solids-not-fat content usually is about 24 to 25%. California is the only state with a composition standard, namely not less than 9.9% milkfat and 24% milk solids-not-fat. The bacterial count must not exceed 20,000 per gram and the coliform count maximum is 10 per gram. The standard for condensed skim milk is not less than 24% milk solids-not-fat.

Fresh concentrated milk may be held for at least two weeks at 35°F, but at temperatures above 40°, deterioration is evident within a few days. The flavor of the reconstituted product generally is good and because of the processing procedures, is free of most volatile feed flavor and odor. When freshly reconstituted, a "harsh" flavor sometimes is evident, because of the presence of minute particles of calcium salts. This defect disappears if the reconstituted milk is held for a day or so at 40°F.

Concentrated milk may be frozen as described in Chapter 12. The Food and Drug Administration (Federal Register, Sept. 9, 1972), describes concentrated milk as a liquid food which has the milkfat and total milk-solids content of evaporated milk. It is pasturized but not processed by heat so as to prevent spoilage. It may be homogenized and vitamin D may be added, provided it is so labeled.

EVAPORATED MILK

Evaporated milk, sometimes called "unsweetened condensed milk," is whole milk from which about 60% of the water has been removed by evaporation. The Federal standard for evaporated milk (December 10, 1973) identifies it as "the liquid food obtained by the partial removal of water from milk. The milkfat and total milk solids content of the food are not less than 7.5 and 25.5%, respectively. It is homogenized. It is sealed in a container and so processed by heat, either before or after sealing, as to prevent spoilage. Vitamin D shall be present in such quantity that each fluid ounce of the finished food contains 25 I.U. thereof. Safe and suitable emulsifiers and safe and suitable stabilizers, with or without dioctyl sodium sulfosuccinate may be used as a solubilizing agent.

Vitamin A may be added in such quantity that each fluid ounce contains not less than 125 I.U."

One pound of evaporated milk is the equivalent of about 2.25 pounds of liquid whole milk. The caloric value averages about 43 calories per fluid ounce.

Evaporated milk is the most widely used form of concentrated milk, although the amount produced has been gradually decreasing. The 1973 production was about one billion pounds, compared to 1.4 billion in 1968. The order of States, according to production, in 1972, was Ohio, California, and Tennessee.

According to the United States Department of Agriculture the consumption in 1972 of evaporated and condensed milk (including canned skim milk) in the United States, Canada, Australia, New Zealand, and Western Europe was 5.2 billion pounds. The per capita consumption averaged 10.3 pounds. The Netherlands was the leader, with 22.7 pounds per persons, followed by West Germany, Canada, Australia, and the United States.

The manufacture of evaporated milk requires the use of milk that meets the sanitary *Standards* set up by the Evaporated Milk Association.[7] The raw milk is graded carefully by chemical and bacteriological examination.

The Evaporated Milk Association classifies milk according to the following tabulation.[7] They recommend the methylene blue test (Chapter 20).

	Standard Plate Count or Direct Microscopic Clump Count		*Methylene Blue Test (Hours)*	*Resazurin Test No color change beyond P-7/4 in Hours*	*Resazurin one hour test*
Class 1	not over	500,000	Over 4½	2¼	Purple
Class 2	" "	3,000,000	"2½	1½	Lavender
Class 3	" "	10,000,000	"1	3/4	Pink
Class 4	Over	10,000,000	Under 1	Under 3/4	Decolorized

Milk falling below class 2 may be accepted for a temporary period. Milk in class 4 must be improved to at least class 3 within 4 weeks. If within the specified period the milk does not conform to the bacterial standard it is not accepted until the producer corrects the responsible conditions.

The milk is weighed and passed through a clarifier or filter to remove any foreign material that may be present and is then standardized to the desired ratio of solids to fat by the addition of skim milk or cream.

The essential piece of equipment needed to make evaporated milk is the vacuum pan (Fig. 15.1). It was devised in England in 1813, but was first used in 1835 for the evaporation of milk. In 1884, John Meyenberg and Louis Latzen, Swiss dairy farmers, established the world's first commercial evaporated milk plant in Highland, Illinois. In 1974, there were 30 plants in the United States, operated by 17 companies.

The milk is heated to a temperature of 190° to 210°F, the exact temperature and time used depending upon the procedures used in the plant. This forewarming is important because it helps to stabilize

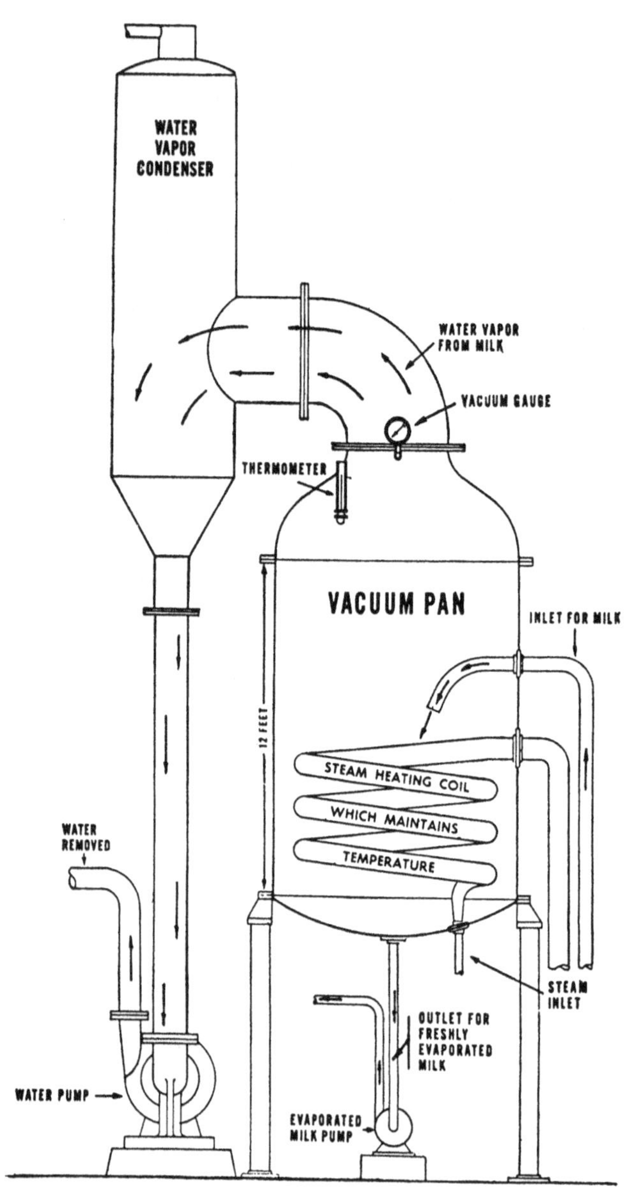

Fig. 15-1. Diagram showing the principle of the Vacuum Pan as used for the manufacture of evaporated milk. (Courtesy of Evaporated Milk Association).

the milk proteins and so prevents the milk from becoming coagulated when it is sterilized at a high temperature after the evaporation process. As prolonged forewarming favors an undesirable increase in color in the finished product, treatment for a short time at a high temperature—248° to 284°F for 25 seconds—is a favored procedure, even though the viscosity of the resulting product may be low.[10] According to one investigation, the increase in color may be prevented or minimized by the addition to the milk of 0.15% sodium hexametaphosphate, 0-1% ascorbic acid, glucose oxidase or 0.01% vitamin A.[8]

The hot, forewarmed milk is drawn into a vacuum pan where it is condensed to the required concentration. The vacuum maintained is such that the milk boils at a temperature between 110° and 135°F in a single-effect vacuum pan such as in shown in Fig. 16:1. Double- and triple-effect evaporators, which operate at a lower temperature, are also in use (Fig. 15:2).

When maximum economy of steam and water is required, the compression evaporator may be used. Instead of taking steam from a boiler, the vapor from the hot milk entering the evaporator is drawn into the compressor (Fig. 15:3). Here it is compressed to a higher pressure with a corresponding increase in temperature. This vapor enters the steam chest of the evaporator and serves as the heating medium for the evaporator. After the steam transfers its heat of vaporization to the milk through the walls of the tube, it condenses and is drawn off as water. The only power needed for the evaporator is that required to run the compressor.

In a conventional evaporator the milk enters the bottom of the pan until the liquid-level reaches sufficiently high to permit circulation. This volume of milk must be kept in the evaporator constantly, otherwise the product will be overheated and adhere to the steam tubes, with possible injury to the finished product. In the *falling-film* or down-flow evaporator the milk enters at the top of the pan and flows evenly over the tubes and is withdrawn within a period of seconds.

After evaporation, the milk is homogenized at a pressure of 2,500 to 4000 pounds in order to prevent the separation of fat in the finished product. The homogenized milk is cooled and pumped to storage tanks, where, if need be, it is standardized with water, skim milk, or cream, in order to obtain the correct content of fat and total solids. If the vitamin D content of the milk is to be increased, a vitamin D concentrate usually is added to the milk before it is homogenized or just before it is canned.

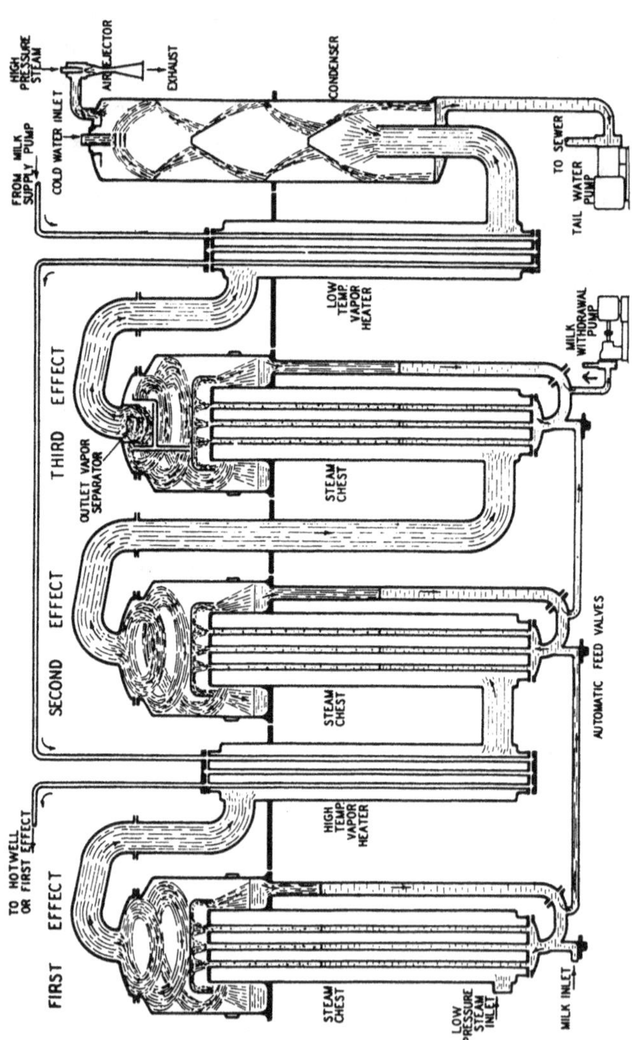

Fig. 15-2. Diagram of Three-effect Evaporator with Vapor Heater and Condenser (Courtesy of Henszey Company, Inc.)

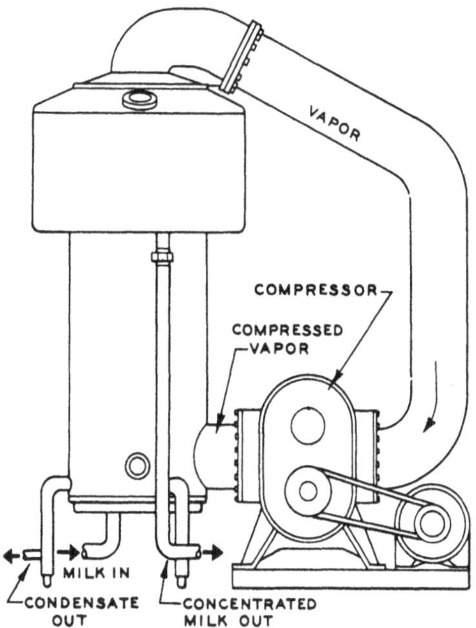

Fig. 15-3. Compression Evaporator (Courtesy of Henszey Company, Inc.)

Experience has shown that some lots of evaporated milk may coagulate during sterilization, even though they had been preheated. In order to prevent this the "salt balance" in the milk is adjusted as needed. The milk is tested by sterilizing a small amount in cans placed in a pilot sterilizer and giving it the same heat treatments as used for sterilization. To some of the cans there is added a small, measured amount of an inorganic salt. Disodium hydrogen phosphate or sodium citrate or both, calcium chloride or sodium bicarbonate may be added as a stabilizer salt, in an amount not to exceed 0.1% by weight of the finished evaporated milk. *Carrageenan*, a colloid derived from the seaweed, chondrus crispus (Irish moss), is now used widely since it efficiently inhibits the separation of milk fat during storage of the finished product.

The small amount of carrageenan that might be used permits the use of smaller amounts of stabilizing salts. So little carrageenan is used that it has no thickening effect on the product. The amount of mineral salts added lies within the actual variation of the salts that occur naturally in milk. If the pilot sterilization of the can of untreated milk gave satisfactory results, the bulk of the milk is processed without further treatment. If some of the cans contained coagulated milk, the smallest amount of stabilizer needed to prevent coagulation is determined and added to the bulk of the milk to be sterilized. Often the forewarming operation is conditioned upon the results obtained when the previous batch of milk was processed.

After the evaporated milk has passed all tests, it goes to the can-filling machine. Here it is forced into the can through a small hole in the top cover. The weight of the milk can be adjusted within 1/25th of an ounce. Usually, *baby size* cans that hold 6 ounces (5⅓ fluid ounces) and *tall* cans that hold 14.5 ounces (13 fluid ounces) are filled, but a *confectioner's* size of 8 pounds is also used. The filling machine solders automatically the small hole through which the milk enters. The cans pass to the can-tester, where they travel fully submerged, through a tank of hot water. A defective can is detected by the presence of air bubbles and is removed. Some plants use a vacuum device to test the cans. A vacuum cup picks up the sealed can and by means of a conveyor belt carries it to the stabilizer. Imperfectly sealed cans drop from the vacuum cup and are not carried to the sterilizer.

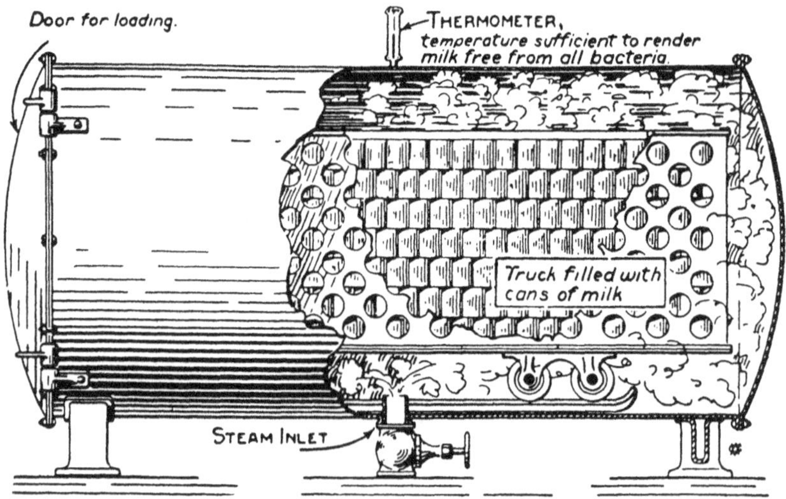

Fig. 15-4. Structure of Batch Sterilizer: The truck is filled with sealed cans of evaporated milk and rolled through the door into the sterilizer. After the door is closed, steam is admitted to heat the milk to sterilizing temperature. (Courtesy of Evaporated Milk Association).

In 1973, the average lead content of evaporated milk was 0.11 parts per million, the maximum content was 0.22 ppm. The Food and Drug Administration limit is 0.5 ppm. The lead content in the solder does not materially add to the lead content of the milk.

Evaporated milk may be sterilized by a batch process, (Fig. 15:4) but the continuous sterilizer is in common use. These methods are similar to those used in the canning of fruits and vegetables. In the batch process, the cans are placed in a container or metal basket which revolves in a steam-tight container, similar to a large pressure cooker. The agitation insures uniform heating. The milk is heated gradually, by means of steam, to 245°-250°F at the rate of about 5° per minute. It is held at the maximum temperature for about 15 minutes. The sterilizer then is filled with cold water and the cans are cooled quickly to about 90°F.

Virtually all evaporated milk is sterilized by the continuous process (Fig. 15:5). The cans first enter a preheater wherein they pass along a spiral track. Here the temperature of the milk is raised gradually to 208°-210°F in about 10 to 15 minutes. The heated milk expands and slightly bulges the ends of the cans (unless they are leaky or not air-tight). As the cans leave the preheater, they travel over a pair of rails so spaced that the bulged cans pass over them but any cans that are not bulged fall through and so are automatically rejected.

Fig. 15-5. Continuous Sterilizer for Evaporated Milk *a*) Preheater *b*) Pressure Sterilizers *c*) Pressure Cooler (Courtesy of the FMC Corporation)

A second intermediate preheater often is used before the cans enter the sterilizer. If they were exposed to the sterilizer directly from the first preheater, a tenacious film of overheated milk could form on the sides of the cans.

From the preheater, the cans enter the stabilizer, where they are heated by steam under pressure, at a temperature of 262°F for two to ten minutes. During this time the cans travel along a spiral track through the sterilizer. The agitation prevents the milk from baking on or adhering to the sides of the can. The travel along the spiral track corresponds to the revolving basket used in the batch sterilizer. After they are sterilized, the cans of milk enter a cooling unit wherein their temperature is lowered by means of cold water to about 85°-90°F in about 11 minutes. Within the cooling unit an air pressure of about 10 pounds per square inch is maintained in order to prevent mechanical strains from rupturing the cans as they cool. During cooling, the pressure within the cans is decreased and the bulged ends contract to their normal position. If a can is leaky, the air pressure within the cooler keeps the ends bulged. This difference in size is again utilized to reject any defective can as it passes over a track when they leave the cooler. The width of this track is such that normal cans drop through, but one with bulged ends will span the track, continue to roll on, and so be separated from the other cans.

The continuous sterilizer is used in the manufacture of other canned liquid products besides evaporated milk, such as chocolate-flavored drinks, liquid diets, formulas for infant feeding, and low-fat milk. From 300 to 500 cans a minute are processed, depending on the product and the size of the container.

During sterilization, milk may acquire a jelly-like consistency, known as *liver*. The cans with sterilized milk made by the batch process, if jelled, are put in a mechanical shaking device which by vigorous shaking removes the *liver* and gives the evaporated milk a smooth texture. The agitation given the cans in the continuous sterilizer is sufficient to prevent the formation of *liver*. After being shaken the cans are labeled and conveyed to packing and storage rooms.

The heat treatment and the high temperature used for sterilization give the evaporated milk a darker color then normal milk has. The increase in color is not due to the increased concentration of the milk solids. It is caused by the Maillard or browning reaction, due to an interaction between lactose and milk protein in the presence of moisture and heat.

Evaporated milk which has been canned by the aseptic process is commercially available (see Chapter 11). As the milk is essentially sterile when it comes from the HTST pasteurizing unit, it need not be further sterilized when aseptically canned. Such milk retains some of the flavor of fresh milk and undergoes little change in color.

Sterilized, canned, evaporated milk will keep indefinitely. In time the fat will rise, either as a thick layer or in globules. The homogenization and the heavy body of the milk will delay the rising of the fat but do not prevent it. A decided change in color may occur as the milk ages over a period of months; the original creamy or brownish color gradually acquires a yellowish to greenish hue. Old evaporated milk loses much of its original viscosity and usually has a thin body.

Small crystals, consisting mostly of calcium citrate, may form in cans of evaporated milk if held for a long period at room temperature, especially if calcium chloride or sodium citrate had been added to the milk before sterilization. When stored at a low temperature, around 35°F, for a period of years, granules consisting largely of a compound of calcium and phosphorus may separate.[4]

The average composition of evaporated milk is given in the following table:

TABLE 15:1
Average Composition of Evaporated Milk

Fat	7.55%
Protein	6.80
Lactose	9.80
Ash	1.45
Water	74.40

Flavor and Nutritive Value

In a comparison of the flavor of the evaporated milk made by different processes from the same lot of milk it was noted that the one made by the aseptic process had the best flavor initially. After two month's storage, its flavor was about the same as that made by the HTST method. Evaporated milk made by the conventional (batch method) procedure had the poorest flavor throughout the test.[9]

Although evaporated milk has undergone changes in its physical and chemical characteristics, it has high nutritive value. The curd formed in the stomach is soft and spongy, similar to that formed by breast milk, in contrast to the hard, compact curd formed by most raw or pasteurized milk as shown in Chapter 13. The fat globules are finely divided by homogenization and therefore are easily digested. The changes produced in the protein by the heat treatment make it assimilable by infants that in many cases cannot tolerate ordinary milk.

It was found that little change occurs in the amino acid content of evaporated milk, even after two years storage. The only major change noted was a loss of glycine from 1.3% in the protein of the fresh milk to 0.5% after storage. A nutritive value of the milk was essentially the same as that of the fresh milk.[5]

The need for addition of carbohydrates to milk in the preparation of infant formulas is a matter of controversy among pediatricians. Evaporated milk diluted with only water and without added carbohydrate is used in the nurseries of some maternity hospitals; in one large hospital this has been done for over 20 years.[6]

Vitamin Content of Evaporated Milk

Evaporated milk is an excellent source of vitamin A. The heat treatment given the milk destroys about one-fifth to one-third of its thiamine content. There is little, if any, loss of riboflavin as this vitamin is not affected by the heat treatment. Evaporated milk is a poor source of ascorbic acid and usually contains less than one-half of the vitamin originally present in the milk. Milk for infants' formulas should be supplemented with ascorbic acid from sources such as citrus fruit juices.

In Canada, a vitamin C enriched evaporated milk is made. Sodium ascorbate usually is used in its manufacture. The legal requirement is that the milk must have not less than 70 mg. of vitamin C per 100 ml. of evaporated milk. It was found that the vitamin C content of the evaporated milk remained above the minimum legal requirement even after storage at 70°F for one year.[2]

Because milk normally contains but a small and variable amount of vitamin D, the Federal regulations require manufacturers to fortify evaporated milk with not less than 25 I.U. of the vitamin per fluid ounce. This will yield a milk with 400 I.U. per quart when the evaporated milk is diluted with an equal volume of water.

SWEETENED CONDENSED MILK

According to the Federal Register (September 9, 1972) sweetened condensed milk "is the food obtained by the partial removal of water only from a mixture of milk and safe and suitable nutritive sweetener. The finished food contains not less then 8.5% by weight of milkfat, and not less than 28% by weight of total milk solids. The quantity of nutritive sweetener used is sufficient to prevent spoilage. The food is pasteurized, and may be homogenized."

Evaporated milk is preserved by heat treatment in the can, but sweetened condensed milk is a concentrated milk product preserved with sugar. It was made on a commercial scale in 1857 by Gail Borden in Litchfield County, Connecticut. When the method for evaporating and sterilizing milk was perfected, the use of sugar as a preservative decreased.

Milk for the manufacture of sweetened condensed milk, after passing chemical and bacteriological tests for quality, is forewarmed for 15 to 30 minutes at a temperature of 180°-200°F. This time and temperature will pasteurize the milk and inactivate enzymes. A lower forewarming temperature is used than for evaporated milk, in order to avoid excessive thickening upon storage. The same stabilizers may be used as for evaporated milk but the use of carrigeenan is not considered necessary. A syrup containing 65% by weight of sucrose, or a mixture of sucrose and dextrose or corn sugar, is added to the milk at a temperature around 190°F. The sugar is dissolved in skim milk or in water. The amount of sugar added should form about a 63% solution with the water in the finished product. The sugar-milk mixture is evaporated in a vacuum pan around about 135°F to a ratio of 2.5:1, that is, 2.5 pounds of the mixture of milk and sugar is evaporated down to 1 pound.

As the sucrose present does not form a saturated solution it will not crystallize from the condensed milk. The volume of water left in the concentrated product however is not sufficient to hold all the lactose in solution and this sugar does separate from sweetened condensed milk.

Considerable care is taken in the heating and cooling procedures as they have a great influence upon the body or viscosity of the finished product as well as upon the degree of separation of lactose. The milk is cooled rapidly to 80°-86°F and held there for 15 to 20 minutes. The milk is *seeded* at this time. A small amount of a previous batch of sweetened condensed milk, or a few ounces of powdered lactose, or even some dry skim milk powder, is added under vigorous agitation for about one hour. The material so added forms areas or nuclei upon which a large number of small lactose crystals will form. The greater the number of crystals that can be formed, the smaller will be the size of each individual crystal. The small crystals do not settle rapidly out of the finished product, and being small, they are not very noticeable to the taste. Fairly large crystals give the milk a gritty body and the product is said to be *sandy*.

The average composition of sweetened condensed milk is given in Table 15:2.

TABLE 15:2
Average Composition of Sweetened Condensed Milk Products

	From Whole Milk	*From Skim Milk*
Water	27.3%	28.0%
Protein	8.2	9.1
Fat	8.6	0.95
Lactose	12.2	18.0
Sucrose	42.0	42.0
Ash	1.7	1.95

Large amounts of sweetened condensed milk are used by the baking, ice cream, and confectionery industries, and it is used widely as an infant food. Large quantities are used in South-eastern Asia and other places where the climate is warm. When intended for industrial use (bulk goods) it is packed in barrels and drums; that packed for the retail trade (case goods) is put up in 6-ounce, 14-ounce, and 1-gallon cans.

Sweetened Condensed Cream

A sweetened condensed cream is made for use in the manufacture of ice cream. According to one formulation, 100 pounds of 65% cream, 53 pounds of sucrose, and 10 pounds of non-fat dry milk solids are mixed and pasteurized at 93°C. for 15 seconds. Storage at 10-15°C. is recommended. A storage life of 6-12 months is reported.[1]

Condensed Skim Milk Products

Except for their lower fat content, condensed skim milk products are similar in their properties and uses to the sweetened products. Their approximate composition is given in Table 15:2. Condensed skim milk usually contains between 20-28% total-milk-solids. Sweetened condensed milk contains between 24 and 28% total-milk-solids. In 1972, 900 million pounds of plain condensed milk were made in the United States and 60 million of sweetened condensed skim milk were made.[11]

BACTERIOLOGY OF PRODUCTS

It is generally assumed that all bacteria originally present will be destroyed during the manufacture of evaporated milk. There are cases on record, however, in which bacterial activity in canned evaporated milk has been observed. The presence of microorganisms may be due to faulty processing, spores, or a defective container, or contamination after opening.

Generally, the bacteria in evaporated milk cause coagulation of the product, but the formation of gas and off-flavor may occur. The thickening of the milk during sterilization should not be attributed to bacterial activity. Geling of canned evaporated milk usually is due to the presence of *Bacillus vulgatus* (*subtilis*). Such milk shows no increase in acidity nor formation of gas. Heat-resistant organisms, such as *Bacillus coagulans* and *Bacillus cereus* may at times survive the heat treatment. Milk contaminated with these organisms may develop a cheesy flavor and odor, and increase in acidity if held in storage at high temperature. Microorganisms are rarely found in cans of bitter evaporated milk; they probably do not survive the storage time needed to develop the off-flavor.

Bulged cans, if not due to overfilling, may be caused by chemical action of the milk on poorly tinned cans. Gas-forming organisms very rarely are present in evaporated milk that has been properly processed.

As with any milk product, after the container has been opened, evaporated milk should be kept refrigerated and protected from contamination.

Sweetened condensed milk is not a sterile product. The low content of oxygen in the can and the high content of sugar prevent the growth of bacteria. The sugar acts as a preservative. Microorganisms that may be present decline rapidly during storage; the gram-negative organisms, especially E. coli, decline most rapidly. The micrococci and spore formers, such as B. cerus, may survive a little longer.[3]

Molds, which form hard, colored bodies or *buttons*, are sometimes found on the surface of the product. Occasionally, gas forms, usually caused by sugar-fermenting yeasts. These conditions are not found in the canned product unless sufficient oxygen is present to permit growth of the organisms. This may be the case when the product is handled in bulk by bakers and other industrial users. When buttons are present, the sweetened condensed milk often is bitter or rancid.

REFERENCES

1. Bell, R. W., Concentrated sweetened cream, Proc. XV Inter. Dairy Congress, 979 (1959).
2. Bullock, D. H., Singh, S., and Pearson, A. M., Stability of Vitamin C, in enriched commercial evaporated milk, *J. Dairy Sci.*, 51:921 (1968).
3. Crossley, E. L. and Craham, P. A., Microbiological Changes during the Storage of Sweetened Condensed Milk, XVII International Dairy Congress, Sec. E/3, 167, Munich (1966).

4. Fox, K. K. et al., Composition of granules in evaporated milk stored at low temperatures, *J. Dairy Sci.*, 50:1032 (1967).
5. Kugenev, P. and Medvedeva, M., Amino acid composition of proteins in preserved milk after 2 years of storage, XVII International Dairy Congress, E/F 99 (1966).
6. Meyer, H. F. *Essentials of Infant Feeding for Physicians*; Charles C. Thomas, Springfield, Ill (1952).
7. *Sanitary Stanards Code and Interpretations,* Evaporated Milk Association Washington, D.C. (1968).
8. Stalberg, S. and Radaeva, I., Effect of various substances inhibiting melanoidino-formation in sweetened condensed milk, XVII International Dairy Congress E/F 153, Munich, (1966).
9. Sundararajan, N. R. et al, Changes in flavor on storage of evaporated milk made by three processes, *J. Dairy Sci.*, 49:169 (1966).
10. Webb, B. H. and Bell, R. W., The effect of high temperature and short time forewarming of milk upon the heat stability of its evaporated product, *J. Dairy Sci.*, 25:301 (1942).
11. *Production of Manufactured Dairy Products,* United States Dept. of Agr., F. D. 3-73 Washington, D.C. (1973).

chapter 16

Ice Cream and Related Products

Surveys indicate that ice cream is the favorite American dessert. Most ice cream is consumed in the home. Vanilla ice cream is most frequently served. Vanilla ice cream combined with other flavors rates second, followed by chocolate and strawberry ice cream. In 1968, the consumption in the United States of frozen desserts was 23.78 quarts per persons, of which 15.5 quarts was ice cream. The annual production of frozen dairy products in the United States and some other countries is given in Table 16:5.

A frozen dessert was introduced to Europe by Marco Polo, about the year 1300. Probably a similar dessert, made of flavored snow or ice, was known much earlier. In America, a letter written in 1744 by Wm. Black, tells of a dessert "fine ice cream, with strawberries and milk" served at the home of the governor of Maryland, Thomas Bladen.

The first recipe for ice cream was printed in England in 1769 in *The Experienced English Housekeeper*, by Elizabeth Raffield. In this country, ice cream was advertised for sale in the New York *Post Boy* of June 8, 1786 and has since become one of our most important milk products. Factory production of ice cream began in 1851 by Jacob Fussel in Baltimore, Maryland. Today, home-made ice cream is a rarity.

The basic ingredients of ice cream are cream, milk, sugar, and flavoring. Gelatin or some other stabilizer, an emulsifier, and sometimes egg yolk, are added. As milk and cream alone do not contribute enough milk-solids-not-fat for commercial ice cream, they must be supplemented with condensed or dry non-fat milk solids. The entire combination, known as the *mix*, is prepared in a vat or batch pasteurizer.

The fat in an ice cream mix is derived from cream, milk, sweet butter, butter oil, or frozen cream. It gives the product a rich flavor and improves its body and texture. The milk-solids-not-fat, or *serum solids*, may come from milk, condensed skim milk, cream, and dry non-fat milk and buttermilk or whey solids. The serum solids contribute to the flavor of the ice cream and impart a desirable texture to the product. Sugar not only adds sweetness to the ice cream but it also lowers the freezing point of the mix so that it does not solidify in the freezer.

Cane or beet sugar as well as dextrose (corn sugar) are used. The milk sugar in the serum solids adds but little to the sweetness of ice cream. It is not unusual for 40 to 50% of the sugar content to consist of corn syrup solids and up to 80 percent has been used, but normally not more than 10 to 20% of the total sugar is used. The amount of corn syrup solids that may be used depends upon its "DE" or dextrose equivalent, that is, its sugar content calculated as dextrose and expressed as the percentage of the total dry substance. The product is a solid up to 42 DE, then it becomes a syrup. The sweetness increases with the DE value. A low DE syrup may replace up to 50% of sucrose without causing excessive sweetness or material lowering of the freezing point of the mix. A high DE syrup is preferred when more than 50% replacement is used since it does not give the ice cream the heavy body that a low DE syrup would give. Corn syrup solids are not commonly used in soft ice cream because they lower the freezing point more than does sucrose.
syrup would give. Corn syrup solids are not commonly used in soft ice cream because they lower the freezing point more than does sucrose.

Practically all ice cream, except some chocolate ice cream, contain added permitted color.

Manufacture of Ice Cream

The liquid ingredients are heated in the mixing vat to about 110°F, and then the sugar is added, together with the other dry ingredients. Some ingredients, such as sodium alginate, used as a stabilizer, require a higher temperature for dispersion in the mix, so they are added after the mixture has reached about 160°F. (Fig. 16:1).

Pasteurization is an important step in the manufacture of ice cream. As the fat and sugar present give some protection to bacteria in the mix, it is necessary to pasteurize at a higher temperature than 155°F is held for 30 minutes. HTST pasteurization is widely used, employing plate heaters, tubular heaters, or vacuum pasteurizers. HTST pasteurization is done at a temperature of at least 175°F for 25 seconds. Recent work has shown that UHT, ultra-high temperature pasteurization, has a beneficial effect upon the flavor, stability, and texture of ice cream.[5] In the APV plate heat-exchanger, processing

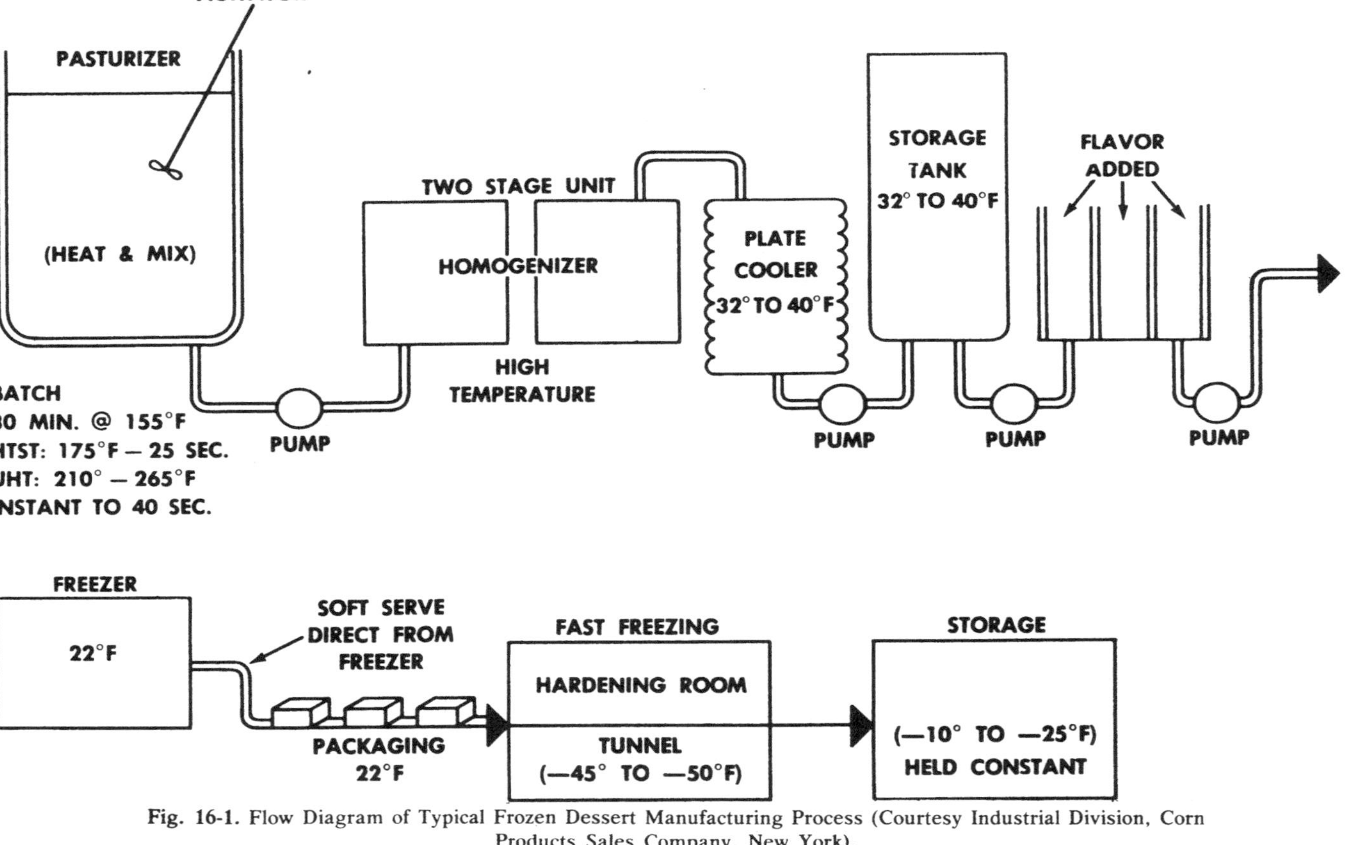

Fig. 16-1. Flow Diagram of Typical Frozen Dessert Manufacturing Process (Courtesy Industrial Division, Corn Products Sales Company, New York).

temperature up to 300°F may be obtained at a rate of 30,000 pounds of mix an hour. At this temperature, pasteurization is practically instantaneous, but usually a temperature of at least 220°F is used, with a 30-second holding time.

After pasteurization, the mix is homogenized at a temperature between 160° and 200°F. This is an essential step because it prevents churning of the mix in the freezer, reduces the time needed to age the mix, affects its viscosity and improves the body and texture of the ice cream. A pressure of about 2500 pounds is used with a single-stage homogenizer; with a two-stage homogenizer, about 2500 pounds on the first stage and 500 pounds on the second for mixes containing up to 14 percent of fat. For mixes containing over 14% of fat, pressures of 1500 to 1800 pounds and 500 pounds, respectively, are used. After homogenization, the mix is cooled to 30-40°F and held in storage at about 34°F until it is to be frozen.

A flow diagram for an automated continuous processing unit is shown in Figure 16:2.

Each ingredient is measured precisely into the weigh tank, from which it enters the blend tank. Here the product is mixed and pumped to the balance tank. A liquid-level sensor maintains a regulated flow of the mix into the HTST pasteurizing unit and heat-exchanger. After pasteurization, the mix goes to the homogenizer and finally to the freezer. In some plants, homogenization of the preheated mix may precede pasteurization. The entire process may be operated by remote control by one person.

Aging the Mix

At one time it was a common practice to hold or age the homogenized mix at a temperature of 32° to 40°F for four to twenty-four hours in order to permit clumping of the fat globules and to increase the viscosity of the mix. This procedure allows the mix to freeze more quickly and gives the ice cream a better body and texture. It is of value when ice cream is made with a fat content of 15% or more and when little consideration is given to the other milk constituents. The modern method of using a stabilizer, emulsifier, and a mix of high milk-solids-not-fat content has made a long aging period unnecessary. The mix may be held for a few hours, or go to the freezer directly from the homogenizer.

The Ice Cream Freezer

Commercial ice cream freezers are of either the batch or the continuous type. In the *batch freezer*, direct expansion of ammonia gas or Freon refrigerant is used as the freezing medium. Within the

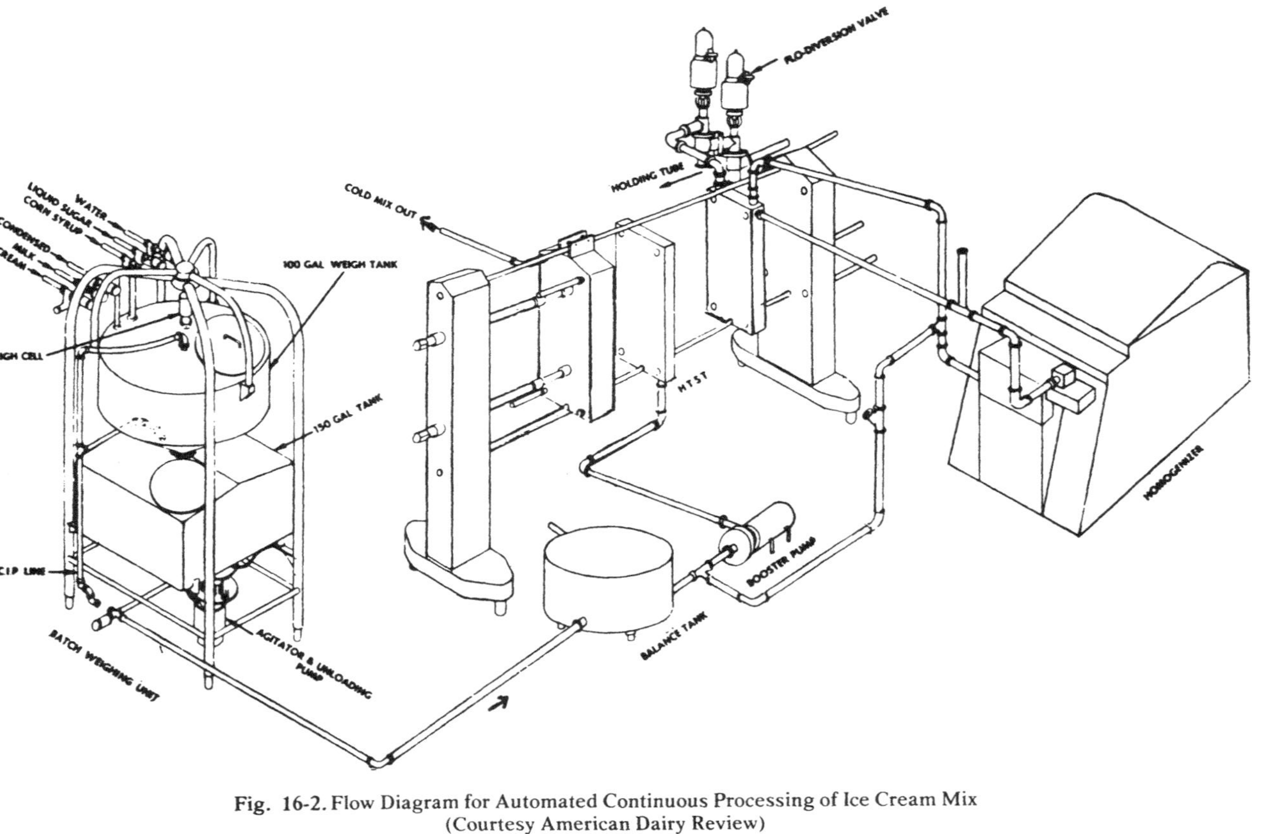

Fig. 16-2. Flow Diagram for Automated Continuous Processing of Ice Cream Mix (Courtesy American Dairy Review)

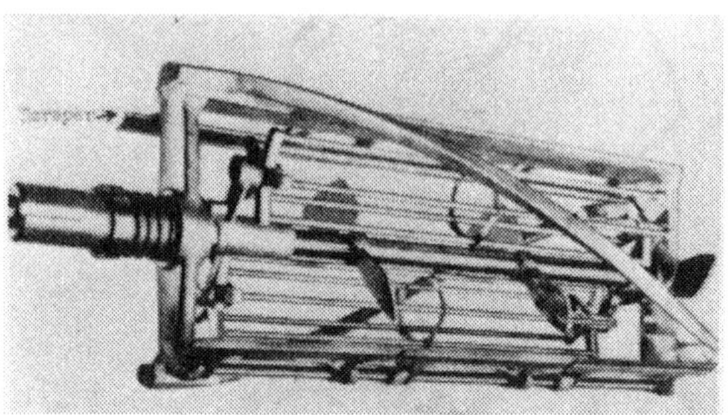

Fig. 16-3. Dasher and Scraper Assembly for Batch Ice Cream Freezer

cylinder, a dasher fitted with scraper blades turns about 200 revolutions a minute and removes the frozen cream from the sides of the freezing cylinder. In seven to eight minutes the mix is frozen to the required consistency at 21 to 23°F. The refrigeration is turned off and a revolving beater incorporates air into the ice cream in order to attain the desired overrun (increase in volume.) Dull scraper blades tend to yield a "wet", coarse ice cream. Withdrawing at a temperature above 23°F will also result in a wet, coarse product.

The *continuous ice cream freezer* operates much like the batch freezer, but differs in that a continuous flow of mix and air are pumped through the freezing unit under controlled conditions. A stiff ice cream, with a smooth body is obtained in about 25 seconds by freezing around 22°F, sometimes as low as 16°F. There is better control of the overrun than with the batch freezer. At about 22°F about ½ of the water present is frozen; at 16°F some 65% of the water is frozen.

The ice cream may be packaged directly as it comes from the freezer. The Cherry-Burrell "Vogt" freezers are made with one, two, or three freezing tubes, each of which has a capacity up to 175 gallons of mix (1050 gallons of ice cream in all) per hour. The Vogt freezing tube is long and narrow, in which the dasher or *mutator* leaves an annular space of about 5/16 inch between its surface and the freezing cylinder, through which space the ice cream travels (Fig. 16:3). (See under margarine.) In the Creamery Package continuous freezer, the freezing cylinder is similar to that of a batch machine. Each cylinder

Fig. 16-4. Ice cream processing area. Three cylinder Crepaco freezer at left. Feeding funnel for adding fruits to the ice cream to the right. Mix vat in rear. Packaging machine in the foreground. Courtesy Crepaco, Inc., Chicago, Ill.

has its own refrigeration and controls, so that it may be operated independently. Three different flavors or products may be made which may be combined or packaged separately.

Fast freezing is always essential as it favors a smooth product, with small ice crystals. In the freezer, about one-half of the water in the mix is frozen, forming the semi-frozen product sometimes served as "soft" ice cream. As the water freezes, the sugar remains dissolved in the remaining liquid which freezes at a lowering temperature as it concentrates.

Overrun

Air is a necessary ingredient of ice cream, because without it the mix would freeze to a hard or soggy mass. The increase in volume effected by whipping air into the mix during the freezing process is known as *overrun*. Ice cream makers commonly consider that the overrun should be between two and three times the total solids content of the mix. The usual overrun for packaged ice cream is about 80%; for soft ice cream it is from 40 to 80%. The usual range for bulk ice cream is from 80 to 100%, but it may reach 150%. Thus, one

Fig. 16-5. Three-Tube Continuous Ice Cream Freezer (Vogt) (Courtesy of Cherry-Burrell Corporation)

gallon of mix makes about two gallons of finished ice cream; and conversely, two gallons of ice cream melt down to about one gallon of liquid. Many States specify a minimum weight for a gallon of ice cream, or require that it contains a certain minimum weight of food solids to the gallon, in order to protect the consumer from a product that has an excessive overrun. An ice cream with an excessive amount of air lacks body and melts too rapidly in the mouth. An ice cream mix that attains the desired overrun rapidly is said to *whip* quickly. In a batch freezer, whipping is accomplished by shutting off the refrigerant as soon as the ice cream is frozen. The beater or dasher is allowed to revolve in order to incorporate the air in the ice cream. Whipping should be completed in two or three minutes in order to prevent the ice cream from melting through lack of refrigeration. In the continuous freezer, whipping is accomplished automatically while the mix passes through the freezer. The formula for the calculation of overrun is:

$$\%\text{Overrun} = \frac{\text{Volume of Ice Cream} - \text{Volume of Mix}}{\text{Volume of Mix}} \times 100$$

When the weight per unit of volume is used, the formula becomes:

$$\%\text{Overrun} = \frac{\text{Weight of mix} - \text{Weight of ice cream}}{\text{Weight of ice cream}} \times 100$$

Hardening

It is necessary to continue freezing the semi-fluid product drawn from the freezer, without further agitation, until it is firm. The quicker it is done, the better the body of the ice cream. In modern practice, the ice cream is packaged as it comes from the freezer. The packages are placed in a room held at -20 to -50°F, or conveyed through a tunnel wherein a blast of high-velocity, equally cold air quickly cools the product. During hardening some 20 to 50 percent of the water remaining in the ice cream is frozen. Even after a long period of time not all the water is frozen; at -15°F about 10% is left unfrozen.

Shrinkage

As the word implies, shrinkage involves the loss of air and a decrease in volume of the frozen product. It is not a serious problem in modern practice. There are many reasons for shrinkage, ranging from excessive original overrun to the effect of barometric pressure. The composition of the ice cream and the temperature at which it is stored also affect shrinkage. Ice cream with a high content of serum solids usually maintains its structure better and shows less shrinkage than a product low in solids. Storage at a low temperature also favors retention of overrun, but if the temperature is too low, the ice cream may be too hard to serve immediately. Pasteurization at about 185°F tends to reduce shrinkage.

Factors that may promote shrinkage include the use of sweetened condensed milk and the presence of free fatty acid in the milk fat. The exposure of ice cream to vapors from dry ice hastens shrinkage. This apparently is caused by the absorption of carbon dioxide, which causes a change in the stability of the protein structure in the ice cream.

Air is forced out of ice cream when it is transferred from the original container to others, such as pint and quart cartons or to cones. It is not unusual for a 5-gallon container to show a total loss of 1 to 1.5 gallons by the time it is dipped out in small lots. The sale of factory-filled half-gallon, quart-, and pint-sized containers is much favored as this means minimizing shrinkage.

Dipping loss is held to a minimum when the temperature of the ice cream is about 5°F or lower and the product is dipped from the highest surface in the container, which should be held stationary.

Stabilizers and Emulsifiers

A stabilizer combines with the water present in the mix to form a gel which improves the body of the product and inhibits the formation of large ice crystals. The stabilizer helps to produce an ice cream with good body, texture, and resistance to rapid melting. Ice cream may be made without the addition of a stabilizer or emulsifier, but they are useful in the commercial production of ice cream.

One of the first stabilizers used was gelatin, which still is used. Gelatin acts slowly and the initial viscosity of the mix is low. Preparations of vegetable gums, made with Irish moss, guar gum, karaya, tragacanth, or sodium alginate, are used as stabilizers. Sodium carboxycellulose and pectin as well as micro-crystalline cellulose especially are favored for the manufacturing of sherbets and ices. Usually less than 0.5% of any stabilizer is used because an excessive amount makes the product too viscous and imparts a sticky or gummy body to the melting ice cream.

Carrageenan is a component of some commercial stabilizers which without it, might cause an ice cream mix to separate or "whey-off". Less than 0.02% of the weight of the mix is sufficient carrageenan to remedy this defect.

About 0.5% of microcrystalline cellulose (MCC), by weight, tends to improve the body and texture of ice cream and ice milk, as well as impart dryness and stiffness, especially when used in a continuous freezer. MCC is recognized as a permissible ingredient in the Federal Standards, but it is not permitted in some States.

An *emulsifier* also influences the body and texture of ice cream but it is added especially to make the product dry and stiff, as well as to reduce the whipping time. The drying action is the result of the formation of an increased number of air cells and a reduction of their size. The greater surface area of the air cells allows the water present in the ice cream to be spread over a much larger surface, and so makes the product appear drier.

The emulsifier also helps to disperse the fat globules throughout the mix so that they do not churn out. The emulsifier does this owing to its structure. One portion of the molecule is soluble in fat, the other in water- the lipophilic and hydrophilic groups. By binding the fat and the aqueous parts of the mix, which normally do not mix, the emulsifier permits the formation of a uniform body in the product. In essence, the emulsifier disperses the fat globules so that they do not churn out. Added milk-solids-not-fat and soluble caseinates act both as stabilizers and as emulsifiers in ice cream. Egg yolk, equivalent to about 0.5% of dry yolk, improves both the whipping properties of the mix and the body and texture of the ice cream. The lecithin content of egg yolk acts as an emulsifier. Mono- and diglycerides of edible fatty acids have been found to be efficient emulsifiers and these are obtainable commercially under various trade names. Polyoxyethylene sorbitan monooleate or tristearate are permitted as emulsifiers but they must not exceed 0.2% of the mix.[3]

Sandy Ice Cream

The crystallization of lactose (milk sugar) in ice cream sometimes gives rise to the defect known as *sandy* ice cream. If more than about 11% of milk-solids-not-fat are present the tendency of lactose to form crystals is increased. *Heat shock*, or the melting and refreezing of some of the ice cream in the dealer's cabinet, is another factor that favors the crystallization of lactose. The crystals are not immediately noticeable in the frozen ice cream but when it melts in the mouth the crystals become evident and give the sensation of particles of sand and grit.

As in the manufacture of sweetened condensed milk, sandiness may be prevented by seeding the mix, just before freezing, with finely divided lactose. Non-fat dry milk and dry whey have also been used for this purpose. The finely divided particles provide nuclei which promote the formation of many small crystals of lactose rather than fewer large ones.

Composition of Ice Cream

As shown in the following table, the composition of ice cream may vary considerably, especially in its content of fat and milk-solids-not-fat.

TABLE 16:1
Approximate Percentage Composition of Ice Cream

Milkfat	*MSNF*	*Sugar*	*Stabilizer*	*Emulsifier*
10	10-11	13 - 15	0.3 to 0.5	0.1
12	11	13 - 15	0.3 to 0.4	0.1
14	10-11	14 - 15	0.25 to 0.3	0.05
16	8-9.5	15 - 16	0.2 to 0.3	—

The quality of an ice cream depends largely upon its total solids and fat content. Cost is a controlling factor in the fat content. An average analysis of commercial ice cream of good quality is as follows:

TABLE 16:2
Commercial Ice Cream (vanilla).

Protein	4.6%
Fat	11.5
Lactose	5.0
Sucrose and Dextrose	15.0
Ash	0.9

Included in the mineral constituents of the ash, are 0.12% of calcium, 0.17% potassium, 0.08% phosphorus, and 0.07% sodium. The vitamin content is given in Table 6.2.

Whey solids often are used up to about 3% of the total weight of the mix. This amounts to approximately one-fourth of the milk solids-not-fat content of the mix. This is about the maximum amount that may be used without having sandiness develop. Whey from cheese colored with annatto may cause the ice cream to develop a pinkish color.

Preparations containing sodium caseinate are used in some commercial products. They improve the whipping quality, texture, and resistance of the ice cream to heat-shock. Up to one percent may be used without producing a flavor defect.

Ice Cream Powder

A dry mix, containing all of the ingredients necessary to make ice cream, except the water, is a commercial product. Large amounts of these ice-cream powders are used aboard ships or where it is not convenient to prepare or store the liquid mix. A representative analysis of an ice-cream powder is as follows:

TABLE 16:3
Composition of
Ice Cream Powder

Sugar	44%
Milk Solids-not-fat	25
Milk Fat	28
Stabilizer	1
Moisture	2

ADVANTAGES AND LIMITATIONS OF THE VARIOUS ICE CREAM CONSTITUENTS

I. Butterfat
   A. Advantages
      1. Increases the richness of the flavor
      2. Produces a characteristic smooth texture
      3. Helps to give body to the ice cream
   B. Limitations
      1. Cost
      2. Fat slightly hinders rather than improves whipping
      3. A high fat content may limit the amount of ice cream consumed
      4. High caloric value
II. Serum Solids
   A. Advantages
      1. Improve the texture
      2. Help to give body
      3. A higher overrun without snowy or flaky texture
      4. A comparatively cheap source of solids
   B. Limitations
      1. A high percentage causes "sandiness"
      2. May cause salty, cooked or other flavor defect.
III. Sugar
   A. Advantages
      1. Usually is the cheapest source of solids
      2. Improves the texture and flavor
   B. Limitations
      1. Excessive sweetness
      2. Lowers whipping ability
      3. Longer freezing time required and the ice cream requires a lower temperature for proper hardening
IV. Stabilizers
   A. Advantages
      1. Very effective in smoothening the texture
      1. Very effective in giving body to the product
   B. Limitations
      1. Excessive body and melting resistance
V. Egg Solids
   A. Advantages
      1. Very effective in improving whipping
      2. Produces a smoother texture
      3. Flavor
   B. Limitations
      1. Excessive amounts may produce "foaminess on melting"
      2. Egg flavor, disliked by some consumers
      3. Cost
VI. Total Solids
   A. Advantages
      1. Smoother texture
      2. Better body
      3. More nutritious
      4. Ice cream not as cold
   B. Limitations
      1. Heavy, soggy or pasty body
      2. Cooling effect not high enough

* Adapted from H. H. Sommer, "Theory and Practice of Ice Cream Making," 1946.

To prepare ice cream, one part of the powder is mixed with 1½ to 2 parts of water and the mixture is frozen in the usual manner. To ensure a satisfactory overrun, some ice-cream powders contain an ingredient such as sodium caseinate, which increases the whipping properties of the mix. In the manufacture of ice-cream powder, the mix is prepared as usual, except that sugar is omitted. After this mix is spray dried, the sugar is mechanically mixed with the dry ice-cream powder.

A number of proprietary compounds are sold for the preparation of ice cream at home. Most of these are mixtures of sugar, vegetable gum or gelatin, starch and flavoring materials which are to be added to milk and cream before freezing. Some preparations contain all the ingredients of ice cream except the water and so are actually a dry ice-cream mix.

Nutritive Value of Ice Cream

Ice cream is an excellent food and a concentrated source of energy, but because of its large amounts of fat and carbohydrate it is an unbalanced food if made the principal part of a diet. Experiments on animals have shown that ice cream of average composition does not supply sufficient protein, mineral salts, or vitamins to maintain normal growth. The addition of more dry milk solids-not-fat to the mix corrects this defect to a considerable degree.

It generally is agreed that ice cream is easily digested, especially because the homogenization and heat treatment of the mix favors the formation of a soft curd in the stomach. The flavor has a positive psychological effect and stimulates the flow of the digestive juice. The widespread use of ice cream in hospitals demonstrates the value placed upon it as a palatable, digestible, and nutritive food. As it is cold, it is acceptable to persons suffering from irritations and infections of the mouth or throat.

Ice cream of average composition furnishes about 615 calories a pint or about 200 calories a serving. If it is assumed that the average overrun of ice cream is 80%, the calorific value of the frozen product is 60% that of the liquid mix from which it is made.

Ice milk with a content of 4.5% fat and 33% total solids, furnishes about 155 calories per 100 grams. A sherbet of average composition would also furnish about 150 calories or 580 per pint.

A method of calculating caloric values is given in Chapter 1.

Varieties of Ice Cream

Plain Ice Cream:

Any ice cream made with only a single flavor may be termed a plain ice cream. Usually the term refers to vanilla ice cream.

Fruit Ice Cream:

This is ice cream made with the addition of fruit or fruit juice.

Nut Ice Cream:

This is ice cream made with the addition of nut meats.

French Ice Cream:

This is an ice cream of high fat content with the addition of 1.5 to 3% of egg yolk solids.

Custard Ice Cream:

Ice cream custard usually is the same as ice cream pudding. More correctly, it is a cooked mixture of milk and egg which is added to the ice cream mix and then frozen. It usually contains more than 10% of fat and not less than 1.4% of egg yolk solids by weight.

Ice Cream Pudding:

An ice cream pudding is a fruit ice cream made with an appreciable amount of egg or egg yolk.

Parfait:

An ice cream of high fat content and usually containing fruit, nuts, and egg yolk, is called a *parfait* or sometimes is referred to as *New York* ice cream.

Mousse:

This is a frozen confection made with whipped cream, sugar, and flavor. An ice cream of very high fat content sometimes is also called a *mousse*.

Spumoni:

Spumoni (Italian, *foamy*) is a frozen confection of high fat content. The bottom and sides of a cup-shaped mold are lined with vanilla ice cream, then the cup is half-filled with chocolate ice cream. The cup is filled with a mousse with added fruit and nuts. After hardening, the cup is dipped in warm water and the spumoni is removed from the mold and cut into wedge-shaped pieces for serving.

Fig. 16-6. Ice Cream in Fancy Molds

Tortoni:

Tortoni is a high fat frozen dessert made from whipped cream and sugar flavored with anise, nutmeg, or almond. The mixture is covered with nut or macaroon crumbs. The tortoni is hardened in small cups or molds.

Soft Ice Cream:

This is a frozen product that is served directly to the consumer as it is drawn from the freezer at 18 to 20°F. The overrun usually is less than that for ice cream.

Ice Milk:

This is a product similar to ice cream but it usually contains only two to five percent of fat. The composition of ice milk usually varies between the following limits:

Fat	2 to 5%
Milk solids-not-fat	10-13
Sugar	14-18
Stabilizer-Emulsifier	0.4

Ice milk usually has an overrun of 50-80%, but is may reach 90%. Owing to the relatively low milkfat and milk solids content the basic product has a bland flavor and so more flavoring is used than for ice cream.

Sherbet:

A sherbet is a frozen confection made from water, sugar, a small amount of milk solids, acid, fruit flavor and color. Enough stabilizer is added to promote a partial gel at room temperature. Ice cream or ice milk mix often is used as the source of milk solids. Citric acid commonly is used as the source of acidity, but other acids permitted in food products have been used. In most areas the acidity of a sherbet must not be less than 0.35%, expressed as anyhydrous citric acid, (*p*H 3.4-3.8).

A sherbet may contain from 25 to 35% sugar, 2 to 5% milk solids, and 0.4% stabilizer in addition to the acid and flavor. The frozen product has an overrun between 30 and 45%.

A representative sherbet contains milkfat 1.1%, milk solids-not-fat 3%, sucrose 23%, corn syrup solids 8%. Whey solids may replace up to 25% of the milk solids-not-fat. A small amount of an emulsifier may be added to impart desirable dryness and to control overrun.

In 1972, 50 million gallons of sherbet were produced in the United States.

MODIFIED ICE CREAMS

Dietetic Ice Cream

Special preparations are available for a frozen dessert acceptable to persons on a restricted diet. Persons with cardiac and circulatory ailments may require a diet low in sodium. A frozen dessert made from spray-dried milk low in sodium may be used in this case, as it would not contain more than 5 to 10 mg of sodium to 100 grams.

Sugar-free ice cream and ice milk are commercial products in which *sorbitol* is used instead of sucrose. Sorbitol is a hexahydric alcohol, or sugar alcohol, which is found in nature in many fruits. Sorbitol and the related product, *mannitol* are made in large quantities by the chemical reduction of dextrose. Sorbitol is widely used in both foodstuffs and industrial products as a humectant, because by preventing the product from losing or gaining moisture, it ensures control of moisture content. Glycerol is not used in dietetic ice cream because it is metabolized like the sugars. As sorbitol does not have the sweetening power of sucrose, a non-nutritive sweetening agent, saccharine, is also used. A ruling of the Food and Drug Administration (April, 1968) states that a food which contains an artificial sweetener must supply at least 50 percent fewer calories than a similar food containing only added natural sweeteners. (One gram of saccharin is equivalent to 300 grams of sugar.)

Until recently,cyclamates (cyclohexyl sufamate, Sucaryl) were used as non-nutritive sweetening agents. (One gram of cyclamic acid or its salts is equivalent to 30 grams of sucrose). The finding that large doses of cyclamates caused cancer in rats resulted in a ruling to prohibit their use in foods and beverages. Similar prohibition of the use of cyclamates was made in Canada and other foreign countries.

In some products, including frozen desserts of low sucrose content, ammoniated glycyrrhizin has been used for its ability to enhance sweetness. A reduction of about 30 percent of the sucrose content may be obtained. Fructose, which has about twice the sweetness of sucrose, has been used in dietetic frozen desserts.

Diabetic Ice Cream

As in the dietetic diet, sorbitol is used as a substitute for sugar. Compared to dextrose and sucrose, it has a slow rate of absorption from the intestinal tract. As a result, sorbitol imposes less of a burden upon insulin demand than does sucrose.[8] It has about 80% of the sweetness of sucrose. The consumption of more than 15 grams of sorbitol at one time, or more than 40 grams a day, may have a laxative effect. The Food and Drug Administration permits the addition of sorbitol to ice cream and frozen desserts, provided that the amount used does not exceed that reasonably required to accomplish the intended physical or technical effect.[3]

In the manufacture of diabetic ice cream, it is advisable to keep the milk fat in the normal range, 10 to 12%. A low lactose non-fat dry milk may be used, as it contains about 36% of lactose compared to 51% for the normal product. Non-sugar flavors such as vanillin, coffee, and maple should be used, rather than fruit syrups which are high in sugar content: see Table 16:4.

TABLE 16:4
Percentage Composition of Diabetic Frozen Desserts

	Ice Cream	*Ice Milk*
Fat	10 - 12	4 - 5
Milk-solids-not-fat	9 - 12	12 - 13
Sorbitol	12 - 15	12 - 13
Stabilizer	0.3	0.4
Non-nutritive sweetening	0.01	0.01

MELTING QUALITY OF ICE CREAM

Ideally, ice cream should melt evenly to the consistence of heavy cream at room temperature. A number of defects may lead to poor melting quality.

If the product resists melting, it usually is due to the use of too much stabilizer or emulsifier. Homogenization of the mix at a low temperature or with excessive pressure also leads to this defect. A high fat content may also favor a resistance to melting.

Seepage or drainage of liquid from the ice cream during melting, called "wheying-off" may occur. A stabilizer which may react with the milk protein could cause this defect. Ice cream of high fat content may also show seepage upon melting.

The presence of curdy particles in the melting ice cream may be due to the use of an unsuitable stabilizer or to improper homogenization procedure. A dry film over the surface of melted ice cream may be due to prolonged storage of the product or to a similar destabilization of the milk protein during freezing of the mix.

An ice cream that melts to a foam usually has an excessive overrun, favored by the use of too much stabilizer or emulsifier as well as the use of a mix that has a high viscosity.

BACTERIOLOGY OF ICE CREAM

With the possible exception of a few flavoring materials, all of the ingredients used in the manufacture of ice cream contain harmless bacteria, often in very large numbers. The mix is pasteurized at a higher temperature than that used for milk. With a batch pasteurizer a temperature of 155-160°F is used, held at that temperature for not less than 30 minutes. The HTST method requires a temperature of 175°F for at least 25 seconds. When vacreation is used, the temperature reached is about 194°F for 2 or 3 seconds. Ultra-pasteurization of the mix calls for flash-heating to 220-300°F and immediate cooling.

A reduction in the microbial content may occur during freezing of the mix. This, in part, has been found to be due to a disintegration of the microbial cells by the abrasive action of rapidly moving ice crystals during freezing.[2]

Flavoring extracts contain, as a rule, only a small number of bacteria, perhaps owing to the action of the alcohol or other solvents used in the extract. The bacteria in the ingredients are destroyed, to a large extent, when the ice cream mix is pasteurized. Bacteria and molds on nut meats may be destroyed by boiling the nuts in a thick syrup for about 15 seconds. Canned fruits are practically sterile if properly processed, but fresh fruits, and some frozen fruits, may contain molds and yeasts as well as bacteria. Because of the whipping and mixing during freezing, the bacteria in the ice cream are evenly distributed throughout the product.

Long periods or storage or holding may permit the growth of some varieties of bacteria but they usually decrease in number during the hardening and holding periods. As a rule, the bacterial content increases when the frozen product softens and is refrozen, as sometimes happens in a storage cabinet.

A few outbreaks of disease have been traced to the use of contaminated ice cream. Such unfortunate events can be avoided if all the ingredients of the ice cream are pasteurized and if care is taken to avoid subsequent contamination.

A number of states and cities have placed legal limits upon the maximum number of bacteria permitted in ice cream and other frozen dairy products. It is feasible to make ice cream with a bacterial content of less than 5000 bacteria per gram. Standards adopted by the various state and local authorities permit more bacteria, varying from 50,000 a gram upwards; some places have no standards at all.

A common source of contamination of ice cream and frozen

products is in the use of scoops for dispensing ice cream in cones, and for packing bulk ice cream into containers by hand as is done in retail stores. It has been pointed out that the use of dippers and scoops in dispensing ice cream differs but little from dipping out pasteurized milk, a prohibited practice.[4]

IMITATION FROZEN DAIRY PRODUCTS

Mellorine and other imitation dairy products are described in Chapter 21.

PRODUCTION OF ICE CREAM AND RELATED PRODUCTS

Recent figures for the production of ice cream and related products are shown in the following table:

TABLE 16:5[6]

Country	*Million U.S. Gallons*
United States,	
Ice Cream	770
Ice Milk	289
Imitation Products (Mellorine)	45.5

Approximate Production in Foreign Countries

Australia	56***	Netherlands*	19.0
Canada-Ice Cream	61***	New Zealand*	13.0***
Ice Milk	3***	Norway*	5.6
Denmark	8.0	Poland	10.5
Finland	6.8	Rep. So. Africa	12.7
France	16.5	Sweden	15.5
West Germany	79	Switzerland*	8.4
India	88	Turkey*	8.4
Ireland (Eire)	6.0	United Kingdom**	51.0
Italy*	80	Yugoslavia	5.0
Japan*	154		

* — includes other frozen dairy products
** — frozen products made with vegetable fats — 36 million U.S. gallons
*** — Imperial gallons

The average consumption of ice cream remains fairly constant in the United States, averaging 15.0 quarts per person. In 1972, ice cream used the equivalent of 11 billion pounds of whole milk; ice milk, two billion pounds. New York produced 66 million gallons, Pennsylvania, 74, and California, 75 million gallons of ice cream.

REFERENCES

1. Anon. Food Processing, p. 81; Jan. 1969.
2. Foley, J. and Sheuring, J. J., Microbial Death During Freezing, J. Dairy Sci., 49:928, 1966.
3. Frozen Desserts: Definitions and Standards of Identity; *Federal Register*, Nov. 29, 1962.
4. Gould, I. A., Larsen, P. B., Doetsch, R. N., Potter, F. E., Maintaining quality in ice cream, *Ice Cream Field,* 52:80-86 (1948).
5. Hatherly, Dave, *Ultra-high temperature processing of ice cream mix;* A.P.V. Company, Buffalo, N.Y. (1961).
6. *Production Index,* International Association of Ice Cream Manufacturers, Washington, D.C. (1972).
7. Production of Manufactured Dairy Products, 1967, U.S. Dept. of Agr. Washington, D.C., July 1968.
8. Wick, A. N., Almen, M. C., Joseph, L., The metabolism of sorbitol, *J. Amer. Pharm. Assoc.,* 50:542 (1951).

chapter 17

Dry Milk Products

Drying is one of the oldest methods used to preserve a foodstuff. The first mention of a dried milk product appears to have been made by Marco Polo, who wrote that the Mongols boiled milk, removed the cream, and dried the remaining liquid in the sun. Buttermilk often was treated in this manner since it provided a better flavored product.

A patent was granted in England in 1855 for a method of dryng milk by evaporation in an open pan; and in 1862 a patent was issued in the United States for a process to dry milk by spraying. These ventures were not practical and it was not until 1883 that the process for making malted milk provided the first commercially successful dry milk product.

Milk, skim milk, and buttermilk, as well as cream, whey, and other liquid milk products may be dried by either spraying or drum-drying. Dry milk, especially non-fat dry milk, makes it possible to supply the nutrients in milk to areas where there is no dairy industry, or where economic conditions place fresh milk out of the reach of many.

METHODS OF DRYING MILK

Drum or Roller Drying

Drum or roller driers for the manufacture of dry milk are either the open (atmospheric) type or the vacuum type (Fig.17:1). In the open type, one or two steel drums or rollers, four to ten feet in diameter, dip into a container which holds the milk to be dried; or, on some machines, the trough formed by the contact of two rollers, acts as the milk container. The drums or rollers are heated internally by steam or hot water and, depending upon the size of the unit, are capable of drying from about 1000 to 6000 pounds of milk an hour. The milk is condensed before it enters the dryer.

In vacuum drum dryers, the rollers are enclosed in a chamber from which the air is withdrawn, thus permitting the milk to be dried at a

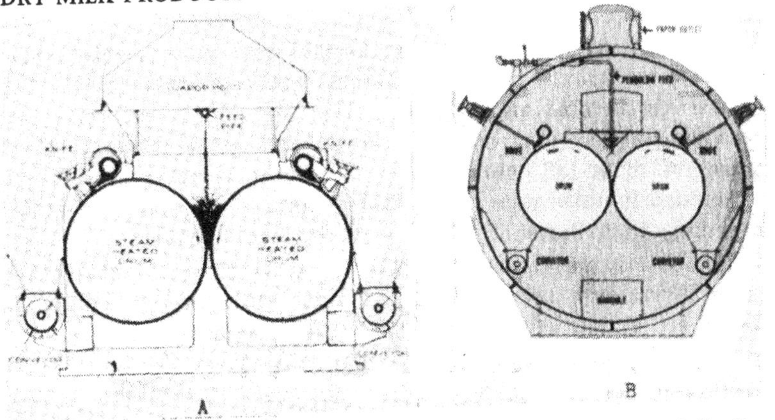

Fig. 17-1. A. Atmospheric Roller Drum Dryer B. Vacuum Double Drum Roller Dryer (Courtesy Blaw-Knox Co.)

much lower temperature than that used for atmospheric drying.

As the drum or drums of the dryer revolve, a thin film of milk adheres to them and this is dried by the time a complete revolution is made. This takes from 6 to 30 seconds, depending upon the degree to which the milk had been previously concentrated by evaporation. The dried milk, usually in the form of a sheet of film, is removed from the drum by means of a knife, called the "doctor blade", and falls into a hopper from which it is transferred to a receiving bin. Usually the product is ground to powder.

The roller process is used mostly for drying skim milk, whey, or buttermilk, especially if the product is to be used for animal feed. The high drying temperature (200°-300°F) required for the atmospheric process yields an insoluble product, whereas a relatively soluble dry milk is obtained by vacuum drying below 212°F.

Spray Drying

Nearly all dry whole milk and about 80% of all dry nonfat milk is made by the spray-drying process. Usually part of the water is first removed, concentrating the milk to 45 or 50% total solids in order to hasten drying and to increase the capacity of the dryer. In the dryer, a fine spray of milk is forced rapidly through a stream of heated air. Commonly, direct gas-fired heated air is used, at a temperature of about 150°F for whole milk and up to 350°F for nonfat milk. Spraying is done by whirling the milk by centrifugal force from the edge of a rapidly revolving disc or by pumping it under considerable pressure through a spray nozzle. The dry milk particles fall to the bottom of the drying chamber while the heated air carries the water vapor away.

Dry milk made by spraying is in the form of tiny, more or less hollow spheres, and usually is very soluble. Spray process units vary considerably in size and are more intricate and costly than the relatively simple roller dryers. The smaller spray driers have a capacity of about 150 pounds of dry milk an hour but some units in current use handle more than 5000 pounds of dry milk an hour, depending upon the solids content of the liquid feed.

The nozzle is an important part of the spray dryer. It controls the rate of flow of milk into the drying chamber and mixes the milk with the hot air, so that it forms particles of the size desired in the dried product. The bulk density, which may vary from about 22 to 30 pounds per cubic foot, the moisture content, and solubility of the powder depend upon the size of its particles. A coarse powder is usually easily reconstituted.

Most of the dried powder falls to the bottom of the drying unit from which it is removed continuously. The smaller particles, which are carried out of the chamber by the exhaust air, are recovered by means of cyclone separators, or by catching them in cloth bags or filters (Fig. 17:2).

The operation of a spray drying process is illustrated in Figure 17:2. Heated, filtered air is blown through duct A through the distributing head B into the drying chamber J. Milk to be dried is preheated at G and pumped at high pressure through spray nozzles located in air inlets, C. The atomized milk is dried and falls to the floor of the drier, from which it is removed by conveyor screw K to the chamber outlet, L. Here it is picked up and carried to the collector M, where it is air-cooled, sifted, and readied for packaging. Powder that does not fall to the floor is carried by the hot air past the baffle D to filter bags E, where it is trapped and automatically shaken at intervals to make it drop to the floor.

Dry Whole Milk

In 1972, 75.2 million pounds of dry whole milk were made in the United States, about one-half of which was exported. Minnesota led in production followed by New York and Wisconsin. Much of the dry whole milk was used in the manufacture of candy and chocolate products. The world trade of dry whole milk in 1972 totaled about 531.3 million pounds. France exported 94 million pounds, followed by Finland, Netherlands, Belgium, Luxemburg and Australia.

Dry whole milk generally is made by the spray process; little is roller-dried. Substantial amounts have been made by the *foam-drying* process developed by the U.S. Department of Agriculture, using either a batch or a continuous method.

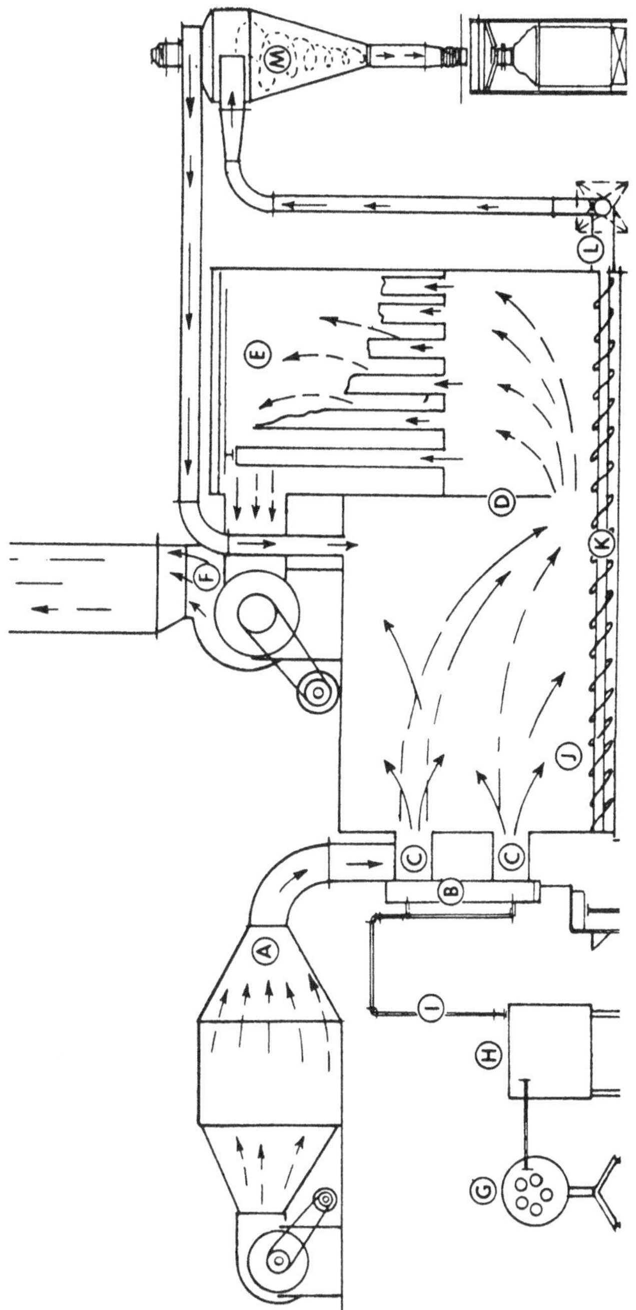

Fig. 17-2. Flow Sheet for Spray-drying of Milk (Courtesy of C. E. Rogers Company)

Foam-Dried Milk[14]

For foam-dried whole milk, milk of 3.1% fat content (26% dry basis) is pasteurized at 162°F. for 16½ seconds and homogenized. The product is concentrated in a vacuum evaporator to about 45% total solids. The concentrate, homogenized at 3000-500 pounds at 135°F. has a fat globule size of two microns or less. Nitrogen or carbon dioxide gas is introduced and the concentrate is cooled quickly and the gas dispersed in a heat exchanger at 35°F. The foam concentrate is then further expanded and dried on a continuous belt in a drying chamber at about 18mm. of mercury absolute pressure.[1] From 5 to 15 grams of carbon dioxide gas is used to foam 100 pounds of 45% solids concentrated milk. When carbon dioxide gas is used as the foaming agent the resulting product has greater density than when nitrogen or air is used and the reconstituted liquid usually has less surface foam.[15]

Fresh dried whole milk is soluble and of good flavor. Dry whole milk has limited keeping quality and so far has defied efforts for

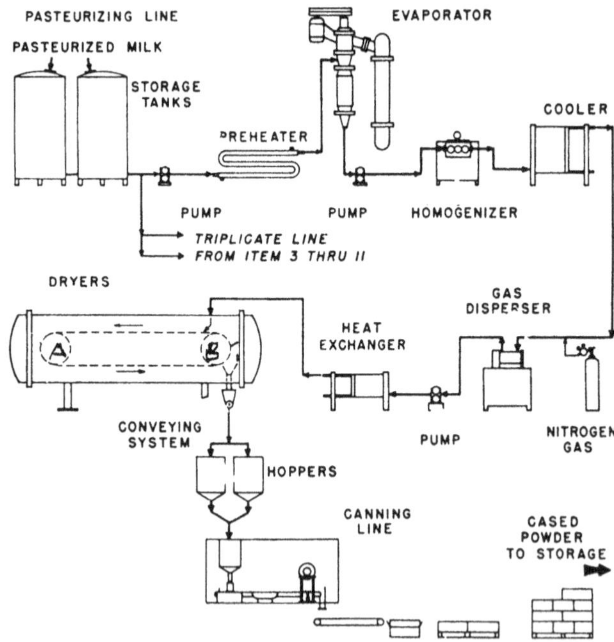

Fig. 17-3. Flow Sheet for Continuous Process Foam-Dried Whole Milk[14] (A - Heating Drum; B - Cooling Drum) Courtesy Food Technology 25:52-64 (1971)

major improvement. The use of anti-oxidants, although helpful, does not ultimately prevent deterioration over a period of time. The use of anti-oxidants was long considered as adulteration, but certain compounds, such as butylated hydroxyanisole and propyl gallate, may be added to instant whole milk powder. Nitrogen or carbon dioxide gas is used in containers of dry whole milk in order to delay spoilage. Vacuum foam-dried milk may be packaged with less than 0.001% oxygen by volume. As a result the product may be held at 40°F. for at least nine months and still have an acceptable flavor. When stored at room temperature, flavor defects may appear within three weeks. Air may be used in the preparation of foam-dried nonfat milk since fat oxidation is not a problem.

An oxidized or tallowy flavor, which arises in the milk fat is a common defect of powdered whole milk. A fishy flavor may develop if the moisture content exceeds 3.5 percent. If improperly heated, lipase may cause a rancid flavor in the powder. During storage, a stale flavor develops and solubility gradually decreases, especially if the moisture content exceeds 3.5 percent.

Dry Cream

Cream may be dried by a spray process. It gives a light cream-colored powder, usually somewhat flaky in appearance. When cream of 18% fat content is dried, the powder contains about 72% of milk fat. Much of the product made is of lower fat content, usually between 40 and 70% fat content. The product usually is made on special order.

During manufacture, the dried cream is moved quickly from the drying zone and cooled promptly to permit fat solidification. Cream

Fig. 17-4. Structure of Foam-Dried Whole Milk obtained by expansion of nitrogen gas in the concentrated milk as the pressure in the dryer is reduced.

with a fat content over 50% may require several days of cooling. The dried cream is sifted through a coarse mesh screen before packaging. If placed in deaerated, hermetically sealed containers dried cream has a shelf-life of several months at room temperature. The use of "coffee-whiteners" made of non-dairy products has materially decreased the demand for dry cream.

Nonfat Dry Milk

Owing to the unfavorable reaction that many persons have to the term *skim milk*, the dairy industry refers to dry skim milk as *nonfat dry milk*. This term was incorporated into the Federal Pure Food, Drug and Cosmetic Act by Congress in 1944. The term is descriptive of the product and does not minimize its high nutritive value.

Nonfat dry milk for use in bakeries is a "high-heat" powder, made from skim milk that had been forewarmed to 190-200°F for 30 minutes. A "low-heat" powder is used in the manufacture of ice cream and cottage cheese, and for household use. In this case the skim milk is forewarmed to not over 160-170°F. After being forewarmed the skim milk usually is concentrated to 35 to 45% of solids before drying. To be classified as a low-heat product, non-fat dry milk must contain not less than 6 mg. of undenatured whey protein nitrogen per gram as measured by the modified Harland-Ashworth Test[7], for the high heat product the whey protein nitrogen must not exceed more than 1.5 mgs. per gram. A modified qualitative procedure for this test is described in Chapter 20.

Much nonfat dry milk is used by the baking industry, not only in bread but also in cakes, crackers, doughnuts, and pie fillings. The lactose of the milk reacts with the protein constituents[6] during baking and gives the bread crust a desirable, uniformly brown color. For the same reason, bread that contains milk solids yields a better colored toast than does plain bread. Owing to its ability to retain moisture, milk bread usually will remain fresh much longer than bread made without milk solids. Milk adds to the nutritive value of bread because it supplies mineral salts and proteins that are lacking in wheat flour. Up to 20% of the weight of flour used may be substituted by non-fat milk solids. Bakers generally used 6% or less. When 6% of the weight of the flour is substituted by non-fat dry milk solids, the bread will contain about 50% more phosphorus, 16% more protein, and about 300% more calcium than white bread made without milk solids. When 6% of nonfat dry milk is used in bread it represents the total amount of nonfat solids that would be present in the bread if the entire liquid content were fluid nonfat milk.

Much nonfat dry milk is used in the manufacture of prepared

mixtures for pancakes, waffles, and biscuits. In some of these products the shortening is a spray-dried mixture of vegetable fat and skim milk.

Nonfat dry milk is an important ingredient of ice cream mix and the powder sometimes is added directly to the freezer in order to increase the milk solids in the ice cream without causing it to become *sandy*.

Infant foods, cocoa and chocolate drinks, malted milk, and other beverages often contain nonfat dry milk. Meat products, such as sausages and frankfurters may contain nonfat dry milk because it helps to retain moisture and gives a plump appearance to the product. For every pound of dry nonfat milk added there is an

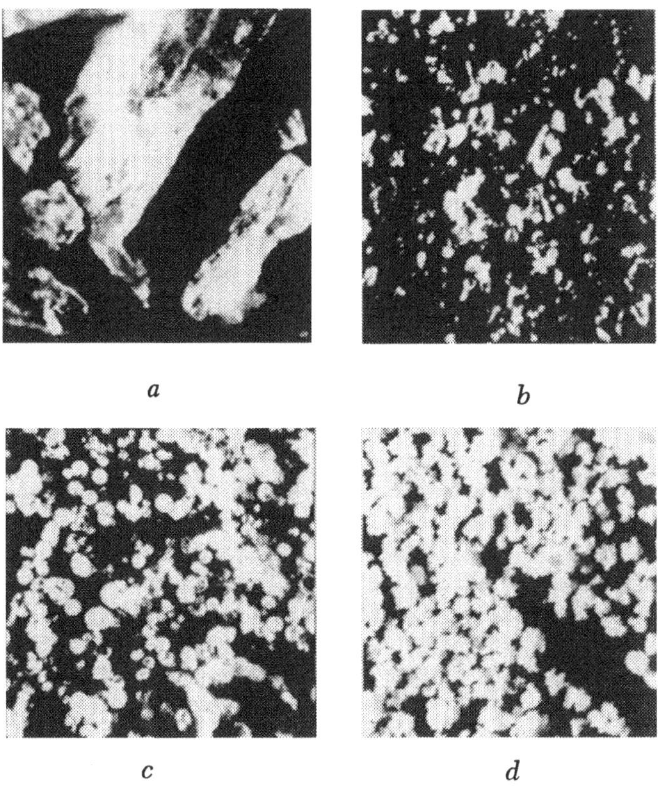

a *b* *c* *d*

Fig. 17-5. Appearance of Non-Fat Dry Milk under Microscope (x 8)
(Courtesy of Blaw-Knox Company)
a)roller-dried *b*) Vacuum Drum dried
c) Spray-dried *d*) Instantized

absorption of one pound or more of water. The Federal standards permit not more than a total of three and one-half percent of nonfat dry milk, cereal or starches to be present in the finished sausage. Either spray dried or roller dried nonfat milk is used. In 1972, about 33 million pounds were used for meat processing.

Many cat and dog foods, as well as feeds for poultry, calves, and other domestic animals contain nonfat milk solids, usually made from milk of inferior quality as well as from buttermilk and whey. In 1972, a total of 5.6 million pounds of nonfat dry milk was used for animal feed.

Some soap and cosmetic preparations contain dry milk solids, and mixtures of starch and dry non-fat milk solids have been used as dusting and cosmetic powders.

The output of nonfat dry milk for human food use was 1.2 billion pounds in 1972. Over one-third of all the nonfat dried milk was produced in Minnesota.

The appearance of the different nonfat dry milks under the microscope is shown in Fig. 17:5.

Instant Nonfat Dry Milk

The difficulty of dissolving nonfat dry milk was long a limiting factor in the use of it in the home. The introducton of "instant" nonfat dry milk remedied this situation and today virtually all nonfat dry milk sold at retail is of the instant type. Although it is sold under many brand names, almost 80% of the product sold for household use is made by only a few companies. In 1972, 224 million pounds of nonfat dry milk was packaged for home use, compared to 184 million pounds in 1960.

Instant nonfat dry milk is spray-dried skim milk which has been processed in such a manner that the powder particles are clustered or *agglomerated*. In this process, individual particles of powder are bound together in durable clusters. The lactose in the product changes from a concentration of about two parts alpha and three parts beta lactose to about three parts of alpha and two parts of beta.[3]

Because they are relatively large and porous, the penetrating moisture breaks up the clusters when they are added to water and allows each individual particle of powder to disperse quickly and dissolve. Nonfat dry milk is used principally in the home for cooking and, when reliquefied, as a beverage. The dry equivalent of 1 quart of fluid skim milk is 3.2 ounces of powder.

There are several methods of making instant nonfat dry milk and some related products, such as instant malted milk powder, and

TABLE 17:1
Some Grading Requirements for Dry Milk Products*

Maximum Allowance of	*Nonfat Dry Milk*					*Dry Whole Milk Gas Packaged*			*Dry Buttermilk*	
	Spray Dried			Atmospheric with Roller Dried		Spray			Spray and Roller	
	Extra	Standard	Instant	Extra	Standard	Premium	Extra	Standard	Extra	Standard
Fat %	1.25	1.50	1.25	1.25	1.50	26.00[a]	26.00[a]	26.00[a]	4.50[a]	4.50[a]
Moisture %	4.00	5.00	4.50	4.00	5.00	2.25	3.00	3.00	4.00	5.00
Titratable	0.15	0.17	0.15	0.15	0.17	0.15	0.15	0.17	0.18	0.20
Acidity %									c	c
Bacterial										
Estimate-Not Over	50,000	100,000	30,000	50,000	100,000	30,000	50,000	100,000	50,000	200,000
(per gram)						b	b			
Scorched	Disc B	C	B	C	D	A	B	C	B	D
* Particles, mg.	15	27.5	15	22.5	32.5	7.5	15	22.5	15	32.5

* From *Standards for Grades for the Dry Milk Industry. Including Methods of Analysis:* published by American Dry Milk Institute, Chicago, 1971
a-minimum content
b-Copper, not over 1.5 ppm; Iron, not over 10 ppm.
c-minimum titratable acidity, 0.10%
Coliform count - Not over 10 per gram.

Fig. 17-6. Ground level view of pasteurizing equipment and evaporator chambers. Portion of Mojonnier Dryer shown on the right. Courtesy Mojonnier Bros. Co., Chicago, Illinois.

chocolate milk powder. In one method, the "double pass", previously dried nonfat milk is agglomerated by making a slurry of the powder with water. The moisture content is raised to about 10% and the slurry is redried under carefully controlled conditions to permit lactose crystallization.

In the Cherry-Burrell A-R-C-S system (Agglomerates, Redries, Cools, Sizes), previously dried powder is fed into the hopper, from which it passes to the agglomerating tube. A high-velocity stream of humidified air picks up the powder and moistens it sufficiently to cluster. Large clusters separate from the air stream in the Wet Collector at the far end of the agglomerating tube. The wetted, clustered powder passes from the wet collector into the re-drying section. High-velocity, filtered air, at 270° to 300°F. is forced through the re-drying section along with the clusters. These are dried to the desired moisture content and discharged into the dry collector.

The re-dried clusters pass from the dry collector onto the shaker-cooler, a nylon-covered vibrator table with filtered, cooled air forced down over the clusters. Any fine particles are collected in the exhaust collector. Stainless steel mesh screen at the discharge end allows small clusters to by-pass the sizing rolls.

The clusters from the shaker-cooler pass into the stainless steel sizing hopper. They pass through stainless steel rolls to make pellets of uniform size for packing in barrels or for passing into a surge hopper for small unit packaging. The equipment can process powder at a rate up to 3000 pounds per hour.

In the Blaw-Knox *Instantizer* a stream of powder is surface moistened by falling between a pair of steam jets. The agglomerated particles then drop through a stream of turbulent air heated to 240-280°F, which removes the moisture picked up from the steam jets. The agglomerated powder is carried by a belt through a gyrating sifter where it is sized and readied for packaging. Fig. 17:7.

The "single pass" procedure starts with nonfat milk and instantizing is an integral part of the drying process.

Instant nonfat dry milk has about one-half the density of the conventional powder and so requires a larger container to hold a given weight of the product (grading standards are given in Table 17:1).

A foam spray-dried nonfat milk was developed by the U.S. Department of Agriculture. Nonfat milk is concentrated to 55-60% solids. As the concentrate is discharged from the spray-nozzle, compressed air, at about 2000 psi, is introduced. Other gases, such as nitrogen, or carbon dioxide, may be used. The foamed milk has an

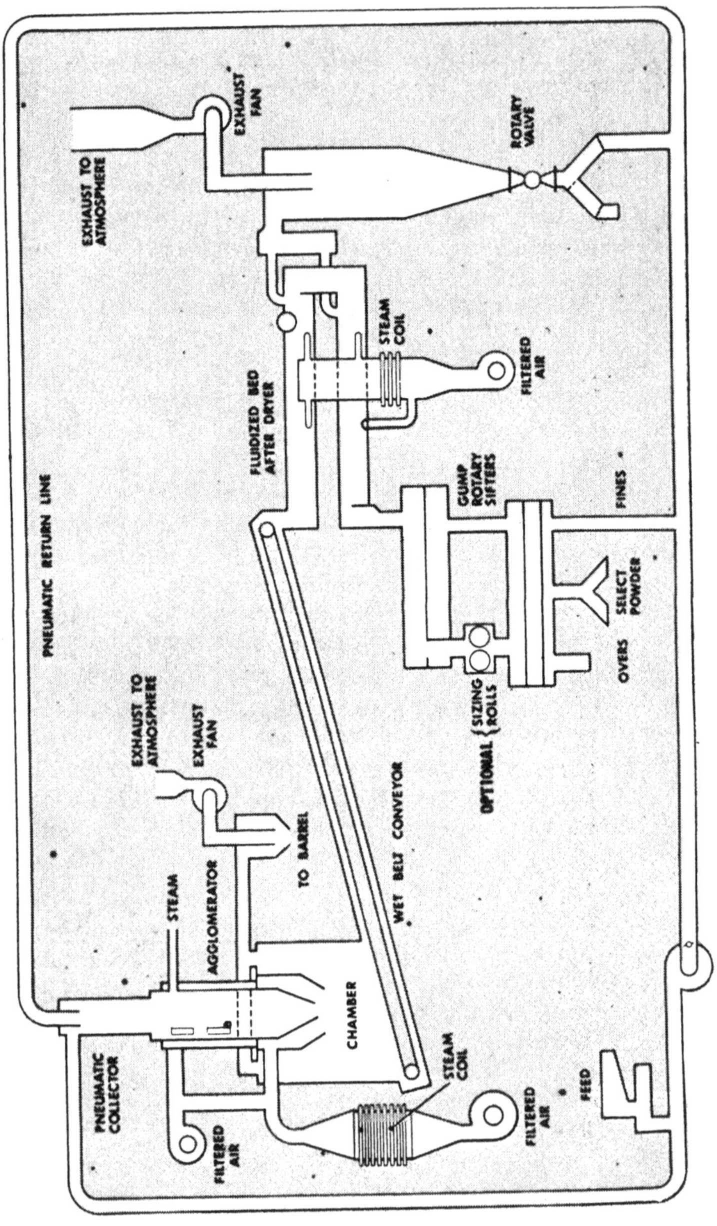

Fig. 17-7. Blaw-Knox Instantizer
Courtesy Blaw-Knox Food & Chemical Equipment Inc. Buffalo, N.Y.

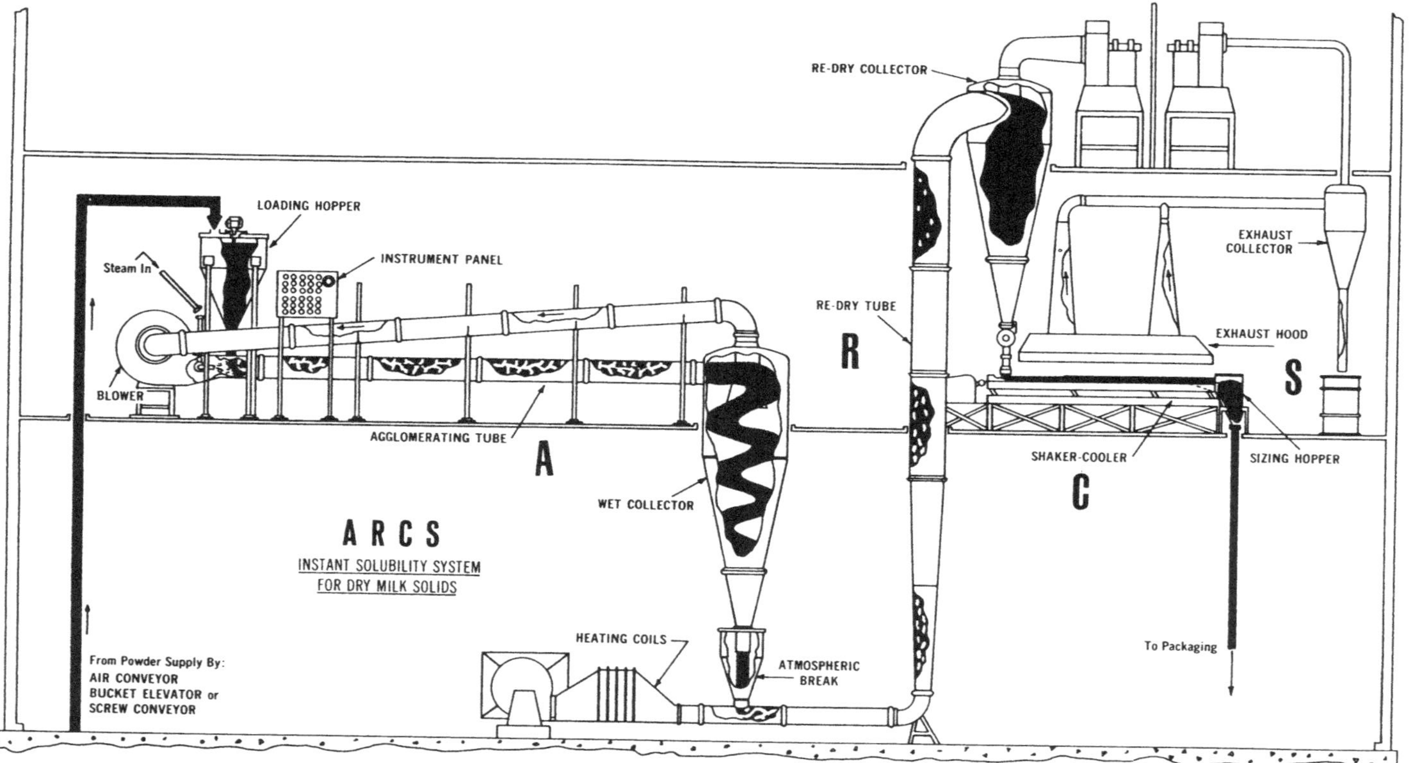

Fig. 17-7. Flow Sheet for Instant Milk Agglomerator (Courtesy Cherry-Burrell Corporation)

increased surface area, which favors a large droplet which dries rapidly. The foam-sprayed milk is easily dispersable and does not require a separate instantizing treatment.

Dry Buttermilk

Dry buttermilk may be made by the spray process from sweet cream buttermilk, but sour cream buttermilk usually is drum dried. In either case, the product is more hygroscopic than is nonfat dry milk, especially that made from sour cream buttermilk. The keeping quality of the product is improved if the dried buttermilk is held in bins for a day before it is packaged. This allows time for the lactose to crystallize and prevents excessive caking when the dry buttermilk is packaged. Further drying of the dry product with hot air improves its quality and delays the discolorization that may otherwise occur on storage.

Dry buttermilk is used by bakers and in the preparation of dry mixes, such as pancake flours. A limited amount is used by confectioners and by manufacturers of ice cream. When butter is used as the principal source of fat in an ice cream mix, the use of dry buttermilk improves its whipping properties, probably on account of the phospholipid material contributed by the dry buttermilk.

In 1972 about 0.4 million pounds of dry buttermilk were utilized for animal feeds and 51.2 million pounds for human consumption. Of this latter amount, about 14.6 million pounds were used by bakers and 13.5 million pounds in prepared dry mixes.[3]

Malted Milk

> The United States Department of Agriculture defines malted milk as: The product made by combining whole milk with the liquid separated from a mash of ground barley malt and wheat flour, with or without the addition of sodium chloride, sodium bicarbonate and potassium bicarbonate, in such a manner as to secure the full enzymic action of the malt action, and by removing water. The resulting product contains not less than 7.5% of butter fat and not more than 3.5% of moisture.

Malted milk was the first dry milk product made commercially in the United States. It was invented about 1883 and appeared on the market about 1887. The manufacture of it was undertaken by William Horlick, at the request of physicians who wanted a baby food prepared from milk and cereals.

Manufacture:

The manufacture of malted milk involves two steps:

1) Preparation of a malt extract:
2) Mixing and drying of the extract with milk.

Barley malt is generally employed but some use has been made of wheat malt. The barley is malted by soaking it in water and allowing germination to proceed to the point where its diastatic activity is at the maximum. *Diastatic activity* is a measure of the amount of diastase present, the enzyme that has the power to convert starch into sugar.

The finished malt is dried, mixed with wheat flour, passed through a crushing device, and hot water is then added. The diastase present converts the starch in the cereal grains into maltose and dextrin. When this action is complete, the mixture is filtered, and whole milk is added in the ratio of about 1 pound of milk to 1.25 pounds of malt extract. A little common salt, and usually some sodium bicarbonate, is added—the latter to neutralize some of the acidity of the mixture. Finally, much of the water is evaporated and then the mixture is dried. Much malted milk powder is available in instant form.

Between 1960 and 1966, the average yearly production in the United States was about 23 million pounds. In 1970 it dropped to 19.4 million pounds. All of it was made in Wisconsin.

Uses:

Large amounts of malted milk are used in soda fountain drinks and other beverages, confections, and baby foods. It is a popular food for invalids.

Nutritive Value:

Malted milk is a rich source of readily available carbohydrates. The cereal proteins originally present are partially digested during the malting process. Malted milk is a good source of mineral salts, vitamin A, and the B complex. It furnishes about 145 calories per ounce.

The average composition of malted milk is given in Table 17:2.

Chocolate Flavored Malted Milk

Chocolate flavored malted milk products often are mechanical mixtures of cocoa, chocolate, sugar, and malted milk. Sometimes considerable amounts of dry non-fat milk solids are added. A wide variation in composition is found in the different brands of chocolate flavored malted milk. In most cases, the high sugar content makes them an unbalanced food from the nutritional viewpoint. An analysis of one widely sold product showed it to contain 40 to 50% of sucrose, 15 to 25% of malted milk, 15 to 25% of cocoa and chocolate, about 1% of salt and flavoring materials, such as vanillin and cinnamon.

TABLE 17:2
Average Composition of Dry Milk Products

Product	*Protein*	*Fat*	*Lactose*	*Ash*	*Water*	*Citric Acid*	*Lactic Acid*
	%						
Dry Whole Milk	26.4	26.3	38.0	6.0	2.3	1.0	
Dry Nonfat Milk	36.0	0.7	51.0	8.1	3.0	1.2	
Dry Buttermilk	34.5	5.0	48.0	8.0	3.0	—	
Dry Cream	11.0	72.0	14.0	2.4	0.6	—	
Dry Whey {Sweet, Rennet}	12.5	1.0	72.0	8.0	4.0	—	2.5
Dry Whey {Cottage Cheese Whey}	12.9	0.3	67.0	8.5	3.5	—	7.8
Malted Milk*	14.0	8.0	20.5	3.5	3.3	—	

*Other sugars, maltose and dextrose — 50.7%

Dry Whey

Dry Whey is defined as

> The product resulting by spray drying sweet, fresh cheese whey which has been pasteurized either before or during the process of manufacture at a temperature of 143°F for 30 minutes or its equivalent in bacterial destruction and to which no alkali or other chemical has been added.[17,18]

A wide range of subjects pertaining to whey is given in detail in reference 19.

Sweet whey (pH 6.1) such as obtained from the whey of Swiss cheese and Cheddar cheese is relatively easy to dry compared to the acid whey obtained from cottage cheese and cream cheese. Sweet whey, with a pH of above 6.2 is clarified to remove suspended solids and passed through a separator to recover the milkfat present. The whey is then preheated to about 170°F and condensed to 45% total solids. The plastic concentrate is cooled and held about one-half hour to allow lactose to separate or it is seeded with lactose to promote crystal formation before being spray-dried. A flow sheet for the De Laval non-hygroscopic whey drier is shown in Fig. 17:9. The nearly dry whey powder is carried from the drying chamber to a moving belt where by the use of infra-red heaters additional drying takes place and the powder becomes non-hygroscopic. The powder is further dried by passing through two additional stages in which the moisture content is reduced to 3% or less. The whey powder is cooled in the final stage before being packaged. This drying equipment may also

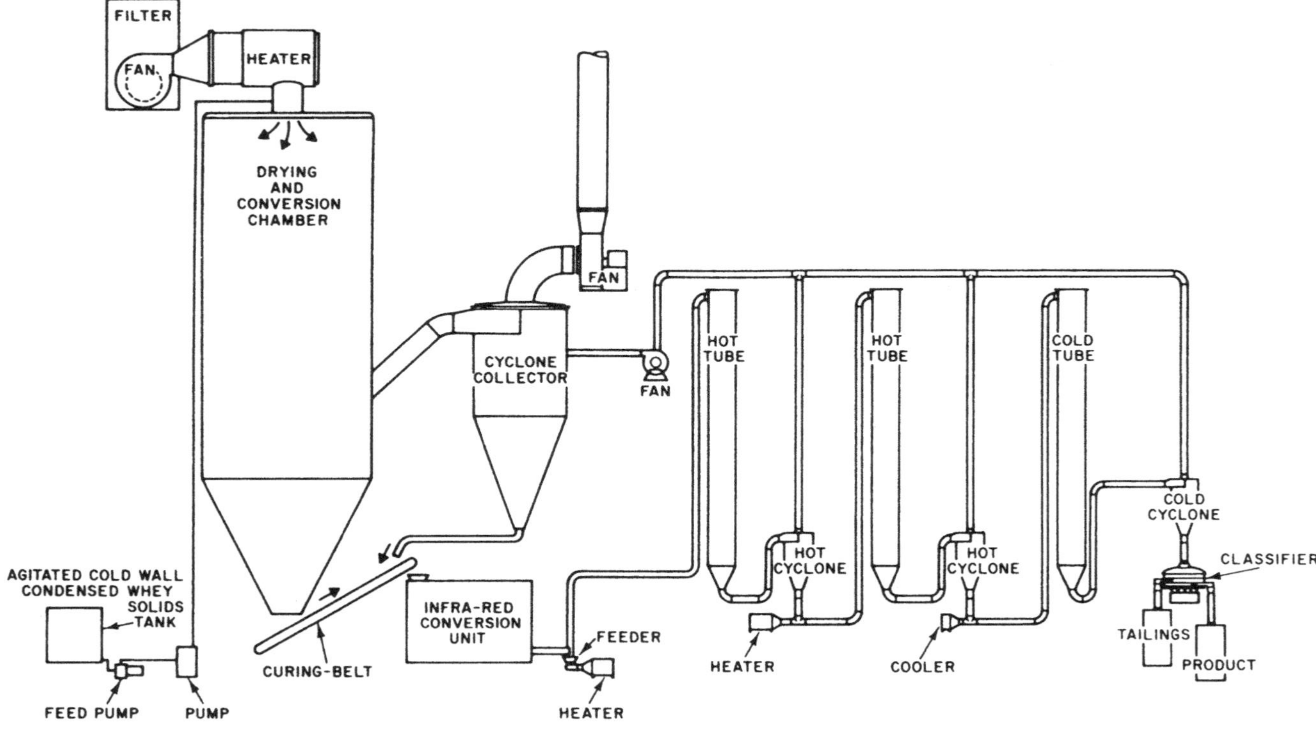

Fig. 17-9. Non-Hygroscopic Whey Dryer (Courtesy of The De Laval Separator Co.)

be adapted for drying non-fat milk. Relatively little whey is roller dried.

The acid-type wheys (pH 4.5) become hygroscopic during conventional drying and tend to coat or clog the equipment. A successful drying procedure was developed by the U.S Department of Agriculture.[5] The whey is heated to 196°F., concentrated to 40-45% total solids and immediately spray-dried. Air, nitrogen or carbon dioxide gas under high pressure is injected as the concentrate reaches the nozzle of the spray-drier. The foamed whey, upon drying, forms tiny, hollow spherical particles which are non-hygroscopic and free-flowing. When air is used to foam, the density of the powder is about one-half that of conventionally dried whey. Increased density is obtained when whole carbon dioxide gas is used as described under foam-drying of whole milk.

Processes have been developed to remove water, salts and lactic acid from whey, after which the product is concentrated and dried. The product is used in the preparation of baby foods. It is claimed that about 10 million pounds of demineralized whey is used per year for infant foods in the United States.

In the electrodialysis method whey concentrated to 35% total solids is circulated through a series of plastic membranes at the opposite ends of which an electric potential is held. This causes the anions, such as sodium, to flow towards the cathode and the chloride and other cations move to the anode. The salt content is reduced by about 95%. The demineralized whey is concentrated to about 45% total solids and then dried in a spray-drier.[8]

Another process is known as *reversed osmosis*.[11,19] In this method, whey under high pressure, is in contact with cellulose acetate membranes which permit the passage of water, salts, and lactic acid, but the protein and lactose is held back. The concentrated solution of demineralized whey is then spray-dried.[11]

Dry whey for animal feed may be prepared by drumdrying. Special techniques must be followed to obtain a satisfactory product, because the large amount of water present and the high content of lactose make it difficult to dry. The high content of lactose and riboflavin in dry whey makes it a valuable component of poultry feed.

Dry whey is used by bakers in making bread, rolls, cookies,and pie crusts. Like nonfat dry milk, it contributes to a desirable color of crust. This color is due, not to direct caramelization of the lactose, but to the browning (Maillard) reaction between lactose and protein in the presence of heat and moisture. Dry whey is also used in ice cream and frozen desserts, in which it may be substituted for as much as 25% of the serum solids. It may make up all of the serum solids in sherbets. Dry whey is an important ingredient in many prepared foods, such as canned soups, sauces, gravies, instant whipped potatoes, and cheese foods.

The essential amino acid content of whey protein in grams per 100 grams of whey protein is as follows:[19]

Leucine	15.5	Threonine	5.5
Lysine	8.2	Tryptophan	2.5
Methionine	4.3	Tyrosine	3.7
Phenylalanine	4.0	Valine	5.5

Dried whey production in 1972 was 377 million pounds; condensed whey output was 175 million pounds.[12]

Dry Ice Cream Mix is described in Chapter 16.

COMPOSITION AND NUTRITIVE VALUE

The over-all nutritive value of a low-heat spray-dried milk product is essentially that of the product before drying. Numerous authorities have pointed out that milk is not the sole item in the diet of adults and children, and its contribution to nutrition must be considered in relation to the rest of the diet. When viewed from this standpoint, changes that may occur in the drying, or otherwise processing, of milk lose much of their significance, even though some nutrient had undergone change or destruction during processing.

Nonfat dry milk, as a result of removing milk fat, contains but little of vitamins A and D. Loss of about 10% of thiamine occurs during spray-drying, and up to twice as much in roller-dried milk. There also is a considerble loss of ascorbic acid during drying—about 20% for spray-dried whole milk and 30% for roller-dried nonfat milk. The loss of vitamin B_6 in spray-dried milk may be as high as 30 percent. In 1973 the United States Food and Drug Administration established standards for fortified nonfat dry milk. Each quart reconstituted according to label directions must provide 2000 International Units of vitamin A and 400 International Units of vitamin D.

Changes that occur in the milk proteins are related to the degree of heat treatment during preheating and drying. Thus practically all of the lysine is available in spray-dried nonfat milk, but only about 68% when it is roller dried.[4]

As the preparation of instant nonfat dry milk involves only a physical change, no difference would be expected in its nutritive value. This is confirmed by a report that shows that the conversion of low-heat nonfat dry milk to the instant form does not alter its content of thiamine,riboflavin, niacin, pantothenic acid, or vitamin B_6, all of which are present in amounts essentially equal to those in fresh milk. As measured by the growth of rats, drying does not change the nutritive value of the milk protein.[6]

The composition of various dry milk products is given in Table 17:2; requirements for grading and bacteriological standards are listed in Table 17:1.

When indirect, steam-heated air is used for drying, no nitrate is added to the dry milk. When direct gas-fired hot air is used, the nitrate present in dry milk may range from one to three ppm; the nitrite from 0 to 1.5 ppm. This is much less than the 10 ppm. permitted in potable water by the U.S. Public Health Service. Normal, fresh cow's milk contains less than 1 ppm. of nitrate or nitrite.[9,13]

BACTERIOLOGY OF DRY MILK

The high temperature used in preheating milk before it is concentrated or dried inactivates the enzymes in milk and destroys most of the micro-organisms present, including pathogenic bacteria. Some bacteria, such as spore-formers and micrococci, do survive, especially if the milk is spray-dried. As any bacteria that may be present are embedded in the particles of dried milk they are thus protected against complete dessication. The cooling effect of evaporating moisture from the milk particle also serves to protect any organism present. Methods of drying that employ a holding or circulatory system for the hot milk may increase its bacterial count by favoring the growth of thermophilic organisms.

Even though viable organisms may not be present, milk infected with staphylococci can be the carrier of enterotoxin in the spray-dried product. The outbreaks of gastroenteritis in Puerto Rico in 1956, described in Chapter 7, apparently were due to staphylococcal enterotoxin in spray-dried milk.

Early in 1966, a marked increase in Salmonellosis was noted by the U.S. Public Health Service. Investigation indicated that the use of certain brands of instant dry nonfat milk was the likely cause.[10] The presence of salmonella organisms in the dry milk pointed to recontamination after the original pasteurization of the milk. Processing in the agglomerating operation does not always provide a pasteurizing environment and may even be favorable for contamination and growth of various bacteria including the salmonellae. Only extreme care in every phase of plant operation can give assurance of a wholesome food product.

As with other milk products, the bacterial quality of dry milk products reflects the care with which the milk was produced and the sanitary conditions surrounding its manufacture. Milk dried for use in food for man often is of Grade A quality. Standards of quality for the various products have been defined by the United States Department of Agriculture and by the American Dry Milk Institute.

REFERENCES

1. Aceto, N. C., Craig, J. C. Jr., Eskew, R. K. and Talley, E. B., Storage aspects of continuous vacuum foam-dried whole milk, XVII Intern. Dairy Congress, E/F 189-196, Munich (1966).
2. Bokian, A. H., Stewart, G. F. and Tappel, A. L., *Food Res.* 22:69 (1957).
3. Census of Dry Milk Distribution and Production Trends, American Dry Milk Institute, Inc., Chicago, (1972).
4. Gupta, J. D., Dakroury, A. M., Harper, A. E., Rate of disappearance of ingested protein, *J. Nutr.* 64:447 (1958).
5. Hanrahan, F. P., Method of spray-drying liquid food products. U.S. Patent 3,222, 193 (1965).
6. Hodson, A. Z., Nutritive value of instant nonfat dry milk, *Food Technol.*, 10:221 (1956).
7. Kuramoto, S., Jenness, R., Coulter, S. T., and Choi, R. P., Standardization of the Harland-Ashworth test for whey protein nitrogen, *J. Dairy Sci.*, 42:28, (1959).
8. Mann, E. J., Utilization of whey, *Dairy Indus.* 7, 552 (1963).
9. Manning, P. B., Coulter, S. T., and Jenness, R., Determination of Nitrate and Nitrite in Milk and Dry Milk Products, *J. Dairy Sci.*, 51:1725, (1968).
10. Marth, E. H., Salmonellae and salmonellosis associated with milk and milk products. A Review. J. Dairy Sci., 52:283-315 (1969).
11. McDonogh, F. E., Whey concentration by reverse osmosis, *Food Eng.*, 24:27 (1968).
12. Production Of Manufactured Dairy Products, U.S. Department of Agr., Washington, D.C., July (1973).
13. Reineccius, G. A. and Coulter, S. T., Examination of nonfat Dry Milk for the presence of Nitrosoamines, J. Dairy Sci., 55:1574-6, (1972).
14. Sinnamon, H. I., Aceto, N. C., and Schoppet, E. F., The Development of Vacuum Foam-Dried Whole Milk, Food Technology, 25:52-64 (1971). Also see Agr. Sci. Rev., 9:35 (1971)
15. U.S. Patent 2 832 686 (April 29, 1958).
16. U.S. Standards for Dry Buttermilk, *Federal Register*, 19, 955 (1954).
17. U.S. Standards for Dry Whey, *Ibid.*, 19:3349 (1954).
18. Weisberg, S. M. and Goldsmith, H. I., Whey for foods and feeds., *Food Technol.* 23:52-56 (1969).
19. Whey Products Conference, Eastern Regional Research Laboratory, U.S. Dept. Agr., Publ. 3779, Philadelphia, Pa., (1973).

chapter 18

Butter

Butter is a mixture of milk fat, buttermilk, and water, usually with salt and added color. Its composition varies somewhat according to the method of manufacture and whether it was made from sweet or sour cream.

By an act, approved March 4, 1923, the Congress of the United States defined butter as follows:

> Butter is the food product usually known as butter and which is made exclusively from milk or cream, or both, with or without common salt, and with or without additional coloring matter, and contains not less than eighty percent by weight of milk fat, all tolerances allowed for.

The recorded history of butter dates to the Hindu Veda, written over 3500 years ago. At that time the Hindus valued cows according to the amount of butter that could be obtained from their milk. Among the ancient Greeks and Romans butter was used as a medicine. The Romans preferred butter with a rancid flavor, rather than the fresh product. They also used it in ointments, applied it to the hair, and the soot of burned butter was supposed to have curative properties for sore eyes. The last observation is of some interest in view of the fact that when included in the diet, butter is a good source of vitamin A, which has a beneficial effect in some eye disorders.

In Ireland and Scotland about 400 years ago a quantity of butter was indicative of the owner's wealth. As with other evidence of wealth, the butter, contained in casks, was buried with the owners upon his death. At times, the butter apparently was buried for safe keeping. In more recent times, when the butter was discovered, it became known as "bog" butter. When unearthed, the butter was rancid and inedible.

Up to about 1870, butter was farm-made. The first creamery to manufacture butter in the United States was built in Manchester,

Iowa, in 1871. The cream used during the early days was usually of poor quality and the butter was inferior to that made on the farm. Today, butter of excellent quality is made by a large number of relatively small plants which usually make more than one milk product. As in other lines of manufacture, the tendency is for the small plants to decrease in number, while plants making one million pounds or more per year of butter have increased.

Butter and dry nonfat milk are the two most important manufactured dairy products. They provide an outlet for surplus milkfat and nonfat solids, products which may be shipped or stored under proper conditions.

Production of Butter (fat content) by major producing countries. (In 1,000 metric tons) (One metric ton = 2204.6 lbs.):

TABLE 18:1

Country	*1973*[1]
Canada	112
United States	350
Belgium	80
Denmark	115
France	450
Germany, West	410
Netherlands	135
Sweden	45
Germany, East[2]	195
Poland[2]	130
USSR[2]	920
Australia	170
New Zealand	220
Others	698
World total	4,030

[1] Includes farm butter
Creamery butter only
U.S.D.A. Fgn. Agr. Service, FFO - 4/74

Manufacture of Butter

Upon the arrival at the plant, the cream is graded for quality, as described in Chapter 13. If too sour, the cream is neutralized by the addition of an alkaline compound, such as sodium bicarbonate, lime, magnesium oxide or hydroxide, or mixtures of these compounds.

Although much butter in the United States is made from either sweet or neutralized cream, the use of ripened cream is common. Practically all butter made in Denmark is made from ripened cream. The acidity and flavor-producing compounds in ripened cream tend to improve the quality of butter made from it. Some defects that may be present in the cream such as feed, neutralizer, and cooked flavor may be disguised in the ripened cream. Butter from sweet cream has a longer storage life than that made from ripened cream. For this reason, most butter made in New Zealand and Australia for export is made from sweet or neutralized cream.

When used in the manufacture of butter, the acidity of sour cream is reduced to 0.15-0.2% as lactic acid. When sweet or neutralized cream is ripened by the use of lactic acid cultures an acidity of 0.45-0.6% is developed and the cream is used at this point (see Chapter 14).

The cream usually is standardized to the desired fat content, preferably to 30-33% of fat. Too long a time is needed to churn cream of lower fat content and some fat remains in the buttermilk. Too rich a cream causes a loss of fat because unchurned cream tends to adhere to the sides of the churn.

Pasteurization may be done at 163°F for 30 minutes but HTST pasteurization at 183°F for 25 seconds is a common procedure. Pasteurization at very high temperatures tends to increased loss of fat during churning, perhaps by binding fat to heat-coagulated protein which is retained in the buttermilk. Much cream for buttermaking is pasteurized by "Vacreation" or other vacuum process, as this procedure is very effective in removing undesirable odors and flavors (see Chapter 11). In Denmark, cream is pasteurized at the higher temperatures in order to meet the requirement that buttermilk returned to the farm for animal feed must have a negative Storch test (q.v.). Butter in the United States must be made from pasteurized cream in order to have an official U.S. grade.

The Butter Churn

The churn is the oldest form of dairy equipment. In prehistoric times it was a bag of skin, hung on the back of a horse or camel; as the animal moved, the motion imparted to the bag caused the milk or cream in it to form butter.

The conventional butter churn is a cylinder, rotated on a horizontal axis. As the churn rotates it raises the cream by means of shelves attached to the sides of the churn. After the butter is formed, the continued raising and dropping of the butter from the shelves imparts the necessary "working." Churns in use today have capacities up to 7000 pounds of butter.

The wooden churn, being porous, is difficult to sterilize and gradually wears out with use. The areas where the shelves are attached are favorite spots for the growth of microorganisms. Wood

Fig. 18-1. Unloading a Stainless Steel Churn (Courtesy J. A. Gosselin Company, Ltd.)

was long considered the standard material for construction of churns because it is one of the few common materials to which butter will not adhere during manufacture, and which will not impart a flavor to the product or be corroded by the milk or salt used.

The *all-metal churn* overcomes many of the difficulties inherent to the wooden churn. Butter will stick to most metals, but aluminum alloyed with magnesium as well as stainless steel are giving satisfactory service (Fig. 18:1). The introduction of the continuous butter-making machine has, in many areas, outmoded the conventional churn.

Churning of Cream

The churn is filled about one-half full with cream. The cream is at about 50°F or less, a temperature at which there is some partial solidification of the milkfat. The buttermaker selects the temperature of the cream since the speed and completeness of the churning process depend upon it. If color has not already been added, the cream may be colored at this point. The churn is then revolved a few times in order to release the pressure caused by the escape of air and gases from the cream. A vent in the churn is provided for this purpose.

The cream is churned until granules of butter, about one-eighth to one-fourth inch in diameter, are formed. This takes 40 to 45 minutes. Butter forms quickly at the end of the churning process, and is said to *break* as it separates from the buttermilk. This point becomes apparent to the buttermaker because a glass observation port on the churn washes clear at each revolution of the churn at this time.

A number of theories have been proposed to explain the change which occurs when cream is churned into butter. The *foam* theory holds that the milkfat gathers in the foam which forms by the inclusion of air during churning. The fat globules continue to pack together until the foam collapses at the break point. According to the *phase-inversion* theory, the churning process causes the fat which is present in cream as a fat-in-water emulsion to change at the break point into butter, which is a water-in-fat emulsion.

Butter is washed while in the granule stage in order to remove most of the buttermilk. When cream of good quality is used, this step is sometimes eliminated. With inferior or neutralized cream, the washing procedure removes some of the substances that produce off-flavor and improves the keeping quality of the finished butter. In order to prevent contamination of the butter, it is necessary to use water of good quality and it usually is chlorinated or pasteurized before being added to the churn.

Several factors influence the texture of the butter being churned. In cream produced in the spring of the year, when cows are on green feed,the fat tends to have a relatively low melting point and it yields a soft butter if the churning temperature is not lowered. Cows on dry feed give a fat with a relatively high melting point, which calls for a higher churning temperature to obtain butter of the proper consistency. If cream is to be churned soon after cooling, the fat globules may still be partially liquid and the cream should be churned at lower than normal temperature to avoid high loss of fat. A hard butter is obtained by rapidly cooling the cream and holding it at a low temperature before churning.[9]

It has been observed that vacuum de-aeration of cream yields a harder butter than cream that had not so been treated.[10] The vacuum treatment disrupts some of the fat globules. The smaller globules make for a more uniform structure in the butter which tends to produce the somewhat harder body.[4]

The temperature of the wash water and of working the butter are more important in controlling the consistency of the butter than the churning temperature, which more directly controls the churning time. Usually the temperature of the wash water is about 5°F lower than that of the buttermilk. If the butter breaks soft, colder wash water is used, which makes the butter easier to work, improves its body and decreases the possibility of a *leaky* butter, that is, one which will not retain its moisture content. In some areas where hard, winter fat is obtained, it is a practice to store frozen cream, produced in the summer months and adding it at the churning time to the cream produced during the winter.

Working of Butter

The working of butter is essentially a kneading process in which the butter granules are forced into a compact mass. At this time any excess water or buttermilk is removed or, if necessary, additional water may be incorporated in order to control the moisture content of the butter. When salted butter is being made, the required amount of salt is added at this point. The working process distributes the salt uniformly throughout the mass of butter.

Properly worked butter has a firm, waxy body which is not greasy and contains no visible droplets of water. The addition of salt favors an increase in the size of water droplets in the butter, and therefore salted butter must be worked more thoroughly in order to divide and evenly distribute the moisture. In well worked butter and in unsalted butter, the small droplets of water give the butter a more opaque, whitish appearance than if larger droplets were present. During the working period, the buttermaker's skill largely determines the quality of the butter being made.

The worked butter usually is removed from the churn manually with a large wooden paddle. In another method, a large metal tray is wheeled to and inserted into the churn, which is then rotated to lift the mass of butter and then let it drop into the tray (see Fig. 18:1.). The Canadian-made Blanchet butter printer can remove the butter directly from the churn by means of stainless steel rolls, which push the butter into the mold and at the same time continue a working action on the butter.

At the completion of the churning process the butter granules consist largely as separate globules of fat. During the working these globules are crushed and liquid fat is released and forms the continuous fat phase typical of butter. The free fat forms up to 95% of the continuous phase; the remaining fat globules and the finely divided water droplets constitute the dispersed phase.

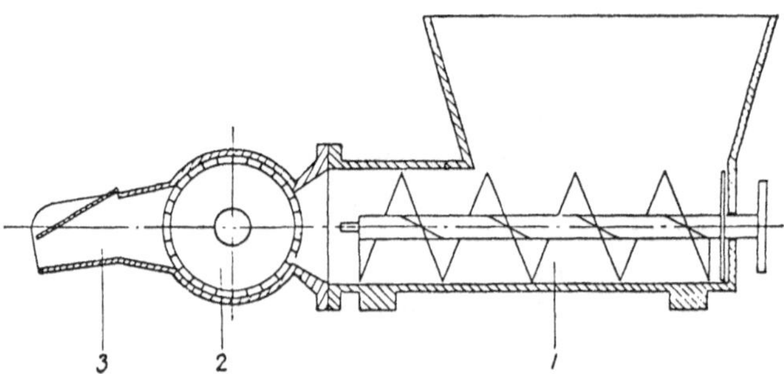

Fig. 18-2. Butter Homogenizer. 1. Hopper with augers. 2. Rotor. 3. Discharge Funnel. (Courtesy American Dairy Review)

The dispersion of finely divided droplets is a factor in improving the texture and appearance of butter. An additional treatment is sometimes given to the butter, especially that made by the continuous process. This is called "homogenization" and it is used to further divide the water droplets. This procedure also reduces the hardness of the finished butter and eliminates the cause of leaky butter.[14] The type of equipment used, such as the "Microfix" made in West Germany by Benz and Hilgers, is shown in Fig. 18:2. The butter from the hopper is forced into the mixing chamber by means of screws or augers. Here a rotor or homogenizing head cuts the butter into thin layers by means of rotating blades which also carry the butter under pressure, to the discharge extruder.

Homogenizing the butter immediately after churning and working has little effect on the hardness, but if it is done after twenty-four hour storage at 40°F. the hardness may be reduced about one-third.[3]

Continuous Buttermaking

Until about 1937, the only way to make butter was from relatively small lots of milk or cream in a churn. Progress in continous operation of pasteurizers, evaporators, and in ice cream freezers led to the development of continuous buttermaking. The first commercial installation was made in Australia about 1937 with a "New Way" butter-maker.[8] The operation of the "New Way" machine was very similar in many ways to methods developed later in the United States which are described in following paragraphs.

It is claimed that one-half of the butter made in the Soviet Union is made by a continuous process.[1]

Continuous buttermaking was further developed in Germany during World War II and two of the methods are still in use.[12]

The modern continuous butter-maker is capable of producing up to 5000 kg. of butter per hour.

Fritz Process:

In the *Fritz* process, used in Germany and to some extent in France, pasteurized cream of 40-45% fat content enters a cooled cylinder at 45-50°F, wherein it is churned into butter in about 1.5 seconds by high-speed dashers. The butter granules pass to an auger-like screw which pushes them out at one end and lets the buttermilk drain away. The butter is worked as it is being extruded from the machine. It is claimed that the moisture content of the butter can be controlled within 0.2 percent. The manufacture of salted butter by this process has been difficult because the added salt is not evenly distributed owing to the short time the butter is in the machine.

The widely used French *Simon Contimab* continuous butter churn is based upon the Fritz process. Although salted butter may be made by blending dry salt into the butter as it is being extruded, a more uniformly salted butter may be made by using concentrated brine, which is prepared in a special tank and is pumped into the buttermaking machine while the butter is being worked (Fig. 18:3).

Fig. 18-3. Churn for Continuous Butter-Making (Made by Simon Frères, Cherbourg, France; courtesy of Len E. Ivarson, Inc., Milwaukee, Wis.)

One procedure to facilitate the addition of salt is to use salt that will pass a 200 mesh screen. In Sweden, a patented method called the PSM method (Passch-Silkeborg Miskinfabrikken) adds the salt to the cream to be churned. To reduce the amount of salt in the buttermilk, cream of 80% fat content is salted and then diluted with water before churning.[7]

Alfa Process:

In the Alfa process, used in a limited extent in Germany and Sweden, butter is made by a method of phase inversion. Pasteurized cream of high fat content—about 78-80%—enters a cylinder or "Transmutator" wherein, by a combination of agitation and cooling to 48-50°F, the oil-in-water form of emulsion in cream is reversed to the water-in-oil form characteristic of butter. For salted butter, the salt may be added to the cream as it enters the transmutator. It is claimed that butter can be made from sour cream by the Fritz process, but in the Alfa process sour cream tends to clog the separator.

A modification of the Alfa process, called the *Meleshin* method is used in the Soviet Union.

The German *Westfalia* continuous butter churn, the "*Buttermatic*" is widely used[18] (Fig. 18:4). Its operation is characteristic of Fritz-Process buttermaking. Sweet cream of 30-50% fat content or sour cream of 30-40% fat content may be churned. In order to avoid reducing the capacity of the machine use of 40% cream is favored. This also minimizes loss of buttermilk. Before processing, the cream is held at the intake temperature for 6-8 hours. Depending somewhat upon local conditions, the temperature of the cream to be churned is determined by the equation 93°F minus the percent fat in the cream.

Cream at a temperature of 46-55°F. is fed into the primary churning cylinder, where it is partially churned by a 4-bladed dasher, rotating at up to 2400 rpm. The product then passes to the secondary churning cylinder where a 2-bladed dasher rotates at about 42 rpm. The butter is pushed to the end of the cylinder from which it enters the Texturizer. At this time most of the buttermilk flows into a collecting pan under the secondary churning cylinder from which it is pumped to a receiving tank.

In the Texturizer, two augers, rotating in opposite directions at from 30-85 rpm work the butter and transport it to the extrusion end of the cylinder. The speed of the augers controls the moisture content of the butter being worked. In the blending section of the Texturizer the moisture and salt are distributed evenly by a series of perforated plates and rotating blades.

Provision is made to wash the butter, if desired, and so reduce the

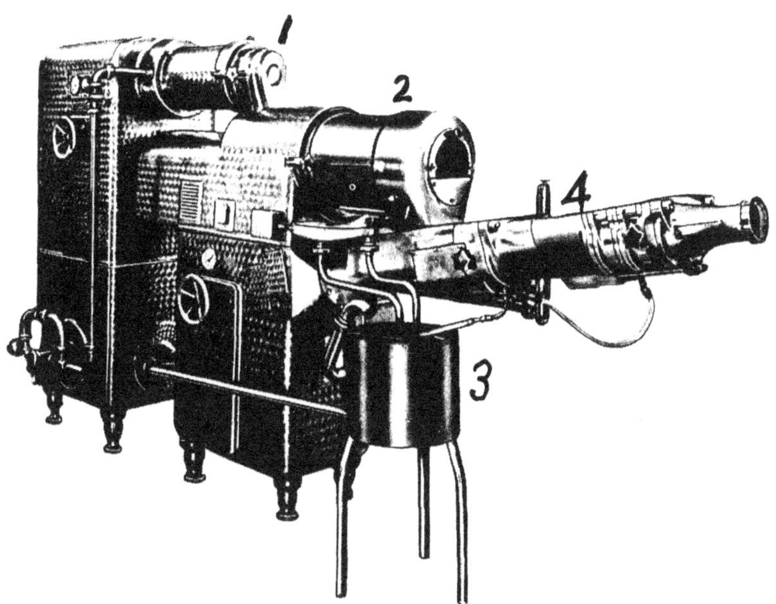

Fig. 18-4. Westfalia "Buttermatic" Automatic Buttermaking Machine. (Courtesy, Centrico, Inc.)
1. Primary Churning Cylinder 2. Secondary Churning Cylinder
3. Buttermilk Vat 4. Texturizer

curd content to less than 1%. The finished butter extrudes from the Texturizer in ribbon form, ready for delivery to the packaging equipment.

Cherry-Burrell Process:

The Cherry-Burrell ("Gold'n Flow") process of continuous buttermaking is used to some extent in the United States and Canada.[6]

Raw cream, at a temperature of 60-70°F is pumped from holding vats through a pressure filter and into a high-speed destabilizing unit. This unit consists of a chamber, always held full of cream, in which a perforated blade travels at high speed, packing the fat globules together by mechanical force. Cream with a high fat content is more readily destabilized than low-testing cream. The destabilized cream passes to a special separator which has three discharge ports, one for the fat concentrate, another for the skim milk, and the third for the discharge of the curd which normally collects in the bowl of a cream separator. The fat concentrate, which contains from 88 to 94%

fat, is discharged into a tank from which it is pumped to a Vacreator (vacuum pasteurizer) in which it is continuously pasteurized at 200°F and then cooled by evaporation in the vacuum to about 110°F. The pasteurized concentrate is discharged into Composition Control Units, which are vats of up to 600 gallons capacity, wherein agitators and baffles mix the concentrate uniformly with the standardizing materials that may be added. The required amounts of neutralizer, salt, water, color, and starter distillate are added at this point, bringing the mixture to the desired content of fat, water, salt and curd (Fig.18:5).

After it is standardized, the concentrate is chilled and worked into butter. The concentrate is held around 120°F to ensure that the fat is in liquid form before it enters the chiller in order to avoid a mealy body in the finished butter. The butter chiller is similar in principles of operation to a Vogt Freezer. It has ammonia-cooled cylinders in which agitators work the butter and scrape it from the walls of the cylinders, thus partly working the butter and giving it a firm body. As conventional butter contains 4-5% of air, the butter chiller is equipped with a device that introduces nitrogen gas into the product and gives it the body and texture of churned butter. From the chiller, the butter enters a "Texturator" around 40°F. Here a time lag occurs, which allows the fat to crystallize before it is forced through a perforated plate that works the butter. As it leaves the Texturator, the butter is ready for packaging in bulk or for printing and wrapping, at a rate of about 2000 pounds an hour.

Butter made by this process has its moisture incorporated in very small droplets. As it is therefore lighter in color than conventionally churned butter it requires somewhat more added color. The use of the Vacreator improves the flavor and keeping quality of the butter, especially if the cream used was not of the best quality. The consistency of the butter differs from that of the churned product, the body being very compact and somewhat harder.

In an electron-microscopical study of their structure it was found that butter churned by the conventional process has a granular structure whereas butter made by the Cherry-Burrell continuous process has a homogenous structure similar to that found in margarine.[11]

A problem that may be encountered in a continuous buttermaking process is the variation in the fat content of the different lots of cream to be churned, as well as temperature changes in the cream supply. A vat, capable of holding up to 10,000 gallons of cream has been devised. Throughout the day's run this would insure a source of cream that is of uniform composition and temperature.

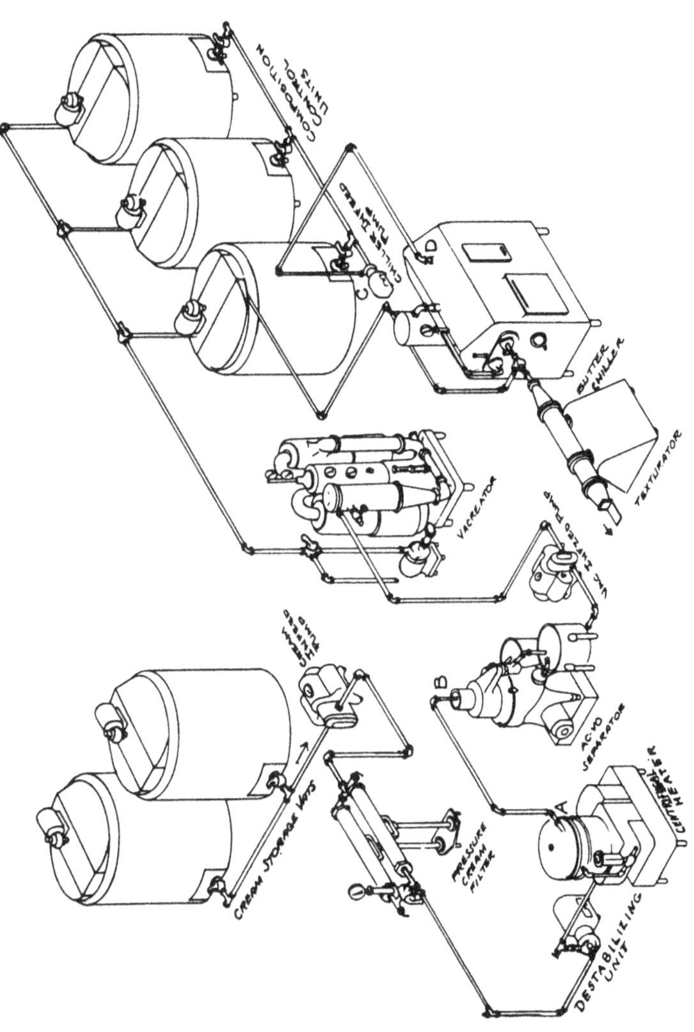

Fig. 18-5. Diagram of "Gold' n Flow" Continuous Buttermaking (Courtesy of the Cherry-Burrell Corporation)

Overrun

The weight of butter obtained from a given lot of cream exceeds the amount of fat in the cream. This is due to the presence of water, salt, and curd in the butter. The amount of butter made in excess of the milk fat present in the cream is called *overrun*. As the minimum legal United States fat content of butter is 80%, the maximum amount of butter that can be made from 100 pounds of milk fat is

$$\frac{100}{80} \times 100 \text{ or } 125 \text{ pounds.}$$

This is an overrun of 25%, but it is not obtainable in actual practice. Efficient creamery operators approach an overrun of about 23%, based upon the amount of milk fat going into the churn.

Composition of Butter

A chemical analysis of butter usually gives the percentage of fat, water, salt, and curd present. The curd consists of protein, mineral matters, the lactose derived from the buttermilk retained by the butter. The average composition of butter, based upon published analyses of more than 10,000 samples from various parts of the United States is as follows:

TABLE 18:2
Average Composition of Butter
(Percentage)

Type	*Fat*	*Moisture*	*Salt*	*Curd*
Salted	80.47	16.54	2.15	0.84
Sweet	81.00	18.05	—	0.95

The variations in the composition of butter are slight because of uniform methods of manufacture and the effort to keep the fat content near the legal minimum requirement.

Salt is added to butter in order to improve its keeping quality as well as to meet the market demand. The amount of salt used varies with the region; most markets prefer 1.5-2%, but in the south about 3% of salt is added. Sweet butter contains no added salt. The salt is present dissolved in the water phase, thus making a solution varying between 5 and 16% of salt content. This concentration is sufficient to inhibit the growth of some bacteria, especially Pseudomonas species. Unsalted butter accordingly is more subject to the development of rancidity and other defects than is salted butter.

It is not unusual for creamery operators to pay a bonus to the butter-maker for each churning of butter in which the fat content does not exceed a certain maximum, say about 80.3%. Milk fat is by far the most expensive fat in common use as a food, but adulteration with other fats is uncommon. Any adulteration would soon be detected by the tests to which butter is subjected by various governmental health and food inspection agencies.

Color of Butter

The natural color of butter is due almost entirely to its carotene content. Carotene is the yellow pigment found in many vegetables and fruits as well as in grasses and forage plants. In the animal body it is converted into vitamin A (see Chapter 6). In naturally deeply pigmented butter, traces of other yellow-colored compounds may be present. Some cows, notably those of the Jersey and Guernsey breeds, can transfer more carotene from the feed to the milk than can cows of other breeds. Under ordinary conditions naturally colored butter is light in winter and a deep yellow in the early summer. This seasonal variation is caused by the difference between the high carotene content of fresh pasture feed during the spring and the carotene-poor feeds usually available in the winter months.

As the seasonal change in color is objectionable to the consumer, who prefers a uniformly colored product throughout the year, butter usually is colored artificially. The shade used varies locally, but in general, the northern and eastern states prefer a light, straw color, the southern states prefer a deep yellow, and in the western states an intermediate color is most popular. In Canada, a richly colored butter is preferred.

Years ago, coloring for butter was made from carrots, saffron, marigold, or egg yolk. In any case, the added coloring is one that is fat- or oil-soluble because it must be retained by the milk fat. The coloring is added directly to the cream before it is churned. Until May, 1959, the certified food colors, FD&C Yellow AB and OB (Nos. 3 and 4) were much used to color butter. This practice was terminated when it was found that these colors could not be produced with any assurance that they would not contain a carcinogenic impurity, beta-naphthylamine. Butter is now colored with annatto or with beta-carotene obtained from carrots or made synthetically. The certified colors, FD&C Yellow No. 5 and No. 6 may be used.

Annatto is a yellow dyestuff obtained from the seeds of the annatto tree, *Bixa orellana*, cultivated in Brazil, India, and the West Indies. The color is oil-soluble and is extracted with vegetable oils. The preparation is used to color butter and other fatty foodstuffs. An alkaline solution of annatto, which is soluble in water, is used to color cheese.

Although the dye known to chemists as dimethylaminoazobenzene is usually called "butter-yellow," owing to its color, it never is used to color butter or other foodstuffs. Because of its popular name, some misinformed persons have believed that this dye is used to color butter.

Nutritional Value of Butter

The nutritional value of butter depends almost entirely upon its content of fat and vitamins. The nutritional value of milk fat has been discussed in Chapter 2. The protein, lactose, and mineral constituents of milk are almost entirely lacking in butter so that it is the most limited of dairy products in the variety of nutritive materials present.

Butter is a heat- or energy-producing food which furnishes about 3,400 calories per pound. Its digestibility usually is stated to be 97.8%, the value found for milk fat. As butter is an excellent source of vitamin A, it often is the chief source of this vitamin in the diet. A survey of the content of vitamin A in butter made throughout the United States showed that a pound of butter made between July and September contained an average of 18,000 International Units of the vitamin. Butter made in March may contain as little as 9,500 units.[6] The amount of vitamin D varies greatly. These factors are discussed in more detail in Chapter 6.

Butter Products

Whey Butter:

In cheese factories, where a large amount of whey is available, it is passed through a cream separator in order to recover the fat. The whey cream is churned and made into a very acceptable grade of butter. Whey butter, as it is called, is not distinguishable from butter made from ordinary cream.

Whipped Butter:

Whipped butter is butter that has had its volume increased through the incorporation of air or inert gas, such as nitrogen. Butter may be whipped much as in the making of whipped cream. It is used as a spread on bread, hot cakes, waffles, and baked potatoes. Various devices are used to whip butter, such as dough-mixing machines, batch- and continuous ice cream freezers, and special machines. Butter, usually at 60 to 70°F, is whipped until it increases 50% in volume, although it is possible to get an overrun of 100% or more.

There is no evidence that the incorporation of air in good butter will cause any deterioration in keeping quality, provided that the overrun does not exceed 80 percent.

Butter Oil-Ghee:

A kind of clarified butter, more properly called *butter oil*, is used in many parts of India and Egypt. It is called *maslee* in the Middle East, *Ghee* in India and *samma* in Egypt. Generally it is prepared from buffalo milk which constitutes 55% of the total milk production in India. Buffalo milk contains no carotenoids and so lacks a yellow color.

The making of ghee is mostly a home industry. The fresh milk is boiled for about one hour and then is allowed to stand in a warm place. Sometimes a little curdled milk (called *dhye*) is added to hasten coagulation. The curdled mass is skimmed off and churned for about one-half hour. Toward the end of this time, water at about 140°F is added and the churning is continued until butter is formed. The fat is heated and strained, so that a clear butter oil is obtained. The method of manufacture removes much of the curd and moisture usually present in butter, and also imparts a pronounced cooked flavor and odor to the product. It has good keeping properties, which is important under the climatic conditions of the Orient. Large amounts of it are used there for cooking and for making confections. When made from cow's milk it often serves as a medicine, especially if it is old and rancid. The use of an anti-oxident, such as butyl hydroxyl anisole (BHA) is permitted. Ghee often is adulterated by the addition of a vegetable oil.

Butter oil may be prepared both by the mechanical procedures described for the preparation of dry milk fat and by a chemical procedure. Various anionic surface-active agents are especially effective in breaking the emulsion and liberating the fat from milk to cream.[13] The method has been used for laboratory purposes but not commercially.

On a small scale, butter oil is made by a "rendering process." The butter is melted slowly and the fat is allowed to separate by gravity from the serum. The serum is drained off and the oil is drawn off into containers.

Ghee differs from butter oil in that it is prepared by heating up to 140°C, whereas butter oil is prepared by melting butter at about 80°C.

Dry Milk Fat:

Dry milk fat is the most concentrated form of milk fat commercially available. It contains about 99.8% fat, 0.1% moisture, and not more than 0.5% free fatty acids. As its moisture content is very low, it may be stored without refrigeration for several weeks without danger of bacterial growth. Dry milk fat is subject to oxidative deterioration, so it should be held in air-tight containers filled with inert gas, such as nitrogen.

For the preparation of anhydrous milk fat the cream is first separated to about 90% fat content and then is diluted with water and passed through the separator a second time. This procedure effects a more exhaustive separation of fat and in the second separation, the added water serves to balance the volume in the bowl of the separator, thus preventing any loss of fat through the discharge ports for skim milk.

The high fat concentrate from the second separation contains at least 99% fat. It is passed to a vacuum pasteurizer, where the remaining water is evaporated and the fat is pasteurized, chilled, and packaged. As dry milk fat sets at 65°F, it must be packaged quickly. To ensure maximum keeping quality, dry milk fat is packed in nitrogen-filled cans and stored in cold storage. A common package is the standard egg can, which holds 45 pounds of dry milk fat.

Dry milk fat may be used as a substitute for butter, plastic cream, and frozen cream. It may also be used to standardize ice cream mix and cream cheese mixes. When blended with nonfat dry milk solids and water, the reconstituted milk can be used where fresh whole milk is not available.

Grading of Butter

The quality of finished butter depends to a large extent upon that of the cream from which it is made. Poor workmanship, however, may cause defects which will detract from the stability and flavor of the finished product. The United States Standards for grades of butter are based first upon the flavor characteristics and then upon the body, color and salt characteristics.[15]

In the scoring procedure used by the United States Department of Agriculture, the best grade of butter is known as *U.S. Grade AA, U.S. 93 Score, or 1st Grade.*

Below this is *U.S. Grade A, 92 Score,* or *2nd Grade.*

Butters of somewhat inferior quality are rated *Grade B, 90 Score,* or *3rd Grade.*

Finally there is a *Grade C,* or *89 Score*, which has definite defects.

The judging of butter is an art that is practiced with a fine degree of distinction. Grade AA butter has a fine, highly pleasing flavor. A slightly normal feed or a definite cooked flavor may be present. Eight types of flavor and slight defects in flavor, such as *cooked, slightly storage, flat,* or *acid flavor*, are allowed for butter that is graded *92*

Score. Any one of sixteen defects in flavor including those impared by an *excess of neutralizer, storage, bitter,* and *old cream,* places butter in the *90 Score* grade. No provision is made for a *91 Score* butter. Butter inferior to *90 Score*, and with evidence of any of fifteen defects in flavor, such as *sour, cheesy, or yeasty,* is placed under Grade C.

Butter that does not meet the requirements for U.S. Grade C, or that has less than 80% of fat, or butter produced under unsanitary or unsatisfactory manufacturing conditions, is not given a U.S. grade.

Body characteristics that detract from the quality of butter include the following designations:

Crumbly: This defect is associated with butter whose fat has a high melting point, as a result of feeding cows dry feed, such as cottonseed meal. Holding cream at a low temperature may also cause this condition. Use of wash water at 10-20°F below that of the buttermilk will help correct it.

Gummy: This defect arises from the presence of fats of high melting point. Cream that tends to give gummy butter requires a relatively higher temperature for churning and washing, as well as a longer working time.

Leaky: Such butter usually is due to the incomplete incorporation of moisture or uneven distribution of salt. It is remedied by proper working.

Some defects of color which lower the quality of butter are:

Mottling: This usually is due to insufficient working. The flow of moisture towards areas of high concentration of salt causes irregular spots of color.

Streaks: These are caused by insufficient working and sometimes by the incorporation of butter from a previous churning.

Defects in Butter

Contamination by Yeasts and Molds:

Butter made from cream that has undergone a yeasty fermentation may retain a characteristic and objectionable yeasty odor. The yeasts responsible usually are lactose fermenting types, such as *Torula cremoris* and *T. sphaerica*. A fruit-like odor and flavor usually is caused by molds, but sometimes it is of bacterial origin. Defects due to yeasts and molds can be minimized by the use of a clean churn and vacuum-pasteurized cream.

The parchment wrappers used in packaging butter may be a source of contamination. To avoid this, the wrappers are run through hot water or a hot salt solution before being used.

Yeasts and molds are discussed in Chapter 7. According to the United States Department of Agriculture the yeast and mold count in butter should not exceed 50 per gram.

Bacteria in Butter

Detailed instructions for the microbiological methods for the examination of butter are given in the Standard Methods for the Examination of Dairy Products (13th edition, 1972).

The enterococcus count is a useful tool to check upon the sanitation of butter manufacturing plant. These organisms are able to survive in butter which should have an enterococcus count of less than 10 per ml. of butter and a yeast and mold count not over twenty per ml.[2]

Churns are difficult to sterilize completely by ordinary plant procedures and they require special attention if all contamination is to be eradicated. If it is not kept clean and sanitized, the churn can become an important source of contamination of the butter, especially with yeasts and molds.

Surface Taint:

Bacteria introduced by the water used to wash butter during manufacture may produce a cheesy flavor. Sometimes it is necessary to chlorinate or pasteurize the wash water in order to destroy the organisms. The flavor defect known as surface taint is associated with a bacterial contamination. The source can be the milk or cream from which the butter is churned, but in general, the water supply appears to be the original source of contamination. The defect is characterized by a flavor and odor of putrefaction, which begins on the surface of the butter and gradually penetrates into its body. Butter made from cream of high acidity or butter that contains over 2% of salt rarely shows this defect. Proteolytic bacteria may be greatly outnumbered by harmless species and yet be present in sufficient number to produce surface taint. Under favorable conditions of temperature, salt, and concentration of moisture, the organism *Pseudamonas putrefaciens* produces surface taint but, undoubtedly, other organisms are also involved.[17]

Flavor and Flavor Defects

The characteristic flavor and aroma of butter, which usually are noted best when the butter comes fresh from the churn, distinguish it from other edible fats. Although its chemical composition may vary but little, the flavor of different lots of butter may vary considerably. To a large degree, the flavor of butter depends upon the quality of the cream used in manufacture, as well as upon the manner and length of time during which the butter is stored before it is consumed. Butter made from sour or ripened cream has a distinctive flavor which is popular in some parts of the United States, while in other sections, especially the west and north-west, butter made from sweet (fresh) cream is preferred.

More than 80 components, present in minute amounts, have been identified in butter. Among these are alcohols, aldehydes, esters of various acids, lactones, and sulfur compounds.

The pleasing aroma developed from heated butter and butter sauces appears to be due in part, to the formation of compounds called lactones.

Flavor-producing substances in ripened cream are discussed in Chapter 14. The activity of desirable flavor-producing organisms, such as are present in ripened cream, does not continue to any extent in the finished butter.

A tallowy flavor, more properly known as *oxidized* flavor, sometimes is present in butter and other products that contain fat. This defect is caused by a chemical change in the fat. In time, oxidized butter may become bleached and actually turn white. Butter made from ripened cream develops an oxidized flavor more readily than that made from sweet cream, but when the cream is neutralized, the butter is less likely to develop an oxidized flavor. A rancid flavor, as has been explained in Chapter 10, is caused by the liberation of fatty acids from the milkfat and should not be confused with oxidized flavor. A fishy flavor, which at one time was believed to be due to the decomposition of phospholipids, is now thought to be due to the presence of oxidation products of the fatty acids.

Body defects are described in the section on *Grading of Butter*.

Flavor Defects of Bacterial Origin:

With the exception of the lactic-acid-forming bacteria introduced by the use of cultures, all bacteria, yeasts, and molds are undesirable in butter. When made from unpasteurized cream, butter may contain any of the organisms present in the original cream or milk. Pasteurization of the cream destroys all pathogenic bacteria, but some harmless organisms generally survive. Some of these are retained in the buttermilk, but others find their way into the butter. Pathogenic organisms are able to survive in butter and the bacteria of tuberculosis and typhoid fever have been isolated from butter made from contaminated cream.

There is no evidence to show that the keeping quality of butter is related in any direct way to the number of bacteria, yeasts, or molds that may be present. In fresh butter of good quality, the predominating organisms usually are streptococci and micrococci.

During storage, the micrococci generally grow more rapidly than the streptococci.

A number of flavor defects in butter are of bacterial origin. A *barny* or *cowy* flavor may be caused by the use of cream of high bacterial content. A *bitter* flavor may be present if the cream had undergone bacterial decomposition. Butter made from sour cream or with the use of excessive amounts of starter may develop a *cheese-like* flavor and odor. This is especially true if large amounts of buttermilk are retained in the butter.

The presence of rod-shaped organisms in butter is undesirable because they often are proteolytic and produce cheesy, putrid, and unclean flavors and odors.

Packaging and Storing Butter

On a wholesale basis, butter usually is packed in parchment-lined wooden boxes or tubs that hold 63-64 pounds. For the retail trade, butter is cut or "printed" into quarter pound, one-half pound, and one pound prints; "chiplets," weighing from one-fourth to one-third ounce each, are made for the restaurant trade. The butter printer or molder extrudes the butter through a die of the desired shape and then cuts the butter into lengths of the required weight, wraps the portions in parchment, and places them in the finished carton or box.

Much butter is made in the months of May to August when a surplus of milk and cream usually is available. It is held in cold storage plants at a temperature of -10 to -20°F, sometimes for nine months or more. Little bacterial growth occurs at this range of temperature and good quality butter does not undergo material change in flavor. Butter made from poor quality or neutralized cream may show flavor defects after prolonged storage, especially if the temperature should rise above -10°F.

Utilization of Butter

Supplies of butter have exceeded consumer demand and its price is determined largely by governmental support prices. Much butter is purchased under dairy support programs and distributed domestically for School Lunch Programs, for welfare purposes, and for donations to foreign countries.

When the price of cheese is high, due to consumer demand, manufacturing milk is diverted to the production of cheese rather than that of butter.

A comparatively recent innovation in the use of butter is in the manufacture of canned and "boiled-in-bag" frozen vegetables, prepared with butter sauce. These products have gained popularity owing to their flavor and convenience in preparation. Unsalted butter of high quality usually is used. The sauces vary in composition according to the vegetable being processed. The viscosity of the sauce is important, in that it must not gel and still must adhere to the vegetables without having the butterfat separate during cooking.

The composition of a typical sauce is reported to be as follows:

	Percent
Water	79.0
Butter	10.9
Sugar	5.7
Salt	2.8
Modified waxy starch	1.2
Stabilizer gum	0.4
Color (carotene)	as needed

REFERENCES

1. Beloussov, A. P. XVI International Dairy Congress, Copenhagen, (1962).
2. Blankenagel, G., D. L. Gibson & C. N. Shih. The enterococcus count of butter for evaluating creamery sanitation. *Canadian Dairy Ice Cream J.* 46:17-19 (1967).
3. Dehn, F., Observations on the working of butter with the Mikrofix, *Molkerei and Kaserei Zeitung* 16: 1885-1891. (1965).
4. Dolby, R. M., *J. Dairy Res.* 20:211 (1953).
5. Dolby, R. M., Jebson, R. S. and Le Heron B. S., Fritz-Process Buttermaking as Applied to New Zealand Cream, XVII International Dairy Congress, C-43-48 (1966).
6. *Gold'N Flow Continuous Buttermaking method:* Bull. G-493; Cherry Burrell Corporation, Chicago, 1954.
7. Hansen, E. A., Continuous butter manufacture-Manfacture of salted butter XVII International Dairy Congress C 39-41 (1966).
8. Hills, G. L., Some theoretical aspects of the New Way Butter Process, *Australian J. Dairy Technology,* 1:2 (1946).
9. Huebner, V. R. and Thompsen, L. C., Spreadability and hardness of butter, *J. Dairy Sci.* 40:839 (1957).
10. Johansson, S. and Swartling, P., The effect of de- aeration on the consistency of butter, XVII International Dairy Congress, C 27-31 (1966).
11. Knoop, A. M., Knoop, E. and Wortmann, Physical Structures of butter and margarine as determined by comparative electron microscopic studies, XVII International Dairy Congress, C 253-263 (1966).
12. *Some Developments in Dairying in Germany:* British Intelligence Objectives Committee, Final Report No. 86, Items 22 and 31; H. M. Stationery Office, London (1946).
13. Stine, C. M. and Patton, S., Preparation of milk fat: III, Properties of butteroil prepared by the use of surface active agents, *J. Dairy Sci.*, 36:516 (1953).
14. Thomasos, F. I. and Wood, F. W., The effect of homogenization on the moisture dispersion of Gold'n Flow and conventional butters, XVII International Dairy Congress, C 83-89 (1966).
15. U.S. Standards for Grades of Butter, *Federal Register*, Jan. 28, 1960.

16. *Vitamin A in Butter:* Misc. Pub. No. 636, U.S. Department of Agriculture, Washington, D.C., 1947.
17. Wagenaar, R. O., The bacteriology of surface-taint in butter: A Review, *J. Dairy Sci.*, 35:403 (1952).
18. Centrico, Inc., Bulletin No. 3446, Englewood, N.J.
19. CP Division St. Regis, Personal Communication, May 8, 1969.

chapter 19

Cheese

The story of who and where cheese first was made is lost in unrecorded history. Earthern-ware cheese pots have been found in the tomb of King Horaha of the Egyptian First Dynasty (3200 B.C.). Egyptian clay tablets dating about 4000 years ago refer to cheese as a tax imposed upon conquered tribes. In Greek mythology, the Gods of Mt. Olympus bestowed cheesemaking as a gift to the human race. A much repeated legend tells of an Arabian merchant who, before starting on a journey, placed milk into a water-pouch made of sheep's stomach. During the day, the combination of the sun's heat and the rennet in the lining of the stomach changed the milk into cheese curd and whey. During the centuries before Christ, cheese was probably introduced into Europe by Asiatic traders. About 400 B.C., the Greek writer Otesia referred to the legend of the Assyrian queen Semiramis, who was fed by birds with cheese stolen from shepherds. In Roman times, cheese was made in Italy, but the important source was the Alpine region of modern Switzerland. Roman patricians obtained many varieties of cheese from different parts of the Empire. The Bible makes several references to cheese, and the people of the Old Testament prized cheese as a valuable food.

Cheese making was introduced into England, between 100 and 300 A.D. by the Romans. Following the decline of the Roman Empire, cheese making was one of the many industries conducted in Europe by the various monasteries. The monks developed special varieties of cheese and even today Trappist Monks in France make Port du Salut cheese and those in Quebec make Oka cheese. Over the many centuries during which cheese has been made, it was also the product of the farmhouse. Not until 1851 was the modern factory method of making cheese introduced by Jesse Williams, who built the first cheese factory near Rome, N.Y. About twenty years later, the factory method of making cheese was introduced in Scotland, where it was called the Canadian Cheesemaking system. Today, the making of

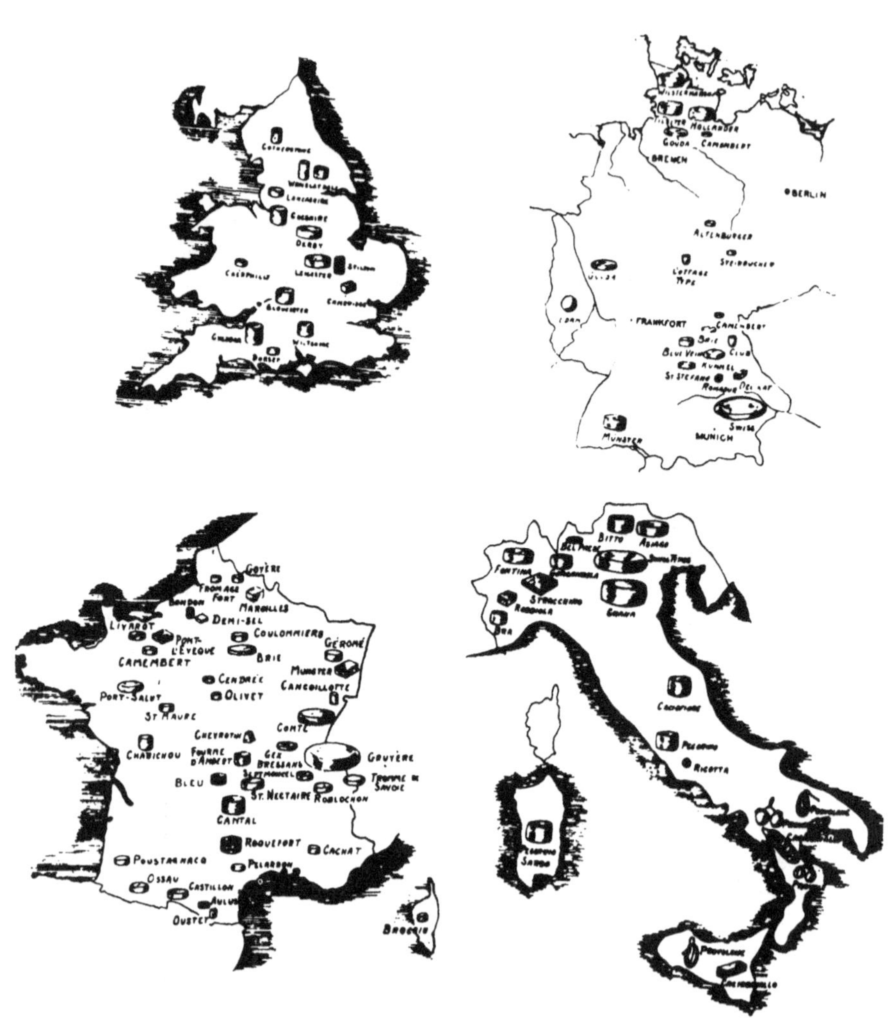

Fig. 19:1. Home Regions of Some European Cheeses: *a*) England and Wales; *b*) Germany and Holland; *c*) France; *d*) Italy
(*New York Agr. Expt. St., Geneva; Circular 187*)

cheese in the leading producing countries is a factory procedure, only small amounts being made on farms for local use. Many of the popular varieties that originated in Europe are now made in the United States. The maps in Fig. 19:1 show the places of origin of some European cheeses.

Cheese is defined as

> "the product made from the curd obtained from whole, partly skimmed, or skimmed milk of cows or from the milk of other animals, with or without added cream, by coagulating the casein with rennet, lactic acid or other suitable enzyme or acid, and with or without further treatment of the separated curd by heat or pressure, or by means of ripening ferments, special molds or seasoning. In the United States the name *cheese* unqualified is understood to mean cheddar cheese."

The making of natural cheese is an art to which only in recent years has precise, scientific control been applied. The basic types of cheese developed in a given locality probably were of accidental origin and

Fig. 19-2. Familiar Varieties of Cheese: Front: Cream Cheese; Processed Cheeses. Middle: Gouda, Edam, Cheese Spread, Roquefort, Limburger, Liederkranz, Brick, Camembert, Port Salut. Rear: Pineapple, Longhorn, Cheddar, American Block, Swiss

depended upon the kind of milk used, the manner in which it was treated and ripened, as well as upon such other factors as the season of year in which the cheese was made, and the way in which it was coated or wrapped. Many cheeses were named for the locality in which they were made; thus, a cheese of the same general characteristics became known by many different names. Natural cheese is made directly from milk or occasionally from whey, in contrast to process cheese, cold-pack cheese, and cheese spread, which products are made by blending or combining one or more different kinds of natural cheese.

CLASSES OF CHEESE

There is no one method by which the different types of cheese may be classified. The handbook *Cheese Varieties and Descriptions,*[5] lists more than 800 names of cheeses and describes more than 400, yet it states that probably there are not more than eighteen distinct types or kinds of natural cheese.

The eighteen kinds of cheese typical of the different processes by which they are made are:

Brick	Neufchatel
Camembert	Parmesan
Cheddar	Provolone
Cottage	Romano
Cream	Roquefort
Edam	Sapsago
Gouda	Swiss
Hand	Trappist
Limburger	Whey cheeses

In another grouping, cheeses can be classified according to texture or according to the manner of ripening, as shown in the following listing:

According to Texture	*By Manner of Ripening*
1. Very hard (grating type)	1. By bacteria
2. Hard	2. By mold
3. Semi-soft	3. By surface micro-organisms
4. Soft	4. By a combination of 1-3
	5. Unripened

A classification of cheeses, with examples, is given in Table 19:1.

TABLE 19:1[5]
Classification of Cheeses

1. SOFT:
   a) *Unripened:*
      1. Low Fat—Cottage, Pot, Bakers'.
      2. High Fat—Cream, Neufchatel (as made in United States).
   b) *Ripened:* Bel Paese, Brie, Camembert, Cooked, Hand, Neufchatel (as made in France).
2. SEMISOFT:
   a) *Ripened principally by bacteria:* Brick, Munster.
   b) *Ripened by bacteria and surface micro-organisms:* Limburger, Port du Salut, Trappist.
   c) *Ripened principally by blue mold in interior:* Roquefort, Gorgonzola, Blue, Stilton, Wensleydale.
3. HARD:
   a) *Ripened by bacteria, without eyes:* Cheddar, Granular, Caciocavallo, Provolone.
   b) *Ripened by bacteria, with eyes:* Swiss, Emmentaler, Gruyere.
4. VERY HARD (grating)
   a) *Ripened by bacteria:* Asiago old, Parmesan, Romano, Sapsago, Spalen.
5. PROCESS CHEESES:
   Pasteurized, Cold-Pack, Related Products.
6. WHEY CHEESES:
   Mysost, Primost, Ricotta.

The soft cheeses contain 45 to 80% moisture and are perishable. The hard and semi-hard cheeses contain from 30 to 45% moisture and under favorable conditions may be stored for a year or more. From one to five months may be needed to ripen soft cheeses, and up to a year for semihard cheeses. Some kinds of hard cheese may require about two years to ripen properly.

The Food and Drug Administration has issued definitions and standards of identity for about 60 kinds of natural cheese, processed cheese, cheese food and cheese spread.[6] Grade designations for cheese are described in Chapter 9.

MANUFACTURE OF CHEESE

In the manufacture of cheese, milk is transformed into a concentrated and less perishable foodstuff. Most of the protein and fat of the milk is retained by the cheese but the more soluble constituents, such as the milk sugar and much of the mineral matter, are lost in the whey. The yield of cheese depends upon the amount of casein, fat and water in the milk. Upon coagulation, the casein forms the network of curd within which the fat and water is held. The relationship of the fat and casein content of the milk to the yield is shown in Figure 19:6. On the whole, the manufacture of cheese is a rule-of-thumb process, wherein the results obtained depend to a great degree upon the skill and experience of the cheese maker. A possible exception to this statement is the modern method for making cheddar cheese from pasteurized milk by carefully controlled schedules and the methods of procedure.

Raw milk generally is used for the manufacture of hard and semi-hard cheeses, but the use of pasteurized milk is increasing because it tends to uniformity of quality in the finished cheese. Most cottage cheese and cream cheese are made from pasteurized milk. Pasteurization destroys not only pathogenic organisms but also other organisms that would cause defects in the finished product. The change in bacterial flora and the inactivation of enzymes caused by pasteurization alter the normal ripening process. As more time is required for proper ripening, the body and flavor sometimes differ from those of cheese made from raw milk.

The Food and Drug Administration and many States require that cheese made from raw milk be held or ripened for at least sixty days before it is offered for sale.[6] It is assumed that pathogenic organisms will not survive the aging process. This is not entirely true, but experience and much experimental work has demonstrated the value of the holding period as a health measure.

Treatment with Hydrogen Peroxide

Although the use of hydrogen peroxide is not permitted in the United States as a substitute for pasteurization (see Chapter 12), it is permitted as a bactericide in milk for the manufacture of Cheddar, Colby, Washed Curd, Granular, and Swiss cheese.[6] Swiss cheese-makers favor the use of hydrogen peroxide. The use of hydrogen peroxide permits lactic organisms to survive and it does not destroy lipase and protease in the milk. Aerobic spore-forming organisms are relatively resistant, the lactic acid bacteria less so, while the coliforms are least resistant to hydrogen peroxide treatment.

The peroxide solution used may be more concentrated but otherwise must meet the specifications of the U.S. Pharmacopeia (to contain not more than 2 ppm of arsenic and not more than 5 ppm of heavy metals). The catalase used to decompose the excess of hydrogen peroxide in the milk must be made from calf livers from animals that have passed inspection by Federal meat inspectors. Its potency must not be less than 100 Keil units (1 ml will decompose 100 grams of hydrogen peroxide under standard conditions). The catalase added must not exceed 20 ppm by weight of the milk.

The milk to be treated is heated to 125°F and the hydrogen

peroxide is added in the specified proportion, not to exceed 0.05% by weight of the milk. The milk is cooled to setting temperature, 86°-95°F, and 5 ml of catalase preparation is added for each pound of 35% hydrogen peroxide used. After 10 minutes, the milk is tested for the presence of any residual hydrogen peroxide by adding about 3 drops of fresh potassium iodide solution (25%) to 10 ml of the treated milk. The presence of hydrogen peroxide is indicated by the appearance of a yellow color. The test is repeated at 5-minute intervals, until it shows that no peroxide is present. If necessary, a small excess of catalase may be added to the milk if a hydrogen peroxide residue persists. The use of too much peroxide tends to form a high-moisture cheese and a bitter flavor.

In-as-much as the hydrogen peroxide treatment is not the equivalent to pasteurization, cheese manufactured with it in the United States must be held for sixty days, just as if raw milk were used. New York State does not permit the use of hydrogen peroxide in the manufacture of cheese.

Bactofugation

The treatment of cheese milk by bactofugation is discussed in Chapter 11.

Ripening the Milk

The formation of acidity in the milk to be used for cheesemaking, known as *ripening*, is usually brought about by use of a starter (see Chapter 14). The acidity developed inhibits the growth of undesirable organisms and influences the rate of coagulation. The starter used depends upon the type of cheese being made. For cottage cheese and cheeses that do not need a cooking temperature above 100°F, the starter consists of *Streptococcus lactis* and *Streptococcus cremoris*. When the curd is to be heated to about 125°F, mixed cultures of heat-resistant organisms, such as *Streptococcus thermophilus* and the *lactobacilli* are used. When the desired acidity is reached, most varieties of cheese require the addition of rennet to the ripened milk in order to obtain a curd of the desired characteristics. The small particles of curd formed by acid alone, as in the case of Cottage cheese, are soft and difficult to handle.

At times, the development of acidity in the milk may be very slow or negligible, even though a good starter had been used. This may be caused by the presence of an antibiotic in the milk or by bacteriophage. These factors have been discussed in Chapters 8 and 14.

Rennin

Rennin is a milk-curdling enzyme found in the stomach of suckling animals which usually is obtained from the *abomasum*, or fourth stomach, of calves reared on a milk diet and slaughtered when less than five months old. In older animals it is replaced by the proteolytic enzyme *pepsin*. Rennin has comparatively little protein-digesting ability. It is most active at *p*H 3.8, whereas pepsin is most active at a much higher acidity, about *p*H 2. One part of purified rennin will coagulate more than five million parts of milk.

The commercial extract of rennin, called *rennet*, is a saline extract, containing a little pepsin, about 10% of sodium chloride, and about 4% of boric acid or some sodium benzoate or propylene glycol as a preservative. Rennet in paste form also contains lipase, and the use of it in making of Italian-type cheeses gives them a characteristic flavor and odor derived from the lipolysis of the milk fat. The paste is made by drying the entire stomach, including any milk that may be present, of calves, lambs or kids. Glands at the base of the tongue of these animals secrete an oral-lipase or pre-gastric esterase, which is lipolytic. The enzyme passes into the milk as it is swallowed by the feeding animal.

The formation of curd depends upon the coagulation of the casein in the milk, as described in Chapter 3. Pasteurized milk often yields a softer curd than when raw milk is used. Other than the formation of the clot, no visible change or chemical change occurs during rennin coagulation.

Slow development of curd may be due to too little rennet or to the use of over-heated milk. In the latter case, the addition of a small amount (about 0.02%) of calcium chloride to the milk usually will restore the calcium-ion balance and permit the normal functioning of rennin.[24]

Rennet is relatively costly and a continuing search is made for acceptable substitutes.[17] Mixtures of pepsin and rennin are used to a considerable extent, since the slow coagulation obtained when pepsin alone is used causes difficulties in the manufacture of cheese. No completely satisfactory substitute for rennin has been found, but at times other milk-coagulating enzymes, such as pepsin, papain, and other vegetable enzymes have been used. In India, where religious beliefs prohibit the use of animal products other than milk, vegetable enzymes, such as the juice of the fig, have been used for making cheese. Such cheese has a different texture than that made with rennet and often has a bitter flavor.[15]

The U.S. Food and Drug Adminstration permits the use of a

non-rennin enzyme in the manufacture of Cheddar, Colby, Granular, and Swiss cheese. The enzyme is formed in a fermentation process by the microorganism *Endothia parasitica*. Commercial experience with the enzyme showed that an undesirable flavor and texture changes occurred in cheddar-type cheeses made with it, upon aging thirty days or more. The use of the enzyme has indicated that it is acceptable in the manufacture of Swiss cheese.[2]

Rennet extract is diluted with water before being added to the ripened milk; usually two to three ounces diluted to about one gallon with water is used for each 1000 pounds of milk. If the cheese is to be colored, the color—usually an alkaline solution of annatto—is added before the rennet.

After the addition of rennet, the milk is stirred for a few minutes to distribute the color and rennet thoroughly. It is then left undisturbed for the curd to form, and this becomes apparent in about fifteen minutes. After about thirty to forty minutes the milk is "set," with a firm curd.

Curd Treatment

The manner in which the curd is handled varies in some degree according to the kind of cheese to be made. A definite time interval occurs during which the acidity of the curd decreases and its body becomes firmer owing to a decrease in its content of whey. Heating the curd favors these reactions. If a soft, high-moisture cheese is being made, the curd is removed from the vat quickly and the whey is drained. For some varieties of cheese, the curd is cut into small cubes, and the whey is allowed to drain off; for other varieties, the curd is cut and stirred in the whey while it is being heated. For Cheddar-type cheese, the curd is heated in the whey and allowed to form a continuous mass, which is then cut or milled into small pieces before further processing.

The manner in which the whey is drained from the curd varies with the kind of cheese. Cream cheese, for example, is prepared by placing the curd on cloths which allow the whey to drain away. Sometimes the curd is placed in forms or hoops put on mats or coarsely woven screens which allow the whey to drain, as in the manufacture of Brick cheese. In the making of Cheddar cheese, the curd is allowed to sink in the vat and the supernatant whey is drawn off. In making Swiss cheese, the curd is separated by placing a cloth under the curd and lifting it out of the vat or kettle.

The rate at which the whey is allowed to drain away is determined by the kind of cheese being made. Acid continues to develop in the curd as long as appreciable amounts of whey are present. With the increase in acidity the curd becomes elastic and can be stretched or, if heated, it can be drawn out into silky strings. This is the basis of a practical test used by makers of Cheddar cheese. A hot iron rod is touched to the curd, and as it is drawn away, the length of the curd fibers at their breaking point is noted; the higher the acidity, the longer the threads. When the threads can be stretched to 1½-2 inches, the acidity of the whey has reached 0.8-1% as lactic acid. The cheddared curd then is considered ready for milling, as described under Cheddar Cheese.

The last portions of whey are removed from the curd by pressing. This operation is also used to mold some varieties of cheese into their conventional shape. The degree of pressure used varies with the kind of cheese.

Treatment of Rind

The manner in which the surface of the cheese is treated also influences its characteristics. For example, Cheddar and Swiss cheese are given a smooth, uniform surface or rind by pressing the curd while it still is warm, and curing the cheese under conditions that allow the moisture to evaporate from the surface. The activity of organisms on the surface is prevented in Swiss cheese by frequently cleaning the rind; for Cheddar cheese, this is done by coating the cheese with paraffin wax and holding it in a cool room with low humidity. The growth of mold on Camembert- and Roquefort-type cheeses, and the growth of yeasts and bacteria on Brick and Limburger cheeses, is encouraged by holding them in a cool, moist environment. Other aspects of the ripening processes, such as the formation of "eyes" in Swiss cheese, are described in the subsequent sections on various kinds of cheese.

In France, some cheeses are given a decorative effect by applying grape skins and seeds to the rind. Some cheeses from goat's milk have ashes rubbed into the rind.

Pigments often are used to color the rind or coating applied to the cheese. The red color of the waxed surface of Edam cheese is a familiar example. In Italy, a mixture of iron oxide and dilute sulfuric acid is used to color the rind of Scanno cheese black. A mixture of charcoal and salt is rubbed into the rind of St. Benoit cheese (France).

In the United States, many varieties of cheese in the form of slices or cuts in consumer-sized packages are allowed to contain not more than 0.2% sorbic acid or 0.3% of sodium propionate, calcium propionate, or a combination of the two. This treatment serves to prevent the development of mold on the cheese, and the label must show the presence of the added chemical.

Salt Content

Most cheeses have some salt (sodium chloride) added to them at some time during processing. In Cheddar cheese, the salt is added to the milled curd; other varieties are salted by rubbing the surface with salt or by placing the cheese in a salt brine. Salt draws the whey out of the curd and serves as a factor in the control of acidity and moisture. The concentrated salt solution, formed by the salt combining with the moisture in the cheese, prevents the growth of undesirable microorganisms in the cheese. The salt content of most cheeses varies from about 1% in cottage cheese and cream cheese to about 5% in Parmesan and Roquefort cheeses. In Switzerland, iodized salt must be used in the manufacture of cheese.

Moisture and Ripening

Cheddar cheese which contains not over 36% of moisture is known as "Long-hold" cheese, and may be ripened for up to a year. "Short-hold" cheese with 37-39% moisture is ripened within four months. It is mild in flavor and does not keep well.

Some kinds of cheese, such as Cottage cheese, are ready for consumption as soon as the curd is collected, whereas other varieties require further processing and holding to acquire their characteristic qualities. Such cheese, when freshly made is said to be *green* or *uncured*. The flavor is bland or only slightly acidic or salty and the body is firm, elastic, and sometimes quite tough.

During the ripening or curing process the cheese undergoes changes which alter its flavor, body, texture and sometimes its color. These changes are brought about by

1) More or less breakdown of the protein to simpler, water-soluble compounds;

2) Hydrolysis of the fat with the formation of fatty acids; and

3) Fermentation of lactose, citrates, and other organic compounds into acids and aroma-forming compounds.

These changes are selectively brought about by the enzymes, bacteria, molds, and yeasts in or on the cheese. The treatment given the curd before the ripening, and the environment in which the cheese is stored influence or control the changes that occur. Each variety of cheese is handled in the manner that has been shown to provide the conditions needed for its proper development. Specific conditions for ripening are described under the variety name.

Flavor of Cheese

A characteristic flavor is associated with each variety of cheese,[14] but this often may be overshadowed by defects in flavor, which range from a lack of all flavor to objectionable ones. As some of these defects are similar to those that are found in milk they may be carried over into any cheese made from milk of poor quality. Defects such as unclean and rancid flavor may occur in the milk. Other flavor defects which are more peculiar to cheese include bitter, acid, salty, fruity, and sulfide flavor.

A bitter flavor usually is caused by the decomposition of protein and it generally becomes more pronounced as the cheese ages. It is favored by insufficient development of acid in the milk from which the cheese is made and by a low salt content in the cheese. A fruit-like flavor often is caused by bacterial contamination from unclean milk or equipment. Low acidity, due to excessive washing of the curd, also tends to favor this defect. In the case of cheddar cheese, a bitter flavor often results from the presence of too much acid formation during the manufacturing process or too little salt content.

It is believed that the amount and relative proportions of a small number of compounds is mainly responsible for the wide range of flavors in cheeses. In cheddar cheese these appear to be mixtures of fatty acids, methyl ketones, and hydrogen sulfide.[9] Flavor development is also discussed in the section on the bacteriology of cheese.

Defects in Body

The evolution of carbon dioxide gas is a normal occurrence in ripening of cheese. The undesirable evolution of gas may occur at various stages in the manufacture of cheese. Raw milk contaminated with coliform organisms is a common source of gas. Hydrogen sulfide may be present in cheese that has undergone abnormal fermentation, especially if the cheese had been ripened at too high a temperature. A gassy cheese is characterized by the presence of many round holes which usually are regular in shape but vary in size. They differ from the holes present in a normally open-textured cheese, which are mechanical openings due to failure of the curd particles to adhere to one another during pressing of the cheese.

Except for a few varieties of cheese, the growth of mold in or on the cheese is undesirable. Such growth discolors the cheese and may spoil the flavor. As the moist environment in which most cheese is ripened is favorable to the growth of mold, considerable effort is made to keep curing rooms clean and sanitized.

Nutritional Value of Cheese

Most kinds of cheese contain approximately the same ingredients, but these vary in amount according to the type of cheese. Changes in the nutritive value due to heat treatment of the milk used in making cheese are not significant, because the milk is not heated above the temperature of pasteurization and much cheese is made from raw milk. Cheese is the most concentrated form of nitrogenous food in common use and even if it is consumed in relatively small amount it can provide a significant part of the daily intake of nitrogen. Most analyses of cheese report nitrogenous matter as "milk protein", but this is not entirely correct as some of the original milk protein may be digested or changed into split products during the ripening process.

Unless made from skim milk, most cheese is a good source of milk fat. Much of the flavor of cheese may be attributed to its content of fat and fatty acid. The fat content of cheese varies from almost none—in plain,uncreamed, cottage cheese made from skim milk—to about 45% in some cream cheeses.

TABLE 19:2
Partition of Nutrients In
Making of Cheddar Cheese
(Percentages of Total in Original Milk)*

Nutrient	*In Whey*	*In Curd*
Water	94	6
Fat	6	94
Total Solids	52	48
Casein	4	96
Soluble proteins	96	4
Lactose	94	6
Calcium	38	62
Vitamin A	6	94
Thiamine	85	15
Riboflavin	74	26
Vitamin C	84	6

*Dairy Council Digest, 31:6 (1966).

Although cheese is a concentrated food, it does not contain all of the nutritive constituents of milk. In the manufacture of Cheddar cheese, about 94% of the fat in the milk is retained in the cheese and about 76% of the protein (Table 19:2). Most of the lactose, lactalbumin, and soluble mineral salts remain in the whey. Some special types of cheese, such as mysost and ricotta, are made from whey, so they contain the constituents of whey rather than those of natural cheese. Some special cheese foods and spreads contain added solids from whey or skim milk, which supply nutritive ingredients missing in most cheese, such as lactose and water-soluble vitamins.

TABLE 19:3
Approximate Percentage Composition of Some Varieties of Cheese

	Moisture	*Fat*	*Protein*	*Ash* (Salt-free)	*Cholesterol**	*Salt*	*Calcium*	*Phosphorus*
Brick	41.3	31.0	22.1	1.2	—	1.8	—	—
Brie	51.3	26.1	19.6	1.5	0.10	1.5	—	—
Camembert	50.3	26.0	19.8	1.2	0.10	2.5	0.68	0.50
Cheddar	37.5	32.8	24.2	1.9	0.12	1.5	0.86	0.6
Cottage								
Uncreamed	79.5	0.3	15.0	0.8	—	1.0	0.10	0.15
Creamed	79.2	4.3	13.2	0.8	0.03	1.0	0.12	0.15
Cream	54.0	35.0	9.2	0.5	0.12	0.75	0.30	0.2
Edam	39.5	23.8	30.6	2.3	0.11	2.8	0.85	0.55
Gorgonzola	35.8	32.0	26.0	2.6	—	2.4	—	—
Limburger	45.5	28.0	22.0	2.0	0.09	2.1	0.5	0.4
Neufchatel	55.0	25.0	16.0	1.3	—	1.0	—	—
Parmesan	31.0	27.5	37.5	3.0	0.15	1.8	1.2	1.0
Roquefort	39.5	33.0	22.0	2.3	0.12	4.2	0.65	0.45
Swiss	39.0	28.0	27.0	2.0	0.10	1.2	0.9	0.75
Process, (American)	40	30	23	(Total Salts-5; Carbohydrate 2%)				

*Fat and Cholesterol Content of Domestic and Imported Cheeses, Fisher, Hans, et. al; New Jersey Agr. Exp. Station, Bull. 832A, 1973.

TABLE 19:4
Recommended Nutrition Information For Selected Cheese Products

Percentage of U.S. Recommended Daily Allowances (U.S. RDA)

Types of Cheese	*Calories*	*Protein (Gm)*	*Carbohy-drate (Gm)*	*Fat (Gm)*	*Protein*	*Vit. A*	*Vit. C*	*Thia-mine*	*Ribo-flavin*	*Niacin*	*Cal-cium*	*Iron*
Cheddar, Washed Curd, Stirred Curd, or Colby (each variety has the same label information)	110	7	1	9	15	4	*	*	6	*	20	*
Swiss Cheese	100	8	0	8	15	4	*	*	4	*	25	*
Pasteurized Process American Cheese**	110	6	1	9	10	4	*	*	6	*	15	*
Monterey	100	6	1	8	15	4	*	*	6	*	15	*
Low-Moisture Part-Skim Mozzarella Cheese—Less than 19% fat	80	7	1	5	15	2	*	*	4	*	20	*
Low-Moisture Part-Skim Mozzarella Cheese—More than 19% Fat	90	7	1	6	15	2	*	*	4	*	20	*
Parmesan Cheese	110	10	1	7	20	2	*	*	4	*	30	*
Romano Cheese	100	9	1	7	20	4	*	*	4	*	25	*
Edam Cheese or Gouda Cheese	100	6	1	8	10	2	*	*	4	*	15	*
Blue Cheese	100	6	1	8	10	2	*	*	4	*	15	*
Provolone Cheese	90	7	1	7	15	2	*	*	4	*	15	*
Cream Cheese	100	2	1	10	4	6	*	*	2	*	2	*

*Contains less than 2% of the U.S. RDA of those nutrients.
**Recommended serving size statement: ⅔ ounce slice—(1 ounce = 1½ slices)
¾ ounce slice — (1 ounce = 1⅓ slices)
1 ounce slice — (1 ounce = 1 slice)

Courtesy National Cheese Institute, 1974

In general, cheese made by coagulating milk with rennet is a good source of calcium and phosphorus. The lactic acid varieties, such as cottage cheese and cream cheese, retain little of these elements. About 55 to 62% of the calcium and 50% of the phosphorus of the milk is retained in Cheddar cheese, whereas only about 20% of each is retained by the lactic acid varieties. As salt is added to cheese, the ash content may not give a valid picture of the minerals originally present in the milk.

A tabulation of the vitamin content of cheese is given in Table 6:2. As most of the fat in the milk is retained by the cheese, the fat-soluble vitamins are retained; cheese made from whole milk is a good source of vitamin A and carotene. In the manufacture of Swiss cheese and some Italian-type cheeses the use of benzoyl peroxide is permitted in order to bleach the milk. As this treatment destroys the vitamin A present, the Food and Drug Administration requires the addition of vitamin A to such cheese. Cheese made from such milk must be so labeled, except for Blue and Gorgonzola cheese.

The content of water-soluble vitamins varies with the amount of whey retained by the cheese and the extent to which the curd is handled. The washing, cutting, and stirring of the curd removes or destroys practically all of the ascorbic acid that may have been present.

The approximate composition of some varieties of cheese is given in Table 19:3.

Production of Cheese

The production figures for the more important types of cheese in the United States in 1972 are given in Table 19:5

Non-milk Substances in Cheese

In the previous sections mention has been made of some substances such as rennet, salt, annatto and hydrogen peroxide which may be added to cheese. In addition to these, many other substances have been used in cheese making.[8]

When the milk is pasteurized at a high temperature, rennet action is slow. The addition of 0.01-0.02% of calcium chloride to the milk assists in its coagulation. A bleaching agent, is used in making Blue cheese. In some countries, an oxidizing agent is used to inhibit the growth of coliform and anaerobic organisms in the cheese milk. Formaldehyde is used for this purpose in Italy. Nitrates, chlorates, and bromates have been used but such use is illegal in the United States. Nisin, the antibiotic produced by some strains of Streptococcus lactis, has been used to check bacterial growth in pro-

TABLE 19:5
Production of Cheese in the United States [23]
(1972)

Type	*Million lbs.*
American (Colby, Washed Curd, Stirred Curd	344
Monterey, Jack).	28.5
Blue Mold & Gorgonzola	21.8
Brick	1,300
Cheddar	1,014
Cottage Cheese, Creamed	101
Cottage Cheese, Lowfat	128
Cream Cheese	512
Italian	2.5
Limburger	45.7
Muenster	5
Neufchatel	745
Processed	177.8
Swiss	31
All other types	

In 1973, in 33 major producing countries the cheese production was 12.5 billion pounds. Western Europe produced six billion pounds and the Soviet Union 1.2 billion pounds.

cessed cheese. It is effective against other streptococci and is useful to prevent spoilage by Clostridia.

A number of herbs and spices may be added to various cheese. The use in cottage cheese of pineapple and fruit salad, as well as chives, is common in United States. Cheeses such as, Provolone, may be smoked, either by the traditional process or by direct addition of a liquid smoke flavor. Dioctyl sodium sulfosuccinate, a flavor enhancing substance, may be added in the United States to cream cheese, Neufchatel, cream cottage cheese, and processed cheese spread. The application of color and other substances to the rind of some varieties of cheese is described under the variety name.

In the United States, up to 0.3% of sorbic acid or potassium sorbate may be added as a mold inhibitor to more than forty varieties of cheese. It is applied to slices of the consumer packaged cheese, either by dipping or spraying the cheese. The cheese may be packaged in wrappers treated with the fungicide.

BACTERIOLOGY OF CHEESE

The need to maintain clean utensils and surroundings during the manufacture of cheese was recognized many years ago. A rhyme attributed to Thomas Tusser* states, among other admonitions that:

> When cheeses in dairy, have Argus his eyes,
> Tell Cisley the fault in her huswifery lies.

The method used for the manufacture of a cheese has a great influence upon the number and type of organisms present. If the cheese is made from raw milk, any pathogenic organism present in the milk may survive. Several epidemics, especially of typhoid fever, have had their origin in the consumption of contaminated cheese. Most States now require cheese to be made from pasteurized milk, unless the cheese is aged at a temperature lower than 35°F for at least 60 days before sale during which time it is assumed that any pathogenic organism will die.

In the finished cheese, the bacteria are fixed in their position by the curd and so are limited in their area of growth. In each type of cheese the ripening process, to which its flavor is due, is associated with the growth of characteristic bacteria or molds. In an acid-curd cheese, such as cottage cheese, in which little or no ripening occurs, the flavor is due largely to lactic acid fermentation.

During the ripening process the protein in the cheese is broken down to simpler nitrogenous substances, many of which have a distinct flavor and odor. Certain non-nitrogenous compounds, such as volatile acids and esters, are formed when the fat is attacked; acetic, propionic, capric, caprylic, and lactic acids are among the acids found. A rancid flavor, due to the presence of butyric acid, often is found in old cheese. Small amounts of biacetyl and related compounds are present in some cheeses, especially those of the cheddar type.

Bacterial activity has no important role in the flavor development in unripened soft cheeses, such as Neufchatel and Cream cheese nor does it have any important part in the ripening of mold-ripened cheese. In Roquefort cheese the action of the lactic acid organisms is inhibited by the high salt content of the cheese as well as by the low temperature at which the cheese is cured. Other bacteria, such as *L. bulgaricus*, may be present, but they are not involved in the ripening of cheese of the Roquefort type.

The bacteriology of a specific cheese is discussed under the name of the cheese.

---

*Tusser, Thomas, *Five Hundred Points of Good Husbandry*, London, 1638, p. 82.

CHEDDAR CHEESE

Cheddar cheese was the first factory-made cheese and it is now the most universally made cheese in English-speaking countries. In the United States it is called *American Cheese*; in Canada, it is *Canadian cheese*, in England and most of the British Commonwealth it is referred to as *Cheddar Cheese*. It was named after the village of Cheddar, near Bristol, in Somerset, England. The date of origin is unknown, but it is called the "best cheese in England" in the *Legacy of Husbandry*, published in London in 1655.

Cheddar is a hard cheese, made from cow's milk. Raw milk is generally used, but in the United States at least 75% of the Cheddar cheese is made from pasteurized milk. "Cheddaring" is the name given to a step in making the cheese and *Cheddar* also is the name of the most common style, as shown in Table 19:6.

TABLE 19:6
Common Styles of Cheddar Cheese

Style	*Diameter* (Inches)	*Height* (Inches)	*Pounds*
Cheddar	14.5	11.5	70-78
Twin or Flat	14.5	5.3	32-37
Daisy	13.5	4.3	20-23
Longhorn	6.0	13.0	12-13
Young American	7.0	7.0	11-12
Picnic			
Junior Twin Commodore	9.75	5.0	11-12

Prints: Usually about 14 inches long, 11 in diameter, and 3.5 thick, weighing about 20 pounds. Large prints weigh up to 80 pounds, and one-pound prints are popular.

Manufacture of Cheddar Cheese

The milk is first heated to about 87°F and 0.5 to 2.5% of its volume of starter is added in order to obtain a growth of lactic acid-forming organisms. When the milk has acquired the proper acidity—about 0.2%—enough rennet (about 3 ounces per 1000 pounds of milk) is added to coagulate the milk in about 30 minutes. Usually from 1-3 ounces of cheese color for each 1000 lb of milk is added.

After the curd has formed, it is cut with curd knives, one set of which has wire blades arranged vertically, the other, horizontally. As these knives pass through the curd from one end of the vat to the

Fig. 19:3. Making Cheddar Cheese: Ditching the Curd in "Cheddar-Rite Vat", Courtesy of Kusel Dairy Equipment Co.

other, they cut the curd into small cubes, from one-half to three-eighths inch square.

The particles of curd are kept floating in the whey by gentle stirring with mechanical agitators or wooden rakes while the temperature is raised to about 100°F. The heating is done by passing steam between the double walls of the cheese vat. During this time the proper body and acidity develop in the curd. The whey is then drained off (*dipped*) at which time its acidity is about 0.14% (pH 6.2).

Care is taken not to break the curd particles while stirring, as this favors an increase of fat loss into the whey. The whey usually contains less than 0.3% of fat, but if the curd is too broken the fat content may

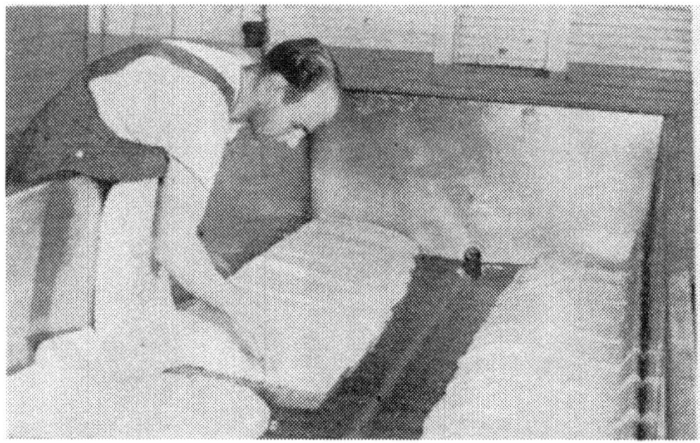

Fig. 19:4. Making Cheddar Cheese: Piling or Cheddaring the Curd by conventional procedure

reach 1%. One-half of the curd is pushed to each side of the vat in order to form a trench or ditch along which the whey escapes. This step is called *ditching* and is followed by the distinctive feature of making cheddar cheese known as *cheddaring*, which means matting or piling the curd (Figs. 19:3, 19:4). The judgment and art of the cheesemaker now become evident since the care taken during these steps determines the quality of the cheese to be made.

The whey is passed through a separator to recover the milkfat present, which represents about 6% of the fat in the original milk. The whey cream is made into whey butter (q.v.).

After the whey has drained away completely, the curd is allowed to stand for about 10 to 15 minutes, during which time it forms a continuous, spongy mass. It is then cut into blocks 7 to 10 inches long, which are turned at intervals of from 10 to 15 minutes over a period of about two hours and then they are piled on top of one another in order to retain heat, increase the pressure, and expel any remaining whey.

After the cheddaring process, the curd is cut by a *milling* machine into pieces two or three inches long and about half an inch square. The cut curd is spread evenly over the bottom of the vat and from two to three pounds of salt is added for each 100 pounds of curd while the curd is being stirred to distribute the salt evenly (Fig. 19:5). The salt hardens and shrinks the particles of curd and so assists in the removal of remaining whey. It also restrains the growth of undesirable microorganisms and checks the development of acidity. The finished cheese will retain about 1.4% of salt.

Cheddar cheese is one of the few varieties of cheese that are salted

Fig. 19-5. Making Cheddar Cheese: Salting the Curd in "Cheddar-Rite Vat". Salting by Mechanical Means is Preferred.

at this stage of manufacture. After the salt has dissolved, the curd is cooled to about 80°F, and is then packed into *hoops* or forms lined with a tubular, seamless strip of cheesecloth called the *bandage*. This cloth remains on the cheese. The filled hoop is fitted with a top, called a *follower*, and with 10 to 15 other filled hoops, it is placed in a press where it is subjected to gradually increasing pressure for 30 to 60 minutes. The cloth bandage is then smoothly fitted over the cheese and it is returned to the press for 12 to 24 hours, during which time the excess of moisture is removed from it.

The cheese is next taken out of the hoop and placed on shelves where it is held for three or four days at a temperature between 50° and 65°F. During this time the surface becomes drier and a rind is formed. After drying, the cheese is dipped in paraffin or cheese wax and placed in the curing room; or it is boxed and shipped to a warehouse for curing

In the United States, much Cheddar cheese is cured by being held for three months to a year at a temperature between 32° and 45°F. In other Cheddar-making areas, the cheese is cured at temperatures up to 60°F with a relative humidity around 90% until the desired flavor is developed, after which the cheese is stored in a cold room. The longer the curing period—which may last up to two years—the sharper and more pronounced is the flavor developed. Curing at the higher temperatures hastens ripening and the development of a sharper flavor but this increases the possibility of incurring flavor defects, such as a sour or a bitter taste. According to Davis, cheese made from raw milk ripens about twice as fast as that made from pasteurized milk. This is a further indication that microorganisms have an important part in protein degradation and flavor development.[8]

The yield of cheese made from a pound of milk decreases as the fat content of the milk increases. This is due to the lower proportion of protein in relation to the fat content of the milk. With less casein present, the cheese has a lower water-binding ability, making for a decreased yield of cheese for each pound of fat in the milk. Milk with 3.3% fat yields about 9 pounds of cheese; milk of 4% fat, about 10.5 pounds of cheese. The relationship of the fat and casein content of the milk to the yield of cheese is shown in Fig. 19:6. When made from pasteurized milk up to 4% more cheese is obtained owing to better retention of moisture.

Experiment has shown that cheese made from milk with increased polyunsaturated fatty-acid content does not develop a normal flavor or body characteristics.

Rindless:

Much rindless Cheddar cheese is made, especially in New Zealand. The curd is weighed into stainless steel rectangular forms, lined with cheese cloth. After the cheese is pressed overnight, the cheesecloth is removed and the cheese is wrapped in a clear, plastic film and

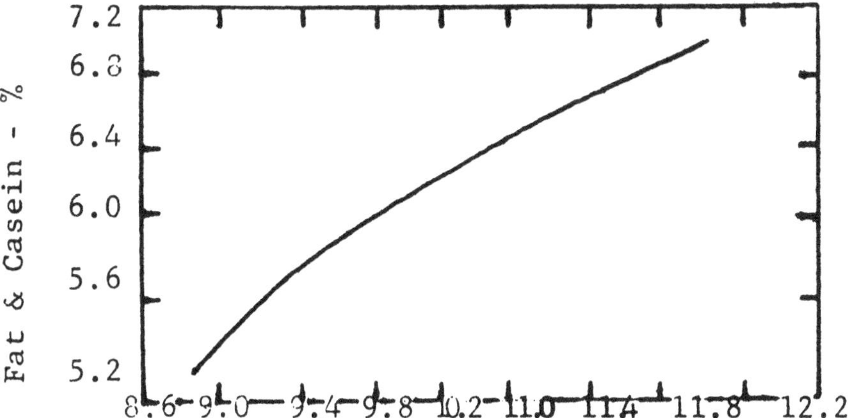

Fig. 19-6. Yield of Cheddar-type Cheese per 100 lbs. milk. (Courtesy, American Dairy Review)

heat-sealed. Such cheese is popular because it is free from mold and there is no waste.

Modified Cheddar Cheese

Several types of cheese are made by modification of the cheddar cheese procedure. In addition to Cheddar the term "American cheese" includes Colby, Monterey, Washed Curd, Jack, and Granular cheeses.

Colby Cheese:

Like many other varieties of cheese, Colby cheese was named after its place of origin, near Colby in north-central Wisconsin. It is not an important commercial variety.

Colby cheese is made in the same way as Cheddar except that the curd is not matted or milled. After about 80% of the whey has drained from the curd, cool water is added to the vat until the temperature of the curd is reduced to about 80°F. The curd is then stirred for about 15 minutes to prevent it from matting, and the whey is drained off. About an hour after the whey is drained the curd is salted and stirred as in making Cheddar cheese, after which it is hooped and pressed. Colby cheese contains from 38-40% moisture and has a softer body than Cheddar cheese. It does not keep as well as Cheddar.

Granular:

Like Colby cheese, Granular (often called *Stirred Curd*) is a modified Cheddar cheese. Water is not added to the curd as in making Colby. After the curd is cut, it is not put through the cheddaring process but is stirred and drained until dry enough to be salted, hooped, and pressed. The cheese has a more open texture than Cheddar and a softer body.

Pineapple cheese is made like Granular, except that the curd often is heated until it is firmer. It is pressed in pineapple-shaped forms, immersed in hot water for a few minutes, and hung in a loose-meshed net to dry and cure. The net forms a design on the curd not unlike the scales on a pineapple.

Monterey:

Monterey (sometimes called *Jack*) cheese, first was made in Monterey County, California, about 1892, by a family named Jack. It is an uncolored cheese, the curd for which is prepared as in making Colby, except that it is cooled to about 86°F. After it is salted, about 10 pounds of curd is placed in a muslin cloth, about 34 inches square. The corners of the cloth are pulled together to form a rounded cheese, the cloth is tied, and the excess cloth is folded over the top of the curd. The cheese is pressed between boards or in hoops. The sides of the cheese are straight if pressed in a hoop, rounded if pressed between boards. After drying for a few days, the cheese, which has a thin, flexible rind, is dipped in melted paraffin. After ripening for about one month it is ready for market. Whole-milk Monterey contains more moisture (up to 44%) than Granular cheese does.

High-moisture Jack is made similarly to Monterey, except that the curd is heated to about 96°F, the whey is drained off, and the curd then cooled to about 72°F. The cheese is cured at a temperature between 40 and 50°F, whereas Monterey is cured around 60°F. It has a moisture content between 44 and 50%.

Monterey cheese, often made from skimmed or partly skimmed milk, is cured for about six months to make *Grating-type Monterey*. This cheese is hard and the moisture content is less than 34%.

Washed-Curd or Soaked-Curd:

This type of cheese is made from curd that is matted and milled as for Cheddar. The curd is then covered with cold water for 5 to 30 minutes. "Soaked-curd" cheese usually indicates a longer washing time than "washed-curd." The curd is drained, salted, and pressed, usually in the smaller (daisy, flat, or print) styles.

The washing process increases both the moisture content and the volume of the curd and forms a soft-bodied, open-textured cheese. The washing also removes some of the whey and reduces the acidity of the curd. It is cured for a period of one or two months, but if made from raw milk, it must be cured for at least 60 days. This cheese has poor keeping quality.

Mechanized Processes for Cheddar-type Cheeses

Work on the development of mechanized procedures for the manufacture of cheese has been undertaken in many areas. One of the first successful applications originated in Australia.[7] In New Zealand, much cheddar cheese is made mechanically on the "Cheddarmaster" equipment.[18] In Australia, a "Cheddarmaster-2" is used to automate the cheddaring process. In the United States and Canada mechanized cheesemaking is gaining favor (Fig. 19:7).

The automation procedures, in general, follow the traditional steps of making cheddar cheese. In some methods, a continuous process is used; in other adaptations a unitized or modular method is followed.

In the Australian procedure, also used in New Zealand, curd from a conventional milling machine is carried upward on a belt to a revolving drum wherein it is automatically salted. The salted curd then drops into a weighing unit wherein it is automatically weighed into cheese hoops which are conveyed to the pressing unit.

The Kusel "Cheddar-Rite" process follows conventional procedures up to the time that about one-third of the whey is drained

Fig. 19-7. Australian Machine for Automatic Cheese-making. Curd from the milking machine is carried up by the belt to the salting drum from which it drops to the weighing unit, prior to entering the hoop conveyor. (Courtesy of C.S.I.R.O., Dairy Research Section, Highett, Victoria).

Fig. 19-8. Cheddar-Rite Machine. Emptying the Curd, Filling and Stomping the Filled Hoops. (Courtesy Kusel Dairy Equipment Co.)

from the vat. The curd slurry is then pumped, or if the vat is elevated, the curd is dropped into the "Cheddar-Rite" vat (Fig. 19:8), capable of holding up to 35,000 pounds. The curd and whey slurry is stirred constantly while it is being conveyed and distributed evenly in the vat. After the whey has been drained the curd is ditched, hot water is circulated under the inside surface of the sides and bottom of the vat to maintain the curd at 100°F.

After final draining, the slab cutting operation and turning is done manually, and is followed by the milling operation. Salt is added and uniformly dispersed by the mechanical forker. After being pushed mechanically to the end of the vat the curd is drawn into a hopper. Prepared hoops are moved into position under the collector by a power conveyer. Here the tare of the hoop and liner is mechanically determined and then the proper amount of curd is dropped into the hoop. The filled hoop is moved to an automatic "stomper" where remaining whey is pressed out and the cheese is closed into its final form before aging (Fig. 19:8).

Continuous cheesemaking procedures developed in France and Holland make use of concentrated, pasteurized milk held in containers moving on a conveyor belt. Starter culture and rennet are added automatically. Warm water (86-90°F) is added to reconstitute the concentrated milk. Coagulation takes place rapidly, depending upon the temperature and acidity of the milk. The curd and whey are separated in a revolving drum, following which conventional cheesemaking takes place.

Bacteria in Cheddar Cheese

At least six types of bacteria predominate in Cheddar cheese. *S. lactis*, which is introduced with the starter, prevails while the cheese still is green. As the cheese ripens, this organism is succeeded rapidly by other bacteria. *Lactobacilli*, such as *L. casei, L. plantarum*, and *L. bulgaricus*, predominate in ripe Cheddar cheese of good quality. About 78% of the non-lactic bacteria are micrococci.[1] *Cocci* are found in cheese of both good and poor grades. Spore-forming organisms are most frequently found in cheese of poor quality. They do not appear to influence the normal ripening process but are associated with the development of undesirable flavors.

The number of bacteria in Cheddar cheese varies greatly, but usually between forty million and six hundred million are found in a gram of cheese and as many as one hundred billion may be present in one gram.

Food-poisoning caused by contaminated cheddar cheese is discussed in Chapter 7.

Some Defects in Cheddar Cheese

An open body is a frequent defect. Gassy openings often arise from coliform contamination or at a time from lactose-fermenting yeasts. Mechanical openings may arise from improper matting, cheddaring, or pressing. Lack of proper development of acidity or inclusion of whey in the curd also may lead to body defects. Improper matting or cheddaring favors an uneven distribution of moisture which may lead to a mottled appearance of the finished cheese.

Oxidation of fat may occur in Cheddar cheese that has been stored for eight months or more. A tallowy flavor develops and the oxidation of fat and carotene gives rise to bleached areas surrounding slits and openings in the cheese. The exclusion of air, as by waxing the surface of the cheese, or wrapping it in plastic film, will prevent this defect.[25]

Small, white spots frequently appear in ripened Cheddar cheese. As they do not detract from the quality of the cheese they are not considered defects. Investigation has shown that these specks are mostly tyrosine associated with calcium lactate.[11] Seams or white lines between the milled curd particles usually consist of calcium orthophosphate. The seam area may become slightly more alkaline and cause a change in the color of the adjacent cheese.

COTTAGE CHEESE

Cottage cheese is one of the most widespread varieties. It is typical lactic acid, unripened, soft-curd cheese. In some households it is made by allowing milk to sour naturally. *Dutch cheese, Pot cheese* and *Schmierkaese* (Smearcase) are other names for this cheese. Most of the cheese is made in fluid milk plants, with increasing amounts being made in butter and by-product plants.

Cottage cheese is made commercially from skim milk, reconstituted skim milk, or nonfat dry milk. Large-grained, low-acid cheese is made by adding rennet to the milk and cutting the curd with knives into 5/8- to 7/8-inch cubes (see Fig. 19:10). This style is sometimes called *large curd* or *popcorn* cheese. When quarter-inch curd knives are used, the cheese is known as *small curd, country* or *farmer's* style. When the cheese contains not less than 4% fat it is called *Creamed Cottage Cheese*. Flavoring materials, such as chives, pimento, pineapple, or vegetables may be added.[6a]

Manufacture

Two types of curd—acid curd and rennet curd—are used in the commercial manufacture of Cottage cheese. With either, the method may be varied to make a curd ready for cutting and heating in 5-6 hours. In the short-set method, pasteurized milk is cooled to a temperature of 88-90°F and from 5-7% of lactic culture is added. The long-set method requires 12-16 hours for the milk to coagulate. The milk is set at 70-72°F and 0.3-1.5% starter is used.

Acid-curd cheese is sometimes made without the use of rennet. When the curd is firm enough, it is cut with curd knives into cubes from one-fourth to three-eighths inch in size and allowed to remain in the vat for about fifteen minutes before it is heated. The small curd requires more careful handling than the large curd cheese. The curd is heated to 120°F during 60 to 90 minutes, and it is stirred gently with the rate of agitation increasing as it becomes more firm. In order to promote proper coagulation, calcium chloride is added if the cheese is made from high-heat treated nonfat dry milk or from milk pasteurized at a high temperature. Curd of high acidity is usually soft and difficult to handle. This condition may be remedied by draining off part of the whey, adding some hot water, and continuing the heating until the desired firmness is obtained. Too little development of acid will produce a tough, rubbery curd, as will too-rapid heating, or too high a temperature during heating.

Rennet-curd cheese is made by adding 1 ml of rennet, diluted with at least 40 times its volume of water, to each 1000 pounds of milk, just before or after the starter is added. The curd is ready for cutting when the clear whey, taken from below the surface of the curd layer, has a titratable acidity of 0.5% or a *p*H of 4.6.

The starters used contain Strep. lactis or cremoris and often include Strep. citrovorans or paracitrovorans, which produce diacetyl which adds to the flavor of the cheese.

About two hours after cutting, the whey is drained off and the curd is washed two or three times with cold water. This removes the acid whey, increases the firmness of the curd particles, and helps to produce a mild-flavored cheese. The washed curd is allowed to drain for about an hour to obtain a drier product which may be more easily creamed. At this time the curd may be salted with 0.5-1% of its weight with a coarse-grained salt which dissolves rather slowly.

Direct Acidification

A process for making cottage cheese by direct acidification of the milk was developed at the University of Wisconsin. Nonfat dry milk solids are added to skim milk to bring its total solids content to about 15%. This is pasteurized and then cooled to 40°F. Skim milk concentrated to about 15% may also be used. About one gallon of food-grade hydrochloric acid is added for each 1000 pounds of milk with constant stirring. The acidified milk is pumped through a series of tubes heated by steam to about 110°F. Coagulation is completed by the time the product leaves the heating unit.

Preparation for Market

Cottage cheese may be creamed and packaged after draining but usually it is placed in a cold room and held at 35°F for up to two weeks. Much cottage cheese curd is shipped to marketing areas where it is creamed and packaged. Cream is mixed with the curd in a vat or with a mechanical mixer. Sometimes the curd is creamed by placing half of the creaming mixture in the can in which the cheese is to be placed and then pouring the rest of the cream over the top of the curd. The can is then stored in a cold room and the cream is allowed to distribute itself through the curd. The fat content of the mixture of cream and milk used to cream the cheese varies with the type of curd. When the particles of curd are large, when a relatively dry product is desired, or when a high fat content is desired in the cheese, a creaming mixture of higher fat content is used. The creaming mixture may contain added lactose, milk powder as well as sodium, potassium, ammonium or calcium caseinate as well as salt.

Creamed cottage cheese furnishes about 100 calories for 100 grams of cheese, the uncreamed cottage cheese furnishes about 77 calories.

The Federal standard for creamed cottage cheese is not less than 4% fat and this is not often exceeded in commercial practice. The moisture content must not exceed 80%.

The high moisture content and open texture of cottage cheese makes it susceptible to contamination by molds and yeasts which may cause characteristic defects of flavor and color. During storage, cottage cheese may develop a gelatinous or slimy coating around the curd particles as well as a putrid odor. This defect is of bacterial origin and it usually arises in the wash water used on the curd or to clean equipment. Species of *Pseudomonas, Alcaligenes, Achromobacter* and *Aerobacter* have been associated with this defect. The use of pasteurized or chlorinated water as well as proper cleaning and sanitizing of equipment will prevent it.

The shelf-life of cottage cheese varies from seven to ten days, but if made under carefully controlled conditions of quality and sanitation, the shelf life may be extended for at least two weeks.

Automated Manufacture of Cottage Cheese

In the automated process for making Cottage cheese, the curd is pumped from the vat into a Drainer Mixer (see Fig. 19:9) where it is drained and then creamed, while the previously creamed batch in the other mixer is pumped to the filler hopper. Continuous processing is ensured by alternating these operations. The low-voltage probe shuts off the pump when the Drainer-Mixer is filled with curd.

Bakers' Cheese

Bakers' Cheese is a soft, unripened, low fat product similar to cottage cheese. A similar German cheese called "Quark" is used in salads and fruit desserts.

In the manufacture of Bakers' cheese about 2% of starter and 35 ml. of diluted rennet is added to reach 1000 gallons of pasteurized skim milk held at 86°F. the temperature is raised slowly to 90°F. at which point the acidity of the milk is about pH 4.5. After the curd is broken, it is drained in cloth bags or the curd may be pumped through a curd concentrator from which it is usually packaged in polyethylene bags.

In making Bakers' cheese, the operator often judges its consistency by kneading it into a ball with his hands. If it shapes readily, and does

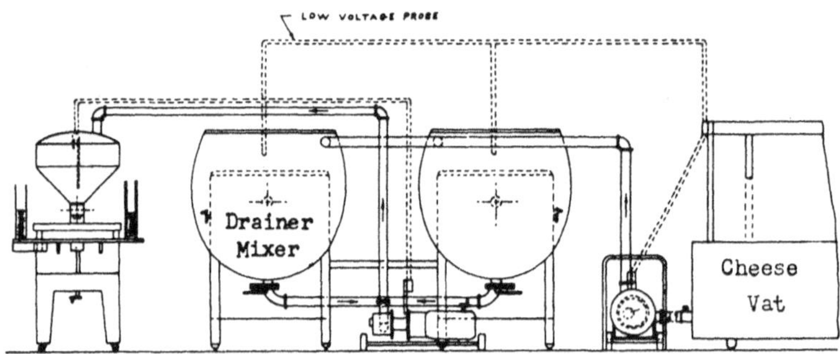

Fig. 19-9. Automated Manufacture of Cottage Cheese (See accompanying text) (Courtesy of the Grace Machinery Company)

not stick to the hands, the cheese is deemed acceptable. A sticky product contains too much moisture. Bakers' cheese contains from 23 to 28% total solids.

PROCESS CHEESE

Process cheese is the product made by mixing and grinding together different lots of cheese with the aid of heat and an emulsifying agent. A number of basic patents relating to the method of manufacture have expired and process cheese is now made in many places.[12] Cheeses of varying degrees of ripeness, those of too low fat content, too high moisture content, or those having an off-flavor or other defect can often be made by processing to give a marketable product of uniform flavor and good keeping quality. Cheddar cheese is the most commonly processed variety, but Swiss, Limburger, Brick, and Camembert varieties may be processed.

Considerable skill is needed to select cheese of the proper age and characteristics of flavor to ensure a uniform product. The use of natural cheese ripened less than ten days tends to give the processed cheese a firm, rubbery body, due in part to increased fat loss during processing. Too much aged cheese in the finished product gives it a weak, grainy, body.

The cheeses selected for processing are run through a grinder into a large steam-jacketed kettle or cooker which may have a capacity of 400 pounds. Water, color, and emulsifying agent are added. The emulsifier is added to prevent the separation of fat during processing as well as to obtain desirable properties of melting and slicing in the end product. Disodium phosphate, sodium citrate or tartrate, sodium pyrophosphate, sodium metaphosphate, sodium aluminum phosphate, and Rochelle Salt or mixtures of these salts, are used as emulsifying agents in amounts not to exceed 3% of the finished product. When Rochelle Salt (Sodium-Potassium tartrate) is used, crystals of calcium tartrate may separate in the finished cheese, causing it to become gritty or "sandy".

The acidity of the cheese may be controlled by the addition of harmless acids, such as acetic, citric, lactic, and phosphoric, provided that the pH of the cheese is not less than 5.3. Sweet cream may be used as an ingredient provided that the fat it adds does not exceed 5% of the weight of the finished cheese.

Federal standards require that the heat treatment during processing be not less than 150°F for 30 seconds. At this temperature the cheese becomes plastic and flows readily. At this stage it is conveyed to a filling machine and finally into the retail package.

The moisture content of process cheese may not exceed by more than 1% the maximum permitted in the natural variety from which it is made. The fat standard is that of the natural cheese. The Federal standards make provision for some variation when two or more types of cheese are used.

Process Cheese Food

The product called Process Cheese Food is made in the same manner as Process cheese from one or more varieties of cheese with the addition of other food products such as cream, milk, skim milk, and whey, as well as fruits, vegetables, and meats. The cheeses must contribute not less than 51% of the weight of the finished product. The finished cheese food must contain not less than 23% of fat and not more than 44% of moisture. In some European countries, the non-milk products must not exceed one-sixth the weight of the total solids.

Process Cheese Spread

Process Cheese Spread differs from the preceding product in that it contains more than 44%, but not more than 60% of moisture. The fat content is not less that 20%. In order to obtain a satisfactory body and texture, water-retaining ingredients may be used, not to exceed 0.8% of the weight of the finished product. For this purpose substances such as guar gum, alginate, gum tragacanth, gum karaya, gelatin, and carboxymethylcellulose may be used. Sweetening agents such as sugar, dextrose, maltose, and corn syrup may also be included.

Cheese spreads have been marketed in aerosol-type dispenser cans which require no refrigeration.

Bacteria in Process Cheese

The organisms present in processed cheese are those that survive the time and temperature used in processing. Aerobic spore-forming organisms, such as *Bacillus mesentericus* and *B. subtilis*, may be present, but anaerobic spore-formers are more important as a cause of spoilage. *Clostridium sporogenes* is frequently present but it does not usually grow because of the relatively high acidity and salt content of the cheese. Processed cheese food and cheese spread may spoil more readily because the non-cheese ingredients may increase the number of organisms present as well as provide more suitable media for their growth.

SWISS CHEESE

Swiss cheese is also called *Emmentaler* cheese, after the Emmental (valley of the Emme), a valley in the Canton of Berne, Switzerland, (Fig. 19:10) where it has been known for centuries. There is some evidence to show that it originated with a Roman invasion of what now is Switzerland about 58 B.C. The cheese now is made in other countries, especially France, Germany, Italy, Russia, and the United States. In Denmark it is called *Samsoe* cheese.

Swiss cheese is a hard, pressed, rennet-curd cheese with a mild, sweet, nut-like flavor. The presence of holes or "*eyes*" is a characteristic of this cheese. The cheese is difficult to make. Special factory equipment is used and the milk, standardized to 3-3.5% fat content, must be of good quality. In Switzerland the cheese cannot be made legally from milk of cows that have been fed with silage. Such milk has been found to subject the cheese to formation of large gas pockets, "blow-holes". The use of pasteurized milk has not proved uniformly successful. Homogenized milk is not used since it may cause defects in the flavor and texture of the cheese. Clarification of

Fig. 19-10. At dawn and in the evening, farmers bring milk by dog cart and more modern means to the cheesemaker.
(Courtesy Switzerland Cheese Association)

the milk at about 80°F. improves the formation of eyes by reducng the "set" or number of eyes that develop.

The milk, is placed in large copper or stainless steel kettles. In Finland, where stainless steel kettles are used, it is considered

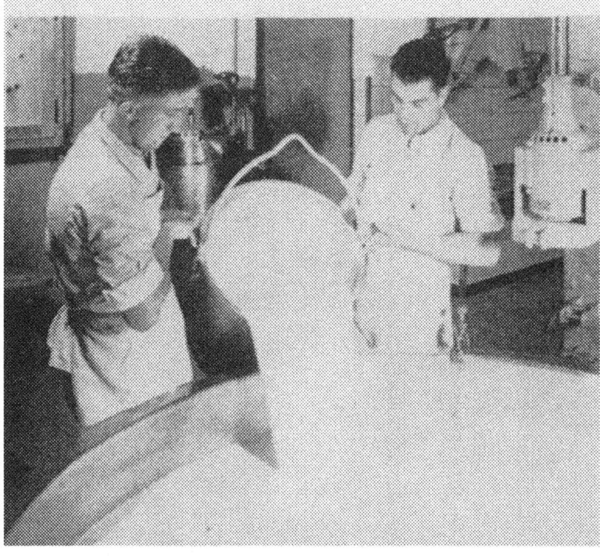

Fig. 19-11. Ripened evening milk and fresh morning milk is poured into the copper vat.
(Courtesy Switzerland Cheese Association)

necessary to add up to 15 ppm. of copper (as copper sulfate). The copper probably activates enzymes associated with proper ripening. This is based upon the finding that this amount of copper is normally present in Swiss cheese made in copper kettles. The milk is heated to 124-130°F. while it is stirred slowly and special starters are added. These consist of a culture of at least two organisms that form lactic acid and one that forms propionic acid. Fig. 19:11.

Lactobacillus lactic and *L. bulgaricus* form acid at the beginning of the process and then *Steptococcus thermophilus* is active when the temperature of the milk has risen to 85°F or more. *Proprionibacterium shermanni*, the eye-forming organism, is also associated with the development of sweet flavor of the cheese. Enough rennet, which in Switzerland is made from calves' stomachs, is added to coagulate the milk in about 30 minutes.

The curd is cut into small pieces about the size of grains of wheat with knives known as "Swiss Harps," which are similar to those used to cut Cheddar cheese (Figs. 19:12, 19:13). It is then heated, with constant stirring, for about an hour at a temperature around 128°F. Swiss cheese whey is less acid than that obtained in the manufacture of cheddar cheese; usually the acidity is between 0.10 and 0.12%, as lactic acid.

Fig. 19-12. Cutting Swiss Cheese Curd with a "Swiss Harp". (Courtesy of Switzerland Cheese Association).

Fig. 19:13. Spooning: The cut curd is turned over with wooden spoons, cut again with the harp and stirred. (Courtesy Switzerland Cheese Association).

The removal of the curd from the whey involves a procedure peculiar to the making of Swiss cheese. A heavy cloth is placed over a steel frame to which the cheese cloth is fastened and the contents of the kettle are set into vigorous circular motion. This motion of the curd is then quickly stopped by means of the stirring implement, which permits the mass of curd to collect into a spinning cone. At this time the frame with its cloth is inserted beneath and the curd is lifted out of the kettle by means of a chain and pulley attached to the hoop (Figs. 19:14, 19:15). The curd weights up to 220 pounds.

The cheese is placed in a press for about a day in order to remove excess moisture and aid in the formation of a strong, uniform rind (Fig. 19:16).

While in the press, the acidity of the cheese increases as a result of continued growth of Streptococcus thermophilus. Later, as the cheese cools, the lactobacilli continue the production of acidity. In a good cheese, the acidity reaches a pH of 5.1 to 5.2. An excess of acid prevents the development of the propionic acid bacteria, with the

Fig. 19:14. A steel frame to which the cheese cloth is fastened is dipped under the mass of cheese, keeping the cloth pressed against the inside of the vat. (Courtesy Switzerland Cheese Association).

result that normal eye formation is reduced and the cheese tends to crack. Proper acidity yields an elastic curd which stretches during eye formation. Propionic acid contribute to the nutty flavor of the cheese.

Next the cheese is moved to a room held at a temperature around 53°F and is salted by immersing it in strong brine and later sprinkling the surface of the cheese with salt over a period of several days. After salting, the cheese is moved to the curing room, where the temperature is around 70°F and the relative humidity is 80-85%. At this time bacterial activity, becomes apparent and in 6 to 10 days the eyes begin to develop. Eyes are formed by the accumulation of carbon dioxide formed by lactate-fermenting organisms. The gas diffuses through the curd and collects in weak spots where the bubble or "eye" is formed. Insufficient formation of gas results in too few or no eyes, a so-called *blind* cheese. If the gas pressure causes the cheese to crack because the curd is too firm or brittle, the defect is called a *glaesler* or glassy cheese.

Ripening requires about six months, during which time an

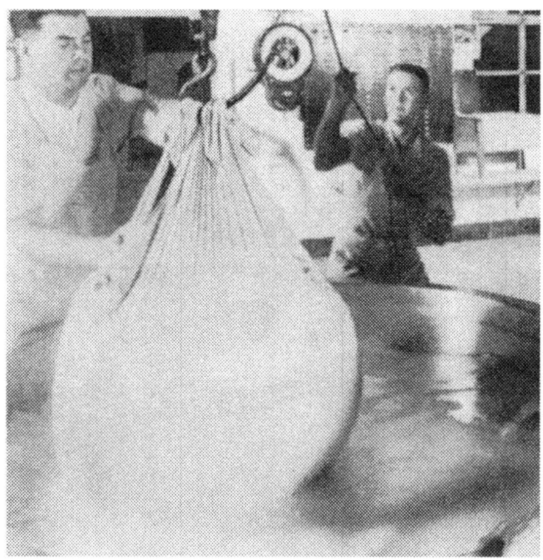

Fig. 19:15. The cheese curd, saturated with whey, is lifted on to the press-table. (Courtesy Switzerland Cheese association).

experienced cheesemaker can follow the development of the eyes by tapping the cheese with his finger, and even can judge their size and location by the sound made. The typical flavor of a good Swiss cheese continues to develop for several months after the eyes have formed and it is associated with the accumulation of traces of acetic and propionic acids. The liberation of amino acids, especially proline, has been associated with good flavor development. The cheese becomes softer as it ripens owing to the activity of proteolytic enzymes. About one-third of the protein nitrogen is changed to a soluble form. Under proper conditions a well-ripened cheese will keep for several years.

Only the cheese made in Switzerland is claimed to have the characteristic body and flavor of Emmenthaler cheese. Cheese of excellent quality made in the United States and other countries does not duplicate that made in Switzerland.

A good Swiss cheese has comparatively few eyes, which range from one-half to one and one-half inches in diameter (Fig. 19:17). A cheese with many small holes, or one with only large eyes near the surface, has not undergone proper development. Gassy fermentations are

Fig. 19:16. Swiss Cheese in the Press (Courtesy of Switzerland Cheese Association, Inc.)

known as *nissler* and *pressler* fermentations. A pressler cheese forms while the cheese is in the press and often is caused by contamination with *Aerobacter aerogenes*; it is characterized by large gas holes, usually near the surface. A nissler cheese has many small eyes or "pinholes," distributed throughout its body. This condition, which develops after the cheese is made, is caused by contamination with anaerobic, spore-forming bacteria, such as the *Clostridia*, especially *Clostridium butyricum*. Cheese in which proper acidity has not developed often shows this defect. Swiss cheese that is made from the curd left in the kettle, the so-called *streble*, frequently forms a nissler cheese.

In Switzerland, the cheese is made by about 2000 family-owned units. Each one makes an average of about two wheels per day. The usual 200 pound wheel requires about 300 gallons of milk. The Swiss normally export cheese with large eyes to the United States. An exported cheese must weigh not less than 143 pounds and bear the cross-bow and the word "Switzerland" in red letters repeatedly on the surface. Usually Swiss cheese are made in very large, "wheels" about six to eight inches thick and up to four feet in diameter, and weighing from 60 to over 200 pounds. For ease in handling and cutting, some Swiss cheese is made in the form of large, rectangular

Fig. 19:17. Swiss Cheese - Called Emmentaler in Switzerland is branded and stamped with the word "Switzerland", and the crossbow when intended for export. (Courtesy Switzerland Cheese Association).

blocks. A Swiss-type cheese with neither rind nor eyes is made in the United States. It is in block form and is coated with a wax or plastic moisture-proof film.

Brick Cheese

Brick cheese is a semi-soft surface-ripened product of American origin. The name may have been derived from the shape of the cheese or from the use of bricks for pressing the curd. In flavor and texture it is between Limburger and Cheddar cheese.

Muenster (q.v.) is similar to Brick cheese. Other similar cheeses include *Oka* made by Trappist monks in Oka, Canada, and in *Port du Salut* in France. The latter cheese when made elsewhere in France is called *Saint Paulin.* In Denmark, Brick cheese is called *Esrom.*

The odor of surface-ripened cheese, such as Brick, Trappist, and Limburger have been attributed to the presence of sulfur compounds, such as methyl sulfide, and hydrogen sulfide.[10]

Fresh milk, preferably pasteurized, is set with a lactic acid starter and coagulated with rennet. The culture contains S. lactis, S. cremoris, S. thermophilus as well as Leuconostoc citrovorum. Yeasts are involved in the surface ripening. The curd is cut as in making cheddar cheese and is heated to about 115°F. After draining, the curd is placed into molds two to three inches wide, five to six inches high and about 10 inches long. These forms, which have neither top nor bottom, are placed on a draining table and the cheese is turned several times during the first day. Often a brick is placed on each form to hasten draining.

The next day, the forms are removed and the cheese is salted by rubbing with salt or floating in brine. It is then cured in a room at about 60°F and with a very high humidity, during which time microorganisms form a reddish-brown growth on the surface of the cheese. After about two weeks, the surface of the cheese is allowed to dry and it is then dipped in paraffin and wrapped in parchment paper. During the subsequent ripening period, which takes about two months, the cheese acquires a soft texture and waxy body.

As with most surface-ripened cheeses, a variety of microorganisms contribute to the characteristics of the cheese. Yeasts are involved, which as they grow reduce the lactic acid present, to about pH 5.6, at which point bacteria predominate, especially Brevibacterium linens. The organism contributes to the proteolysis on the surface of the cheese, the Limburger flavor and the reddish-brown coloration on the cheese surface.

Brie Cheese

Brie cheese originated in France, where it has been made for over five hundred years. It is named for the province of La Brie, just east of Paris, around the city of Meaux. It is a soft rennet cheese, ripened by molds similar to those that ripen Camembert cheese. Brie cheese as made in the United States is partially ripened by bacterial action and it does not have the texture and flavor of the French product. The American made Brie cheese originated in attempts to make Camembert cheese, but the resulting product did not ripen in the French manner.

After the milk is coagulated with rennet, the curd is dipped into small forms and hoops and allowed to drain for 24 hours. The hoops are then removed and the surface of the cheese is sprinkled with salt. The cheese is held in a dry, well ventilated room for about a week and is then put in a dark, moist room or cellar for two to four weeks. During this time the cheese ripens and acquires a soft body, a sharp odor, a surface growth of B. linens and Penicillium camemberti which contributes to a characteristic flavor.

Brie is a perishable cheese and must be held under refrigeration.

Caciocavallo

This cheese was first made in the south of Italy and in Sicily around the twelfth century. The origin of the name is not known; it may originally have been made with mare's milk, or named for a place of manufacture—Monte Cavallo. Some authorities believe the name is associated with the trademark of a horse's head; others state that it refers to the method of drying by hanging the cheese in pairs, one on each side of a pole, as if they were on horseback or *cacio a cavallo*. The cheese is a *pasta filata*, or drawn-curd, made by pulling out the curd in strands from very hot water, as in making Provolone.

To make Caciocavallo, cow's milk is set and curded with rennet paste. After the whey is drained off, the curd is covered again with hot whey in which it is allowed to stay until, when it is placed into very hot water, the cheesemaker can stretch it into long, elastic fibers. The curd is then drained and cut into long slices which are placed in a tub of hot water. When these become plastic, they are pulled by hand into rope-like pieces and draped on poles. These pieces are kept hot by being placed in hot water and are formed into a figure-8 with the upper loop much smaller than the lower one.

The cheeses are allowed to cool and then are salted in brine for several days. They are then tied in pairs and hung over poles in the curing room at a temperature around 62°F and relative humidity around 85%. They are cleaned when they become moldy and the surface is rubbed with oil. Caciocavallo cheese is cured for two to four months for table use and for a year or more for use in grated form.

Camembert Cheese

According to legend, Camembert cheese was first made in the village of this name, near Vimoutiers, in the Orne Department of France, in 1791, by Madame Jeanne Harel. A statue to commemorate her was erected in Vimoutiers in 1928, but it was destroyed in 1944 during the war. Actually, this cheese is much older as it is mentioned in French dictionaries published around 1700. It is now made in many countries, including the United States where it has been made in California since 1880.

To make Camembert cheese, cow's milk is set with a lactic starter and enough rennet is added to make a firm curd in about one hour or a little longer. The curd is sliced, transferred to round metal forms and carefully placed on reed mats for draining in a warm room with high humidity. By the second day, the cheese is taken from the hoops and salted. Ripening is done at about 50 to 55°F., and 85-90% humidity for three weeks or longer. The curd dries until it has lost about half of its original moisture. During this time, mold grows on the surface, especially *Penicillium camemberti*, although *Geotricum candidum* and other molds, yeasts and bacteria, such as *Bacterium linens* may be present.

Camembert cheese made in the United States usually is inoculated by sprinkling it with an aqueous suspension of *Penicillium camemberti*. This is considered unnecessary in France, where the air of the curing rooms contains enough mold spores to inoculate the cheese in a natural way. The growth of microorganisms forms a grayish-white layer on the surface of the cheese which later changes to a reddish color. On the cheese made elsewhere, including the United States, only a growth of P. camemberti takes place, forming a white surface mat.

Enzymes liberated by the molds act upon the protein of the cheese and ripen it by proteolytic action, working from the surface to the center of the cheese. The body of a well-cured cheese has the texture of soft butter. The surface generally is removed before the cheese is eaten because it has a much stronger flavor than the body of the cheese. An odor of ammonia is apparent in an overripened cheese.

The finished cheese is about four and one-half inches in diameter and one inch thick and is packed in flat, round containers.

Cold-Pack Cheese

This product, also called *Club cheese*, is made by grinding and mixing together one or more varieties of cheese, without heat. As in making Process cheese, the soft, unripened cheeses, like Cottage, Cream, and Neufchatel, are not used. A well-ripened cheese often is used, in order to give the product a distinct flavor. As the product is not heated, the cheeses selected must be made from pasteurized milk, or must have been held for at least 60 days at not less than 35°F.

Cold-Pack Cheese Food is a cold-pack cheese with the addition of approved acids, sugar, emulsifiers, spices, or flavoring.

Coon Cheese

Coon cheese, made by a process patented in 1926 by E. W. Coon, is a Cheddar cheese cured for about 3 months at the relatively high temperature of 55-70°F, and at a humidity around 95%. High-quality cheese must be used under these conditions for curing. The product develops a very sharp, tangy flavor and a crumbly body. A blue-green mold is allowed to grow on the cheese. Uncolored wax is applied to the cheese at a temperature of about 240°F. This changes the greenish surface to a black color.

In Canada, a similar cheese is known as *Black Diamond*.

Cream Cheese

Cream cheese is a soft, unripened cheese made in practically the same way as Cottage cheese (*q.v.*). A milk-and-cream mixture, of 11.5-16% fat content, is homogenized and pasteurized, cooled to about 80°F, and set as in making rennet-curd Cottage cheese. The curd is not cut, but it is stirred until it is smooth and homogenous and is then heated to about 120°F for about one hour. Sometimes the curd is heated by mixing water at 180°F into it and draining off the mixture when the curd reaches about 125°F. This method shortens the making time and produces a less acid cheese.

The curd is drained off by pouring it into muslin bags which are then placed on draining racks. In order to hasten draining, pressure may be applied near the end of the draining period, or the bags may be placed in alternate layers separated with a layer of cracked ice.

In another method, the hot curd is passed through a special centrifugal machine which mechanically removes the whey and allows for the addition of salt, stabilizer, and flavoring material to the cheese before it is discharged from the machine.

In order to prevent leakage of whey from the cheese, the Federal Food and Drug Standards permit the addition of not over 0.5% of a stabilizer such as gelatin, algin, or locust bean gum. Cream cheese must contain not less than 33% of milk fat and not over 55% of moisture.

Devonshire Cream

Devonshire (also called Cornish) Cream is a form of clotted or thick cream. It is popular in Devon and parts of Cornwall, England, where it is used as a spread for bread or toast and as a dressing for salads, fruits, and the like. It is made by placing milk in shallow pans and allowing the cream to rise. The pans are then heated at 180°F until a wrinkled skin spreads over the surface of the cream. After cooling and standing about twenty-four hours, the clotted cream is removed with a perforated ladle. Devonshire cream has a slightly granular texture and a cooked, nut-like flavor. The fat content usually is between 60 and 65%.

Edam Cheese

Edam cheese is a semi-hard, rennet-curd cheese chiefly made in the northern part of the Netherlands where the town of Edam is a famous market place for it. The method of manufacture is very similar to that for Cheddar cheese, except that the curd is not allowed to develop any acidity nor is it cheddared or salted. The starter cultures used sometimes may contain rope-forming organisms, such as occur in "taette" milk. While still warm, the curd is placed into spherical molds, often made of wood. The pressed cheese is salted either by rubbing the surface with dry salt every day for a week or by immersing the cheese in a strong brine for several days. After curing in a cool room for about two months, the cheese is colored by dipping it into a solution of red dye. The surface is rubbed until smooth with linseed oil or is paraffined. Edam cheese has a dry, mealy body, with a somewhat salty flavor.

In Holland, the rind of the cheese is not colored if made for home consumption. The presence of a few eyes in the cheese is considered normal.

Feta Cheese

Feta is a so-called pickled cheese (*q.v.*). It is made in Greece,usually from ewe's milk, but milk from goats and cows also is used. The milk, which may be skim milk, is heated to about 90°F and rennet is added. At times, about 0.5-1.0% of lactic starter is added, especially if skimmed cow's milk is used. Rennet paste or kid rennet is used in order to obtain the desired flavor in the finished cheese, which is like a very dry, high-acid cottage cheese.

After it has coagulated, the curd is cut and treated much as in the making of brick cheese (*q.v.*). Instead of being dipped into small forms, the curd is poured into a wooden form about eight inches deep, three to four feet long, and three feet wide, which is placed over a draining cloth on a table or rack.

After the curd is well drained and matted, it is cut into blocks and salted on all sides. The salting is repeated later in the day. The next morning the blocks are cut into 1-inch slices which are salted and placed into paraffined wooden kegs or other containers that hold about 125 pounds of cheese. The space between the pieces of curd is filled with brine or salted whey. The cheese is ready to eat in about one month.

Teleme is very similar, but usually is ripened for about 10 days.

Gorgonzola

Gorgonzola is a semi-hard, rennet curd, mold-ripened cheese of Italian origin, which is similar to Roquefort. Evening milk is set at about 85°F and enough rennet is added to coagulate it in about 15 minutes. This curd is drained over night. The morning milk is treated in the same way. The two lots of curd are cut into pieces and sprinkled with a *Penicillium roqueforti* mold powder,* as they are placed into hoops about one foot in diameter and ten inches high.

Most of the warm, morning curd is formed into a layer on the bottom of the hoop and around and over the cold evening curd. Much of the curd is prepared on farms and sold to central cheese-finishing establishments.

After a few days, growth of mold appears, at which time the cheese is salted by rubbing its surface with salt daily for about two weeks. The cheese is then pierced about 150 times with a metal skewer, about six inches long. The holes let air in, and this permits molds, similar to those of Roquefort cheese, to grow in veins and mottles throughout the interior of the cheese. The surface of the cheese may be coated with a colored mixture of gypsum and tallow in order to prevent growth of mold on its surface.

For curing, which takes from three months to a year, Gorgonzola cheese is stored in a room held at about 40°F with a humidity of about 80%.

Gouda Cheese

This cheese (pronounced "khowda" in Dutch), was first made near Gouda, in the south of Holland. It is a semi-soft to hard cheese, similar to Edam except that it contains more fat. The rind may be dyed red.

Milk with 3-3½% fat is pasteurized at 155°F. and cooled to 86-91°F. One-half to 1% starter is added and 3-5 ounces of rennet per 100 gallons of milk. The curd forms in about fifteen minutes and is cut as for cheddar cheese. The whey is drained off and heated to 125°F. and added to the curds or hot water may be used instead. This reduces the acidity and heats the curds to the cooking temperature. After draining again, the curds are stirred or piled and 6-7 pounds of salt is added per 100 pounds of cheese. After draining, the curd is placed into round molds or hoops and pressed lightly. It is salted by rubbing or holding in brine for 6-10 days. The cheese is turned daily at this time and then once a week for 6-8 months during ripening at 50-65°F. The cheese usually is pressed into the shape of a flattened sphere.

Gouda cheese has about 40% moisture, 30% fat, 26% protein, and 1½% of salt.

---

*The mold powder is prepared by inoculating bread with a culture of *Penicillium um roqueforti*. After the mold has developed on the bread for about one month, the bread is crumbled, dried, and ground for use in making the cheese.

Grana

Grana is the name given to a group of Italian cheeses that have a hard, granular body and texture and a sharp flavor. They are used mostly in grated form, but sometimes, when broken into small pieces rather than sliced, appear on the table. Outside of Italy, these cheeses are known as *Parmesan* (*q.v.*), whereas in Italy they are given different names, according to the place and manner of manufacture.

Grana is a half-skim cheese. The cream from the night's milk is churned into butter and the skim milk is added to the morning's milk for cheese making. In Italy about 25 ppm of formaldehyde is added to the milk occasionally in order to reduce undesirable bacterial activity.[4]

Gruyère

Although Gruyère cheese is named for the town of Gruyère, Switzerland, near the French border, much of this cheese is made in nearby France. The process used is similar to that described for making Swiss cheese, but the product is ripened differently. The cheese is pressed harder and cured at a lower temperature, thus producing smaller eyes. The surface of the cheese is moistened and a secondary fermentation occurs, during which the curd softens from the rind inwards and some protein decomposition occurs, which, with the formation of ammonia, gives the cheese a characteristic aroma and flavor. Process Gruyère is processed from Swiss and natural Gruyère cheeses.

Hand Cheese

Hand cheese is of German origin, and is so called because originally is was molded by hand into various shapes and sizes. Skim milk, often mixed with buttermilk, is allowed to sour and it then is heated to coagulate the curd. The whey is drained off and the curd is molded into cakes or balls. Salt is added, and sometimes also caraway seeds, spices, or beer. The molded cheese is dried and then ripened in a cool, damp place where its surface becomes coated with a heavy growth of mold. The enzymes produced by the surface growth penetrate into the body of the cheese as it ripens over a period of about two months. It has a sharp, pungent odor and flavor which is disagreeable to those that have not acquired a taste for this cheese. In some localities the cheese is known as *Bierkaese* or *Kuhkaese*.

Junket

Junket is a very simple form of soft, rennet cheese. It is often made at home by curdling milk with junket tablets, which contain rennet in a dry form. It is used as a dessert or in the diet of invalids.

Limburger Cheese

This soft, rennet-curd cheese derived its name from the town of Limburg in Belgium. Very little, if any, Limburger cheese is imported into the United States, because domestic manufacture supplies the demand.

Milk, heated to about 93°F is coagulated with enough rennet to form a jelly-like curd in about 40 minutes. The curd is cut, stirred for a short time, and then dipped into rectangular forms, 5 inches wide, 8 inches deep, and 10 to 30 inches long. The next day, the cheese is cut into one- or two-pound bricks. These are rolled in salt and placed in a cool room with a high humidity for ripening, which takes about two months.

The cheese is ripened by bacterial action and its characteristic strong odor and flavor are due to the protein decomposition that takes place. During the first stage of ripening there is a surface growth of *Mycoderma*, salt-tolerant yeasts which use lactic acid as food. With the decrease in acidity, both lipolysis and proteolysis become pronounced. Finally, there is a growth of *Bacterium linens*, which gives the surface a red color and is associated with most of the change that occurs during the curing of the cheese.

The cured cheese is given a protective wrapping in parchment or paper in which it continues to ripen until consumed. The characteristic odor is confined largely to the portion near the surface. A waxed or air-tight coating prevents normal flavor development.

Mozzarella

Mozzarella is a soft, plastic-curd cheese, originally made in Italy from milk of the water buffalo. Much now is made in New York and California. Raw cow's milk is warmed to 88-90°F., about 0.1% of lactic starter usually is added together with the rennet. If made from pasteurized milk, a temperature of about 95°F. is used and about 1-3% of starter containing *S. thermophilus, L. bulgaricus* and sometimes *S. faecalis*. Rennet is added and the resulting curd is heated to 104°F. and matted slightly as used in making Cheddar cheese. After the whey is drained off, the curd is cut into small blocks, washed with cold, potable water and then placed in cloth sacks to drain and cure for three to six days. In the Eastern states, a low salt cheese is preferred (1.5%) but in the west, a higher salt content is used, up to 2%. Cheese with too high a salt content does not melt well. Often the curd is stored by the dealer and made into the finished cheese as required by consumer demand.

The manufacture of Mozzarella cheese is completed by placing the curd into hot water (170-180°F.) for a few minutes. It becomes stringy and is then pulled and kneaded until it is smooth and cohesive. The cheese is molded by hand or in a vacuum chamber into a spherical shape but sometimes into a brick or loaf. A popular size weighs five pounds.

The cheese is eaten fresh with little or no ripening. It is an important ingredient of *pizza*, in which the cheese is sliced and baked in a pie made with tomato sauce and other ingredients. Cheese made especially for pizza-use differs from Mozzarella mainly in having a higher fat and lower moisture content.

Scarmorza cheese is made in a similar method, molded into small egg-shaped pieces, often with protruding ears to which cords are attached for hanging.

Muenster

Muenster is a semisoft, whole-milk cheese that first was made around Muenster in the Vosges Mountain region of western Germany. It is similar to Brick cheese, but contains numerous small mechanical openings and less surface growth. In the United States it is made from pasteurized milk to which a lactic starter is added and sometimes annatto to color. The milk is warmed to about 90°F and rennet is added. About one-half hour later, the curd is cut into small pieces, stirred and heated. As manufactured in Europe, the curd is permitted to set for about two hours before cutting. It is then not stirred for another half hour nor is it heated.

The drained curd, to which caraway seed is sometimes added, is dipped into perforated, cloth-lined forms which are made in two parts. During the first day, the forms are turned several times and then the upper part of each form, with its cloth liner, is removed. The curd remains in the lower half of the form for three or four days more and is turned twice daily. When the cheeses are removed from the forms they are rubbed with salt or immersed in brine.

After being dried, the cheeses are cured at a temperature of 50°-55°F and a relative humidity of about 75%. They are frequently turned, cleaned, and washed with salty water. After drying, the cheese may be rubbed with a solution of annatto in oil to give the rind an orange color. In the United States, Muenster cheeses may be marketed after ripening for ten days to two weeks, but sometimes they are cured for about two months. In Europe they are cured for at least two to three months and a surface-ripening is permitted, as for Brick cheese.

Neufchatel Cheese

Neufchatel is a soft, rennet-curd cheese made in a manner like that used for cream cheese. As made in the United States, the cheese has less fat and more moisture than cream cheese.

In France, after the curd is cut enough salt is added to give it a salt content of 2-4%. The curd is then placed on draining boards for 24 hours and is then put in a drying room for two or three weeks. The cheese is turned daily during which time the surface becomes coated with mold growth, chiefly *Penicillium camberti.* The ripened cheese is packed in small, cylindical or flat form, about three inches long and two in diameter.

Nuworld

This is a cheese that has the characteristics of blue cheese but contains no coloration. By a patented process of ultraviolet irradiation, a colorless, mutant strain of *Penicillium roqueforti* is produced.[13] When used to make the cheese, a soft cheese with a mild, piquant flavor is obtained.[19]

Parmesan

According to an Italian governmental decree issued in 1954, all Parmesan-type cheese made in certain provinces should be called *Parmigiano-Reggiano.* This type of cheese, which was known before 1300, was made first in the vicinity of Parma, Reggio Emilia, hence the name.

Parmesan, as the cheese is called outside of Italy, is a Grana cheese (*q.v.*) made from partly skimmed milk by a method very similar to that followed for Swiss cheese. It is salted by soaking the cheese in brine for two to three weeks.

The curing process takes from two to three years. The first stage takes place in a room held around 60°F, with a relative humidity of about 80°F. Here the cheese is turned frequently, and its surface is cleaned and rubbed with oil. When ready to be moved, the cheese usually is given a coating of black protective material, made of vegetable oils, lamp black, and Fuller's earth. The second period of ripening usually takes place in dealers' curing rooms, held at a temperature of 55°F and a relative humidity up to 90%. The cheese is very hard and is grated for use for soups, macaroni and other foods. A medium-cured cheese is not sliced, but is broken into pieces for table use.

The cheese made in the United States is cured for about 1½ years. A dried, powdered Parmesan cheese product is a commercial product. The cheese is emulsified and then spray-dried. It is used for sauces, dips and pizza blends.

Pickled Cheese

The so-called *Pickled* or *brine-ripened* cheeses are important varieties in the countries of southeastern Europe and the Near East, especially in Greece and the Balkan regions. They are a group of soft, bacteria-ripened cheeses that have been heavily rubbed with salt or soaked in brine or salted whey for one week to one month before they are ready for consumption. The salt acts as a preservative which permits the cheese to be held in the warm, prevailing climate. Typical cheeses of this type are *Feta* and *Teleme* of Greece, *Salamana* in Bulgaria, and *Brandza de Braila* in Rumania. The general method of manufacture is described under *Feta Cheese.*

Provolone

This is an Italian cheese now also made in the United States. It is made in various shapes, each having its own name. *Provolone* comes from the Neapolitan name *prova* for a round cheese. The suffix *-one* indicates the larger of two or more things; thus *Provolone* is a large cheese, *Provoletti*, a smaller one. The cheese is shaped into spheres, cylinders, sausages, and truncated cones. It may be molded in a grooved form or hung with cords which indent the cheese and form grooves as it hangs during curing and storage.

Provolone is a "pasta filata" cheese (plastic-curd). Calf's rennet is used to make a "sweet curd" cheese; lamb or kid's rennet for a "strong" cheese.

In Italy, Provolone is made by a method similar to that for making Caciocavallo (*q.v.*) but it usually contains more fat, and the cheese is smoked after it is salted and dried. In the United States, the cheese is made much like Cheddar, but rennet paste is used rather than rennet extract. After the whey is drained away, water at 180°F is added to the curd, and the curd is mixed until its temperature reaches 135-140°F. The elastic, stringy curd is then cut into pieces and pulled and worked as in making Caciocavallo cheese.

In Italy, up to 0.06% of hexamethylene amine is sometimes added to the water in which the curd is worked. This releases formaldehyde which lessens contamination of the cheese with undesirable organisms.

To meet the requirement of some markets, especially in the Eastern states, the curd is bleached with benzoyl peroxide. In this event, separation of the whey for use in butter manufacture may produce an off-flavor in the butter.

Romano

This cheese dates back to the Roman Empire and originated in the area around Rome. It may be made from the milk of cows, Vacchino Romano; from ewe's milk, Pecorino Romano; or from goat's milk, Caprino Romano.

Romano is similar to Parmesan and other Grana cheeses. The cheese has a high salt content and a characteristic sharp flavor due to enzyme activity during a ripening period of up to one year. The natural yellowish color may be bleached by the use of benzoyl peroxide or neutralized by the addition of blue or green food dye.

Roquefort Cheese

Roquefort is a semi-hard cheese, mold-ripened, and characterized by a mottled or marbled blue-green appearance. It originated in the village of Roquefort, about 70 miles west of Nîmes in southern France. Cheese of this type has been known for over 1000 years. The genuine cheese is made from ewe's milk and it is the only such cheese which is available commercially throughout the world. There are about 500 dairies, in the area of the Larzac hills and Les Causses in the region of Auvergne and to the south, to which ewes' milk is sent for the first stages of cheesemaking. The Locaine sheep of this area are bred for milk production and carry little wool. A ewe produces about 230 pounds of milk annually.

French law, dating to the Parliament of Toulouse in 1666, limits the use of the word *Roquefort* to *cheese made in the Roquefort area from ewe's milk;* other French Roquefort-type cheese is called *Bleu* cheese. The Roquefort name and red sheep trademark is protected by copyright laws and is zealously guarded by the manufacturers. Similar cheese, made in other countries, usually from cow's milk, is known as *Blue* cheese (*q.v.*). Other well-known blue-veined cheeses are the English *Stilton* and the Italian *Gorgonzola.*

Roquefort cheese is made from mixed morning and evening milk. Rennet made from lamb's stomach is preferred by the French cheese makers. The milk is heated to 125-140°F, and then enough cold milk is added to bring the temperature down to 85°-90°, at which point 2-3% of starter is added. Rapid formation of acid favors a firm curd and supresses the growth of undesirable organisms. Enough rennet is added to coagulate milk in one and one-half to two hours, after which the curd is cut and drained. Makers of Blue cheese usually add about 1% of salt to the curd at this time, but Roquefort cheese is not salted.

The curd is placed into hoops about 7½ inches in diameter and 6 inches high; three or four layers of curd are used and each layer is sprinkled with mold powder (see footnote for *Gorgonzola*) as it is put into the hoop.

The hoops are turned several times during the first week and then the cheese is removed for salting and curing. The surface of the cheese is sprinkled with salt two to four times over a period of a week to ten days. The heavy salting contributes to the flavor of the cheese and hinders the growth of undesirable microorganisms. After the salting period, the cheese is pierced with a large number of small holes which permit air to enter and allow carbon dioxide to escape. This promotes a favorable growth of mold in the interior of the cheese. The mold, Penicillum roqueforti, produces a water-soluble lipase which hydrolyzes the milkfat, liberating fatty acids, such as caproic, caprylic, and capric acids which impart a sharp, peppery flavor to the cheese. Ketone derivatives of the fatty acids also contribute to the flavor of mold-ripened cheese.[21]

Roquefort cheese is ripened in the natural or man-made caves of the Roquefort area as shown in Figure 19:8. Their temperature varies from 40 to 50°F, and the relative humidity is high, about 95%, owing to the presence of an underground lake. The ripening period lasts from two to ten months.

French cheeses, similar to Roquefort, include *Auvergne Bleu, Septmoncel, Gex,* and *Epoisses*.

Roquefort produces about 13,000 tons of Roquefort cheese each year.

Fig. 19:18. Vaults in Roquefort Where the Cheese Is Ripened (Courtesy of The Roquefort Association)

An interesting finding concerning Roquefort and Blue cheeses is that the surface of the cheese often contains an enzyme capable of inactivating penicillin.[16]

Sapsago

This cheese is also known as *Schabzieger, Krauter,* and *Glarus Green* cheese. It is a hard, gray-green cheese, shaped into *Stockli* or flat-topped cones. It has been made in Switzerland for about 500 years. A mixture of skim milk and buttermilk is allowed to sour and is then heated to boiling. The whey is drained off, the curd is placed in wooden troughs to cool, and then allowed to ripen over a period of several weeks. The dried curd is ground to a smooth paste and from 4 to 5% of salt is added as well as about 2.5% of dried, aromatic clover leaf. This plant, blue melilot, (*Melilotus Trigonella Coerulea*), is grown especially for this purpose, mostly in the Canton Schwyz. The cheese is packed tightly into molds, pressed, and dried until it is very hard. It usually is eaten in grated form, mixed with butter and spread on bread.

Stilton

Stilton is a semi-hard, blue-veined cheese, made of cow's milk. It was first made in Leicestershire, England, and probably received its name when it was offered for sale at the Bell Inn in Stilton, Huntingdonshire, about 200 years ago. It is now made in other parts of England and in New Zealand, but attempts to make a good quality of it elsewhere, including the United States, have not been successful.

Rich milk is used, which is ripened slightly by the addition of a starter. Enough rennet is added to coagulate the milk in a little over one hour. The curd is cut into slices and placed in cloth bags for draining. When the desired firmness is reached the curd is broken into small pieces and 1.5-2% of salt is added. The curd is poured into perforated, metal hoops which are stood on drain boards. The hoops are pressed very lightly and turned frequently over a period of one week. The cheeses are removed from the hoops, scraped to smooth the surface, bandaged, and placed on draining shelves. When the cheese is firm enough it is moved to a cool, well ventilated room where it is held until the surface is coated with mold. The bandage is next removed and the cheese is allowed to ripen, during which time mold grows in the curd.

Both Penicillium roqueforti and P. glaucum* have been identified

*Penicillium glaucum is believed to be identical to P. roqueforti by many authorities.

on Stilton cheese. Growth of mold is sometimes promoted by piercing the cheese with holes, but usually the light pressing keeps the pieces of curd from completely matting, thus making a space between the curd particles which admits enough air to maintain growth. The cheese has a wrinkled surface, free from cracks, and of a brownish color, derived from the dry molds and bacteria that grow on the surface. Mold starts to grow in the cheese in about six weeks but up to six months is needed to ripen it.

WHEY CHEESES

Nearly one-half of the solids present in the milk used for making cheese, which consist largely of the lactose, whey protein and mineral salts in the whey, are discarded in the cheese-making process. The fat in the whey is recovered by mechanical separation and used in the making of butter, or the whey cream may be used in cream cheese.

Cheese-like products made by coagulating the albumin in the whey with heat and acid are made in many countries.

Mysost

A cheese from whey, called *Mysost* (from *Myse*, meaning whey), is made in the Scandinavian countries; if made from the whey of goat's milk, it is called *Gjetost* (*gjie* meaning goat).

The whey is heated in a kettle and sometimes a starter, sour whey, or even some vinegar, is added. It is evaporated by boiling until about one-fourth of the original volume is left, at which point it forms a brown, caramelized paste. This is called *Prim* and the cheese sometimes is called *Primost* rather than Mysost. The pasty liquid is poured into a trough or tank and stirred as it cools in order to prevent the formation of large crystals of lactose. When cold, the product is poured into greased forms and packaged in metal foil.

Mysost that contains not less than 80% of total solids and not less than 33% fat in the solids is called *helfet* (whole fat), and sometimes is called *floteost*, from *flote*, meaning cream.

Ricotta

Ricotta, meaning "recooked", is a whey cheese which originated in Italy. Like other whey cheeses it is a by-product of the manufacture of other varieties of cheese. It is made from fresh whey, to which about 10% milk or skim milk may be added. The whey is heated to boiling and the coagulated albumin is dipped out and allowed to drain.

Sometimes about 1% of salt is added at this time. Ricotta often is made from milk, in which case it has the characteristic of richly creamed cottage cheese. The curd may be pressed and dried around 100°F. The dried product is suitable for grating.

The fat and moisture content varies widely according to the manner of manufacture. The fat content may range from less than 1% to about 20%; the moisture usually ranges between 70-80%. The protein content is relatively constant-around 11-13%.

Ziger is the German name for Ricotta-type cheese.

REFERENCES

1. Albright, J. L., Tuckey, S. and Words, G. T., Antibodies in Milk-A Review, *J. Dairy Sci.* 44:779 (1961).
2. Anon., First Practical Substitute for Animal Rennet, Food Process. Marketing, 28: 28:63 (1967).
3. Anon., Cheese; 150,000 lbs. a day, *Dairy and Ice Cream Field*, Jan. 1969.
4. Bottazzi, V. and Corradini, C., Grana cheese-Manufactured with milk containing formaldehyde, XVII Inter. Dairy Congress, D 121: Munich (1966).
5. *Cheese Varieties and Descriptions*, Handbook No. 54, U.S. Department of Agriculture, Washington, D.C. (1953).
6. *Cheeses, Processed Cheese Foods, Cheese Spreads and Related Foods: Definitions and Standards of Identity* (Revised); Federal Security Agency, Food and Drug Administration, Washington, D.C. 1962.

6a. Cottage Cheese and other Cultured Milk Products, Emmons, D. B. and Tuckey, S. L., Chas. Pfizer Co., N.Y. (1967).

7. Czulak, J.; Dairy Eng., 78:58 (1961), also Ann. Rep., Dairy Res. Inst., Highett, Vic., Aust., (1964).
8. Davis, J. G., Non-milk substances in cheese, *Dairy Ind.* 26:573-577 and 646-649 (1961).
   Ibid, *Cheese*, American Elsevier Pub. Co., Vol. 1, New York, 1965.
9. Forss, D. G., and Patton, G., Flavor of cheddar cheese, *J. Dairy Sci.*, 49-89, (1966).
10. Grill, H. Jr., Patton, G. and Cone, J. F., Aroma-Significance of sulfur compounds in surface-ripened cheese; *J. Dairy Sci.*, 49:409, (1966).
11. Harper, W. J., Swanson, A. M. and Sommer, H. H., Observation on the chemical composition of white particles in several lots of Cheddar cheese, *J. Dairy Sci.* 36:436 (1954).
12. Irvine, D. M., and Price, W. V., Process cheese Abstracts, Dept. Dairy and Food Ind., Univ. Wis., Madison, (1955).
13. Knight, S. G., U.S. Patent, 2,665, 990, Jan. (1954).
14. Kosikowski, F. V., Cheese Flavor, in *Chemistry of Natural Food Flavors: A Symposium*. Committee on Foods, Department of the Army, Quartermaster Food and Container Institute, Washington, D.C., May, 1957.
    Ibid, *Cheese and Fermented Milk Foods*, Published by the Author, Ithaca, N.Y., 1970.
15. Krishnaswamy, M. Johar, D., Subrahmanyan, V., and Thomas, S., Manufacture of Cheddar cheese with milk-clotting enzyme from *Ficus Carica* (vegetable rennet), *Food Technol.*, 15:482 (1961).

16. Ledford, R. A. and Kosikowski, F. V., Distribution and characterization of anti-penicillin activity in cheese, *J. Dairy Sci.*, 49:621 (1966).
17. Mann, E. J. Rennet substitutes, *Dairy Ind.*, 32:761 (1967).
18. McGilliwray, W. A., and King, D. W., "Cheddarmaster", New Zealand Mechanized Process, XVII Inter. Dairy Congress, D-109, Munich, (1966).
19. Morris, H. A., Jezeski, J. J., and Combs, W. B., The use of white mutants of *Penicillium roqueforti* in cheese making, *J. Dairy Sci.* 37:711 (1954).
20. Olson, N. F., Italian cheese, What effects yield and how to improve it., *Amer. Dairy Rev.*, August (1966).
21. Patton, S. The methyl ketones of blue cheese and their relation to flavor, *J. Dairy Sci.* 31:611 (1948).
22. Peters, I. I., and Nelson, F. E., *Milk Prod. J.*, 51:14, (1960).
23. Production of Manufactured Dairy Products, U.S. Dept. of Agri., Washington, D.C., June, 1973.
24. Pyne, G. T., Calcium salts and rennet coagulation of milk, *Chem. and Ind.*, 302-303; (1953).
25. Riddet, W., Whitehead, H. R., Robertson, P. S. Harkness, W. L., Fat oxidation in Cheddar cheese, *J. Dairy Res.*, 28:139 (1961).

chapter 20

Testing of Milk —Dairy Arithmetic—

Milk is one of the most valuable of all agricultural products; it is important that the producer obtains full value for his products, and that the processor does not lose in the preparation or manufacture of milk products. For this reason accurate testing is an important part of dairy work.

The various public health codes provide standards for the sanitary quality of milk and milk products in order to provide products that are wholesome, of low bacterial content, and free from pathogenic organisms. Routine tests are made to ensure conformance to these standards. The results of these tests are used by regulatory officials to correct improper production and processing, and also to serve as guides to improve the quality of the milk supply. The requirements upon which these standards are based have been discussed in previous chapters, especially Chapter 7. Detailed directions and reasons for the various tests are given in *Standard Methods for Examination of Dairy Products* and the *Milk Code.*[16, 24]

The legal standards which define the products in terms of their content of bacteria, fat, and solids-not-fat have been listed in previous chapters.

BACTERIOLOGICAL TESTS

The Standard Plate Count

The Standard Plate Count generally is used for the routine examination of milk by health departments and control laboratories.[24] The method is empirical, but if carefully done, it gives reliable results on milk that may contain either very few bacteria or millions in a milliliter.

In the method, a small, known volume of milk is mixed in a Petri dish with a jelly-like medium containing agar and nutrient materials that favor the growth of bacteria. Modifications, which compare in accuracy with the Standard Plate Count, may be used, especially with milk that has a bacterial count less than 200,000 per ml. These methods—the oval tube, micro-plate and roll tube procedures—are described later.

Various media, including the *Standard Methods Agar* are available commercially in dehydrated form. The composition of this agar is as follows:

TABLE 20:1
Standard Methods Agar

Pancreatic digest of casein (USP)	5 g
Yeast extract	2.5 g
Glucose	1 g
Agar, bacteriological grade	15 g
Distilled water to make	1 liter
Final *p*H	7.0 ± 0.2 at 25°C

The prepared medium melts when warmed and in this condition it may be mixed easily with the milk sample. The mixture soon solidifies again at room temperature. The covered Petri dish prevents the entry of bacteria other than those present in the milk sample. These are nourished, grow, and reproduce quickly. One bacterium is invisible to the naked eye, but in the medium the rate of growth and reproduction upon incubation is so great that vast numbers soon are present. Thus a group of bacteria growing in a fixed position on the medium forms a mass visible to the eye. Such a mass is called a *colony.* Each colony corresponds to one living bacterium or a group of bacteria originally present in the milk added to the medium. By counting the number of colonies formed, a measure is obtained of the number of bacteria present in the sample. In order to facilitate the counting, a magnifying lens usually is used in the examination of the Petri dishes.

Thorough agitation of the milk is necessary before it is sampled. If the milk is in a large container it is mixed with a sterile stirrer long enough to reach to the bottom of the vessel; if practical, by repeated forceful inversions of the container. Samples may be taken from well stirred cans or vats with a sterile metal tube. If the sample is not to be plated immediately, it is placed in crushed ice or otherwise held at a temperature between 32° and 40°F. All equipment used for the plate count, such as containers, dilution bottles, and Petri dishes, must be sterile to prevent contamination of the sample.

The bacteria in milk or cream usually occur in groups of two to six or more individuals. Immediately before being plated, the sample is shaken 25 times within seven seconds, each shake being an up-and-down movement of about one foot, in order to break apart any clumps of organisms and to distribute them uniformly.

At least two different dilutions of the sample are used, preferably 1:100 and 1:1000. One milliliter of sample added to 99 milliliters of sterile, phosphate-buffered distilled water gives a 1:100 dilution. After thorough shaking, 1 milliliter of this solution is added to another container with 9 milliliters of sterile water in order to make a 1:1000 dilution. If desired, 0.1 ml of the sample may be added to 99.9 ml of sterile dilution water for the 1:1000 dilution. If a low bacterial content is expected, plates of dilutions of 1:10 and 1:100 are prepared. Dilutions of the sample that will yield between 30 and 300 colonies are preferred, because they provide the most accurate results. Only distilled or demineralized water that has been tested and found free of bactericidal and inhibitory substances may be used for the preparation of culture media, reagents and dilution blanks.

The dilutions are shaken in the same manner as the original sample. Then 1 ml is transferred to an empty, sterile Petri dish, using a separate, sterile pipet for each sample and for each dilution. When measuring, the pipet is held at an angle of 45° against the Petri plate or neck of the dilution bottle. The pipet is allowed to drain and the tip touched once against a dry spot on the glass. The cover of the Petri dish is raised only high enough to insert the pipet.

The cover of the Petri dish is marked with a wax pencil or water-soluble crayon to indicate the number of the sample and the dilution used. A convenient way to do this is to place the number of the sample near the bottom of the cover to indicate a dilution of 1:10; in the middle of the cover for 1:100; and near the top for a 1:1000.[8] In another method, a key number is used. Thus "0" indicates no dilution, "1" indicates a 1-10 dilution, "2" a 1-100 dilution, and so on.

From 10 to 12 ml of melted agar medium at 44°-46°C are poured into each plate. An excess of agar favors a spreading growth of colonies on the surface. The interval between the first transfer of the sample and the addition of the agar to the last plate in a series should not exceed 20 minutes. The cover of the dish is lifted only high enough to pour the agar. The sample and the agar are mixed thoroughly by rotating and tilting the dish carefully.

As soon as the agar has solidified, the plates are inverted and placed in an incubator for 48±3 hours at a temperature of 32°C± 1°C. This temperature is preferred because it favors the growth of the bacteria commonly found in milk. Keep incubators in rooms where

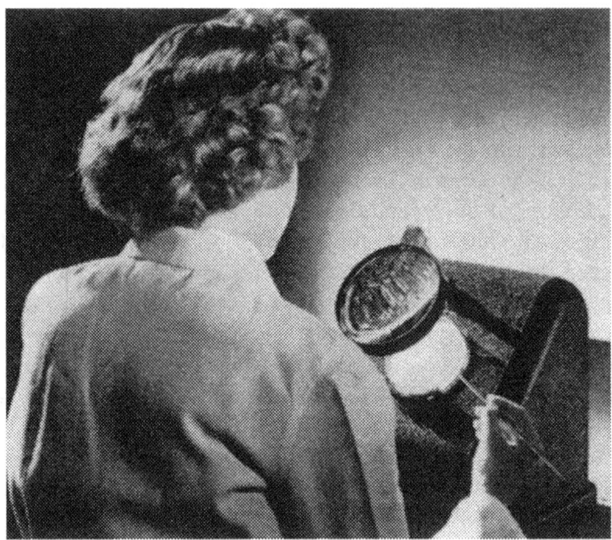

Fig. 20:1. Colony Counter (Quebec Dark Field): Petri dish is placed on illuminated guide glass. Markings on guide enable operator to estimate number of bacteria present on crowded plates by counting a few areas and multiplying by a factor. Operator is removing a colony with sterilized platinum wire loop for further examination. (Courtesy of American Optical Company)

temperatures are within the range of 60.8-80.6°F (16-27°C). After incubation, the colonies on the plate are counted, preferably with a Quebec-type colony counter, and multiplied by the dilution used. This gives the *Standard Plate Count* per ml of sample. If the number of colonies exceeds 300, a fraction of the plate may be counted. Plates with fewer than 30 colonies should not be counted, unless they are the only ones available. In cases of doubt, a microscope should be used to distinguish between colonies and foreign material (Fig. 20:1).

A number of automated procedures have been devised to count the colonies on a petri plate. The Fisher Automatic Colony Counter counts up to 9999 colonies in one-half second. The count appears on a 4-digit display and the contents of the dish are shown on a remote TV monitor (Fig. 20:2).

Counts from agar plates give the *estimated* number of colonies that would have developed if an entire milliliter of milk had been examined under the specified conditions, and if all the bacteria had developed into colonies. As an unknown ratio exists between the colony count and the total number of bacteria, it is incorrect to say that the count shows the number of bacteria per milliliter. The results should be reported as *Standard Plate Count per milliliter*. Only the

Fig. 20:2. Fisher Automatic Colony Counter Courtesy, Fisher Scientific Company, Pittsburgh, Pa.

two significant left-handed digits are used in reporting the count, raising the second digit to the next highest number if the third digit from the left is 5 or more. Thus a colony count of 225 in a 1:100 dilution plate is reported as a *Standard Plate Count of 23,000 per milliliter* not 22,500.

If the colonies per plate appreciably exceed 300, and an estimate of the total number present is made from the count of a representative area, the actual estimate is recorded and not reported as "too numerous to count" (TNTC). Some explanatory term other than the word "standard" is used with the count.

A series of at least four or more samples should be examined before the quality of a given milk supply is judged. The average is determined logarithmically rather than arithmetically. By averaging the logarithms of the counts, an occasional high count has less effect in reducing the quality rating of the milk. If all the counts are high, the logarithmic average approaches the arithmetic one.

Sometimes bacteria are present that do not grow well at the temperature or on the particular medium used, in which case the colonies are small and hard to see. These are called *pin-point* colonies. Another cause of pin-point colonies is the overcrowding of the plates, owing to the presence of so many bacteria that in their competition for food the colonies do not develop well. Obviously the latter condition can be prevented by plating a higher dilution of the sample, yielding fewer colonies. Pasteurized milk may contain thermophilic organisms which form pin-point colonies because they grow best at 40° to 55°C, rather than at 32°C.

The estimate of the number of thermophilic organisms is made as for the standard plate count except that 15-18 ml. of agar is used and the plates are incubated at 55°C±1°. The estimate for psychrophilic bacteria is made as for the standard plate count except that the plates are incubated for 10 days at 7°C±1°.

A comparison of counts made from milk before and after it is pasteurized will give information concerning the ability of the organisms to resist the action of heat. The decrease in the number of bacteria as shown by such counts is a measure of the efficiency of the pasteurization process. Provided that the milk had not become contaminated after pasteurization, a higher count in it than in the raw milk indicates the presence of thermophilic organisms. This is not a common occurrence. Usually about 96 to 99% of the bacteria in raw milk are destroyed by pasteurization.

Some authorities believe that sufficient information concerning the sanitary status of a milk supply may be obtained from the reductase test and an examination for coliform organisms, combined, in the case of pasteurized milk, with a phosphatase test. A common objection to the agar plate count is that it requires much equipiment and results are not obtained until after two days. In Great Britain the plate count has been abolished for the official examination of pasteurized milk and succeeded by the phosphatase test and a modified reductase test.[25] The claim is made that a wide margin of error appears inevitable in arriving at the count and, more particularly, the test takes account of heat-resisting organisms whose presence is of no material significance for the safety or keeping quality of the milk.

Simplified Viable Count

As the Standard Plate Count is time-consuming and expensive, several modifications have been proposed and adopted as official procedures for use on Grade A Raw Milk for pasteurization. The tube methods may be considered refinements of the procedure devised by Burri in 1928.

Oval Tube:[24]

A 0.001 ml or 0.01 ml. sample of milk is transferred by means of a sterile, standardized loop to an oval tube, containing 4 ml of melted medium. A one ounce bottle and 3.5 ml. of agar may be used. After closing with a cotton plug or a stainless steel closure, the agar and milk are mixed by swinging the tube back and forth for about five seconds. The tube is then placed in a flat or slightly tilted position until the agar has solidified. The tube is then placed horizontally, with the agar adhering to the upper side of the tube in an incubator at 32°C for 48 hours. The colonies are counted over a Quebec colony counter and the count per ml is computed. This method may be used for making a count of yeasts and molds in dairy products.

Roll Tube:

In this method milk, or milk diluted in phosphate-buffered water, is used so that it will provide between 10 and 200 colonies to the tube. One-half ml of milk or dilution is placed into a tube containing Standard Methods Agar (made with 2.5% agar) cooled to 44-45°C. After inoculation, the tube is closed and placed in the apparatus, which rolls the tube and directs a spray of cold water to the lower surface. In about two minutes, the agar is firm and the tube is placed in the incubator for 48 hours at either standard temperature. The tube is then placed in the illuminated counting chamber and marked so that one-fourth of its length is counted at a time. The tube is rotated until all segments are counted. As 0.5 ml of milk or dilution is used, the colony count is multiplied by two and by the dilution used in order to obtain the Roll Tube Count per ml.

This method has become popular in Great Britain and Canada, where it is known as the Astell Roll-Tube System. It is not an official method in the United States.

Micro-Plate:

This procedure with a micro-plate is an improvement of the Frost Little Plate Method.[5] One ml of milk is mixed with 2 ml of melted Standard Methods Agar, both at 45°C. Glass slides, marked with circular areas covering 1 sq cm are sterilized in a flame and placed on a slide dryer held at 45°C. With a sterile pipet, 0.03 ml of agar-milk mixture is spread evenly over an area of 1 sq cm. The slide then is placed in a moist chamber held at 32° or 35°C for 12 to 20 hours. After incubation, the slides are dried at just under 100°C, stained with thionine, and then washed with water. After drying rapidly on a slide dryer, the colonies are counted with a microscope.

Advantages claimed for this procedure are that it is relatively simple, that little equipment is needed, and that the results may be had in less than one-half the time needed for the Standard Plate Method. It is not a Standard Methods procedure.

Direct Microscopic Count[24]

The number of bacteria and somatic cells in milk may be estimated in an actual count by means of a microscope. The method is called the *Direct Microscopic Count* or the *Breed Count*, after Dr. R. S. Breed, who developed the procedure.

A measured volume of milk (0.01 ml) is spread over an area of 1 sq cm on a glass slide, dried, and stained. Bacteria and body cells (leucocytes) are stained selectively whereas the milk solids are relatively unstained. The stained slide is examined under a microscope. The estimate is most accurate with high-count milk and decreases as the count level declines.

The examination of a few fields can indicate the quality of the milk—poor milk will show many clumps of bacteria per field whereas high-quality milk shows few or no bacteria per field.

The use of a calibrated wire loop is no longer permitted by the Standard Methods. Either a pipette or a metal syringe, calibrated to deliver 0.01 ml. of sample, may be used. A needle with a bent point is used to spread the milk or cream over the square centimeter. The film is dried at a temperature of 40°-45°C. Excess heat may cause the film to crack or peel.

The film is stained by dipping for two minutes into the *Levowitz-Weber* Stain, which is a modification of the *Newman-Lampert* Stain. The essential difference between these stains is that only 0.6 grams of certified methylene blue is used instead of one gram as in the latter stain. The dye is dissolved in a mixture of 52 ml. of 95% ethyl alcohol and 44 ml. of tetrachlorethane. After it is dissolved let stand for 12-24 hours., then add 4 ml, of glacial acetic acid. The mixture is filtered through a fine filter paper and stored in a tightly closed bottle.

The stained slides are examined under a microscope so adjusted that each field covers a known area of the smear. When the area is known the number of bacteria per milliliter of milk can be calculated. Microscopes are available that are especially adjusted for the direct microscopic count. If the microscopic field measures 0.160 mm in diameter the average clump count per field is multiplied by 500,000 to obtain the number per ml. of sample. When the microscopic field has a diameter of 0.146 mm a factor of 600,000 is used to obtain the count per ml. of sample. In practice, more than a single field is counted, and the average number of clumps in a field is used to determine the count. As shown in the following tabulation, more fields are counted when the milk is of low bacterial content than when many bacteria are present.

Range of individual microscopic counts per field (0.160 mm diameter.)	Number of fields to be counted
Under 300,C00	50
300,000 to 3 million	20
Over 3 million	10

Relation Between Direct and Plate Counts

In the agar plate count, only living bacteria are counted because these are the only ones that can grow. In the microscopic count it is possible that some dead organisms are stained and counted, even though they do not stain so well as living bacteria. In the plate method, a group or cluster of bacteria may develop into a single colony and be counted as if the colony originated from a single bacterium. For reasons such as these, the direct count usually gives results that are from three to four times as great as those obtained by the agar plate method, but no exact relationship exists.

Unlike the plate count, the direct microscopic count is, for the most part, limited to the examination of milk that contains more than about 50,000 bacteria per ml because such a small sample is used that it may otherwise be difficult to find enough bacteria to count. The method is not useful for the examination of pasteurized milk. Such milk may contain relatively few viable bacteria and the killed organisms present may be counted because they stain well for several hours after pasteurization.

The presence of leucocytes and other body cells, which would not be apparent from a plate count, may indicate an abnormal condition of the udder (see Chapter 7). Long chains of streptococci usually are seen in milk from cows suffering from an infected udder.

Recognition of the type of organism present may give valuable information concerning the production and handling of the milk. Large masses of bacteria, especially cocci in clumps of varying sizes, usually denote milk that was handled in improperly cleaned and sanitized equipment. Rod-shaped organisms indicate contamination with soil, manure, or barn dust. If the milk were improperly cooled, short chains of streptococci, usually *S. lactis* or *S. cremoris*, may be found.

In some areas, microscopic counts are made on pasteurized milk offered for sale at retail outlets. Present day practices in processing and distribution have increased the shelf-life of milk. If the milk is not properly handled some organisms, such as psychrotrophic (psychrophilic) bacteria may grow rapidly in the product and result in decreased quality of the milk. Regulatory officials find the microscopic count useful for the examination of such milk.

The microscopic count may be used to examine milk for the presence of thermophilic organisms. An increase in the number of rod-shaped organisms in milk pasteurized by the holding method (145°F for 30 minutes) indicates contamination with thermophiles.

Dye Reduction Tests

Reduction tests with methylene blue or resazurin, are used to check the bacteriological quality and to grade raw milk. The reduction method, using either dye, does not give a measure of the number of bacteria in the milk, but it does indicate the degree of activity of certain types of bacteria and thus it permits classifying the milk as acceptable or non-acceptable for certain grades or uses. As these organisms grow in the milk they consume the oxygen present and by lowering the oxidation-reduction potential they create a condition that causes an indicator dye to change color. The time needed to change the color or reduce the dye is, roughly, inversely proportional to the number of organisms present. Large numbers of *Streptococcus lactis* cause a rapid reduction, but most thermoduric and thermophilic organisms are relatively inactive.

Methylene Blue Reduction Test:

In this test, 10 ml of milk is mixed in a sterile, stoppered test tube, with one ml of a solution of methylene blue thiocyanate of such strength that the final concentration is one part of dye to 250,000 of milk. Tablets containing 8.8 mg. of the certified dye are available. One tablet is dissolved in 200 ml. of sterile, distilled water to prepare the dye solution. The dye and milk mixture is placed in a water bath held at 35°-37°C.

As soon as the sample tube has reached the incubation temperature it is inverted gently three times in order to distribute the cream. After 30 minutes, the sample is examined, and if there is no change in color it is examined at hourly intervals thereafter. If the methylene blue has been reduced to the colorless form within 30 minutes the sample is recorded as having a *methylene blue reduction time* of 30 minutes. If it is not reduced, the tube is again given one complete inversion and examined after one hour. If the color disappears between the one-half hour and the 1.5 hours examination, the result is recorded as a *one-hour reduction time*, and so on for each succeeding hour.

By inverting the tube after each reading, the bacteria are carried into the cream layer and redistributed in the milk. This inversion must be done gently in order to minimize incorporation of the oxygen present in the air in the tube. The surface of the milk in contact with the air remains blue, so, for the purpose of the test, the sample is considered as decolorized when four-fifths of the visible portion is white.

Sometimes milk is graded by the methylene reduction test according to the time intervals indicated in the table that follows. As the table shows, the test is of little value for the examination of pasteurized milk or of high quality milk because they contain much fewer than 500,000 bacteria per milliliter. It is useful for the examination of manufacturing milk, especially that used in the manufacture of cheese and evaporated milk. The classification used by the Evaporated Milk Association and the American Dry Milk Institute for milk by the methylene blue and by the resazurin tests is given in Chapter 15.

Factors that limit the usefulness of the test include:

1. Different rates of consuming oxygen by different species of bacteria.
2. Milk constituents, such as body cells, which may consume oxygen and influence the reduction time when few bacteria are present.
3. The germicidal or bactericidal effect of raw milk, which is not equal for all species of bacteria.
4. Colostrum, mastitis milk, and milk from cows late in lactation, which have shorter reduction times than their bacterial content would indicate.

Resazurin Reduction Method:

The continuing improvement in the milk supply, with the corresponding decrease in its bacterial content, makes the methylene blue reduction time for such milk less suitable for routine work. The indicator *resazurin* is now widely used because it gives results within three hours and also permits the detection of colostrum and of milk from the diseased udders.

Resazurin gives a blue color to fresh milk; as it is reduced it goes through various shades of purple until a distinct pink color is developed. The pink compound is known as *resorufin*, which, unlike methylene blue, is not restored to its original blue color when the milk is in contact with atmospheric oxygen. Upon continued reduction, resorufin is converted to colorless *hydroresorufin*, which may be oxidized to resorufin. The resazurin reduction test usually used is the three hour, or "triple reading," test in which the color in the test samples is compared with a standard lavendar color, midway between the initial blue and the pink stages (Munsell color standard 5P 7/4). The colors are compared at three-hour intervals.

The test is conducted in the same manner as the methylene blue reduction test except that a resazurin solution made with certified tablets is used instead of methylene blue. After the end of the first hour of incubation, any tubes of milk that show the lavender color are removed and the result is recorded. This is repeated with the remaining tubes at the end of the second and the third hour of incubation.

The time required to reach the lavender or 5P 7/4 end point is approximately one-half that of methylene blue on identical samples. Resazurin is sensitive to the reducing action of leucocytes, body cells, and associated substances and so milk containing colostrum or from diseased udders is detected more readily by this test than by reduction with methylene blue. Generally, milk having a reduction time over 2 hours is regarded as ''acceptable'' for milk shipped in cans and over 3 hours for bulk milk.

Sometimes a ''One Hour Test'' is used. According to this test, milk that retains the initial bluish color is put in Class I; for Classes II and III, the milk assumes a lavendar to pink color. Milk that is decolorized within 1½ hours is undergrade and given probational status.

In England a ''Clot-on-Boiling Test'' has replaced the methylene blue test as a measure of keeping quality.[25] The milk sample is refrigerated over night and then held at room temperature for seven hours. It is then incubated at 22°C. for 17.5 to 23.5 hours, depending upon the room temperature. The warmer the weather, the shorter the time of incubation. Five ml. of the milk is placed in a test tube and put in a bath of boiling water for five minutes. The tube is then removed and tilted to expose a thin film of milk. If clots or particles of milk adhere to the side of the tube, the milk is regarded as unsatisfactory.

A unique procedure to estimate the amount of microbiological contamination in a product depends upon the measure of the light output caused when there is introduced into the sample a luciferin-luciferase reagent obtained from fireflies. The reaction is based upon the assumption that living cells contain adenosine triphosphate (ATP) which reacts to produce the light. The procedure has not been used for milk products.

Somatic Cell Estimates

A procedure for a rapid and accurate estimation of the somatic cells in milk involves the use of membrane filtration and the colorimetric estimation of the DNA (deoxyribonucleic acid) present.[31]

Numerous procedures have been proposed for the somatic cell estimate. The ones generally used are the California Mastitis Tests (CMT), Wisconsin Mastitis Test (WMT), and the Direct Microscopic Count.[24]

The Technicon Cell Counter is an electronic devise for automatic counting of somatic cells in milk. The principle of operation is shown in Fig.20:3. When no particular matter is present in the sample, the light beam passes through and is blocked when it strikes the dark

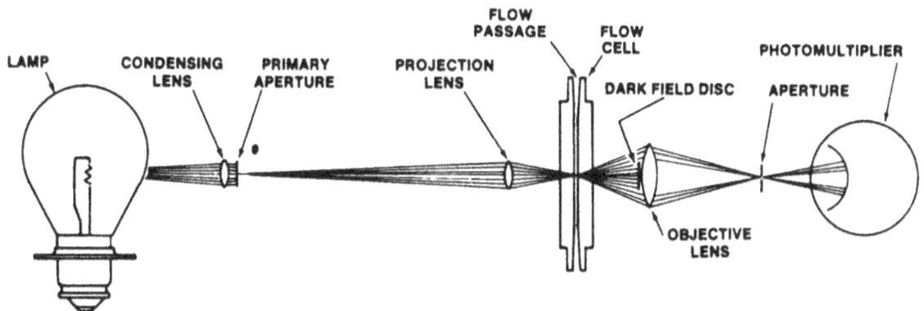

Fig. 20:3. Principles of Operation of the Technicon Cell Counter.

field disc. Any suspended particle interrupts the light beam and scatters the light so that there is a flash of light around the dark field disc. This flash, due to light bouncing off the particles is focused on the photomultiplier tube where it generates an electrical pulse for each interrupting particle. The pulses may be counted or displayed on an oscilloscope.

CHEMICAL TESTS

Sampling for Chemical Examination

An essential requirement for any accurate test is to obtain a sample that is representative of the material under consideration. Careful testing in itself will not make up for inaccuracies in sampling.

In the milk plant it is impossible to obtain a representative sample of the milk in a large vat or weigh-tank unless the milk is first thoroughly mixed, preferably by means of properly placed mechanical agitators. Fresh milk from small containers should be sampled immediately after the milk has been thoroughly mixed with a suitable agitator, or has been poured from one container to another several times.

When it is not practical to test each lot of milk, *composite samples* may be taken. A composite sample is a mixture of single samples taken from different lots of milk or cream. They must be *representative of and in proportion to* the amount of product sampled. For example, if a dairyman delivers 50 pounds of milk one day, 100 pounds the next and 75 pounds the third day, the amount of milk in the sample from each shipment must be in proportion to its weight. Thus, if one dipper of milk was taken as the sample for the first day's shipment, two dippers must be taken the next day and one and one-half dippers for the third shipment. A further explanation of this reasoning is given in the section on dairy arithmetic for computing the weighted average of milk fat in different lots of milk.

If the milk in a composite sample is gathered over a period of time, it is necessary to preserve it lest it spoil before it is tested. This is done by adding to the milk a special preservative tablet containing mercuric chloride or by adding to each ounce of milk a drop or two of a solution of formaldehyde. Experience has shown that the most reliable fat tests are obtained with composite samples that have been preserved with formaldehyde and are not over two weeks old.

The Babcock Fat Test[3]

The dairyman usually is paid for the milk he sells on the basis of its fat content. Sometimes an allowance is made for its solids-not-fat, especially if the milk is to be used in the manufacture of cheese or dry milk. In California, all milk priced at the producer's level must be paid for on its content of fat and solids-not-fat.

The Babcock Fat Test generally is used in creameries in the United States for the determination of the fat content of milk and cream. Probably no other single chemical test of any kind is used so extensively to place a value upon a commodity. A constant error of as

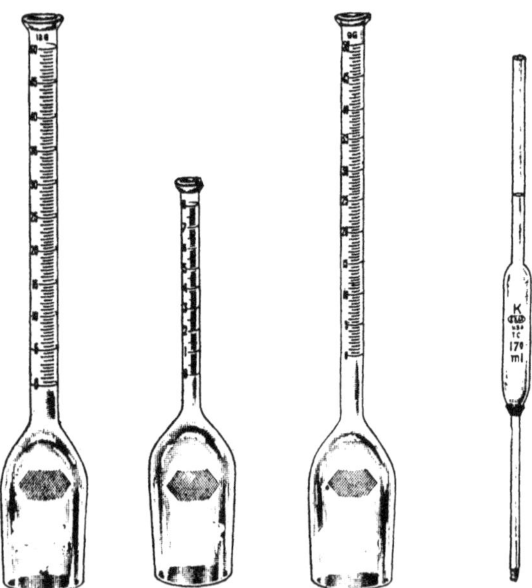

Fig. 20:4. Glassware for Babcock Test: 1. Cream Test Bottle for 18 grams of cream (one-third actual size) 2. Milk Test Bottle (one-third actual size) 3. Cream Test Bottle for 0 grams of cream (one-third actual size) 4. Milk Pipet (one-fifth actual size)

little as 0.1%, one way or the other, in the results of the fat test would mean a loss of millions of dollars to the milk producers of the country or to the creamery operators.

The Babcock test was developed about 1890 by Professor S. M. Babcock of the University of Wisconsin. The procedure is as follows:

Place 17.6 milliliters of sample in Babcock milk test bottle by means of a Babcock pipet (Fig. 20:4). The pipet is calibrated to contain 17.6 ml. but actually delivers 17.5 ml. Insert the pipet in the neck of the bottle, allow the milk to drain into the bulb of the bottle and remove the final drop by blowing into the pipet. Add 17.6 milliliter of sulfuric acid (specific gravity 1.84). Mix the acid and milk by a rotary motion until all traces of undissolved curd have disappeared. Hold the bottle with its neck away from the operator in order to avoid any accidental spatter of acid. Place the test bottles in the centrifuge, taking care that the load is properly balanced. The centrifuge should be heated to 140°F. The correct speed of the centrifuge depends upon its diameter, measured with the carrier cups horizontally extended. The correct speed is as follows:

Diameter of wheel, inches—	14	16	18	20	22	24
Revolutions per minute—	909	848	800	759	724	693

Whirl the test bottles for 5 minutes, then add sufficient hot water (140°F or over) to fill the bulb of the bottle. Whirl another 2 minutes and then add hot water until the fat column approaches the top graduation on the bottle. Whirl one minute more. Transfer the bottle to a water bath maintained at 130-140°F for at least three minutes. The water level in the bath should reach to the top of the column of fat.

Measure the percentage of fat with a pair of sharp-pointed dividers, holding the bottle so that the fat column is at eye level. Place one point of the dividers at the extreme bottom of the fat column, the other at the top of the meniscus.* Then without changing the spread of the dividers, place the lower point on the 0 graduation mark on the bottle; the other point then indicates the percentage of fat in the sample (Fig. 20:5).

The fat column should be read in a diffused or indirect light, such as is obtained from a north window. By proper adjustment of the light, or by seeking the proper place in the room from where to make the reading, the top of the meniscus is easily seen. It will then appear

*The meniscus is the crescent-shape portion on top of the fat column, caused by the attraction of the fat and the glass; the fat rises higher where it touches the glass than in the center of the column where there is no such attraction.

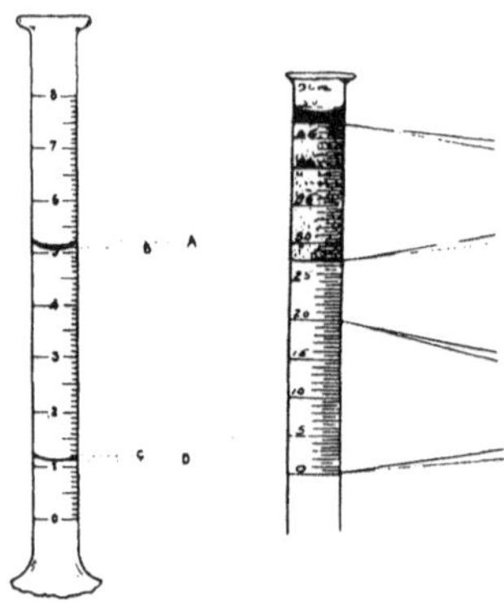

Fig. 20:5. Left, Milk test: Meniscus is the dark line, A-B. Test is read by measuring height of column of fat, A-D, with a pair of dividers. With dividers at same setting, lower arm is placed at zero on graduated neck; upper arm then points to percentage of fat in sample (see figure at right). Right, Cream Test: Colored mineral oil (glymol) is added to fat to destroy the meniscus.

like the letter *D* lying on its side, the top of the meniscus itself appearing as a straight, dark line.

As milk has the average specific gravity of 1.03, the sample delivered by the milk test pipet weighs 17.5x1.03 or 18 grams. The neck of the test bottle is so calibrated that each whole percent represents a volume of 0.2 milliliter. The capacity of the neck of a milk test bottle graduated from 0 to 8% is 1.6 milliliters. The melted fat obtained in the test has a specific gravity of 0.9. A milk sample that contains 4% of fat yields 18x0.04, or 0.72 gram of fat. This fat occupies 0.72÷0.9 or 0.8 milliliter. As each percent of graduated neck of the test bottle occupies 0.2 milliliter, a fat column that occupies 0.8 milliliter represents 0.8÷0.2, or 4% of fat in the milk sample.

A trace of fat is destroyed in the test and some of the fat globules in the milk are so small that they do not rise into the neck of the test bottle. In order to obtain an accurate test on milk, this loss is compensated for by reading the fat column from its lowest point to the extreme top of the meniscus.

Some lots of acid of the proper specific gravity may not give a satisfactory test. In such cases, dilute the acid with sufficient water so that a fat test is obtained in which the fat column is translucent golden yellow or amber in color, free from charred matter or other visible particles. (Note: In diluting sulfuric acid, always add the acid to the water, never water to acid.)

At the time of testing, the temperature of the prepared milk sample should be between 60° and 70°F, and the temperature of the acid should not exceed 70°F.

Dark-colored fat columns or those that contain charred matter are caused by (1) too strong acid, (2) too much acid, (3) milk or acid too warm when mixed, (4) allowing the milk and acid to stand in the bottle before mixing, or (5) by improper or inadequate shaking to mix the milk and acid.

Pale-colored fat columns, often underlaid with white sediment, are caused by (1) weak acid, (2) too little acid, (3) acid or milk too cold, or (4) insufficient mixing.

It is well to adjust the amount of acid added to the milk according to the color of the fat column obtained. Formaldehyde has a hardening action on the curd of milk, and milk so preserved must be mixed with the acid for at least three minutes in order to ensure complete dissolution.

Homogenized milk may be tested in the same manner as ordinary milk, except that the sample is mixed for at least five minutes after adding the acid.

Skim milk and buttermilk may be tested by the Babcock method, provided a special test bottle is used. Owing to the small amount of fat in these products, this test bottle has a very small graduated neck in which the fat is measured.

A modification of the test, known as the *butyl alchohol test* is used. For this test place 2 milliliters of butyl alcohol in a Babcock test skim milk bottle, add 9 milliliters of skim milk or buttermilk and then add 7 milliliters of Babcock test sulfuric acid. Mix thoroughly, centrifuge for 6 minutes, and complete the test as for the Babcock test for milk. Double the fat reading to obtain the percentage of fat.

As cream varies in thickness, it cannot be measured satisfactorily by volume as is done with milk. Cream is weighed in to the test bottle, using either 9 or 18 grams of sample, depending upon the size of test bottle used. The cream test bottle has a larger neck than the milk test bottle because of the higher fat content of cream. Otherwise, cream is tested practically by the same procedure used for milk. Less acid is needed in the test, because of the lower solids-not-fat content in cream. A small amount of water is added to the test bottle, in order to wash down any cream adhering to the inside of the neck of the bottle, as well as to dilute the acid used for the test. Enough acid is used to give the mixture of cream and acid a chocolate-brown color.

Experience has shown that all the fat in cream appears in the fat column and in order to obtain an accurate test, the meniscus must not be included in the reading. In a test of cream, the meniscus is destroyed, just before reading the fat column, by allowing a small amount of colored mineral oil, called *glymol*, to run down the neck of the bottle and rest on top of the fat column. This levels off the meniscus, and the reading is taken at the dividing line between the milk fat and the added mineral oil (Fig. 20:5).

The Gerber or Fucoma Fat Test[24]

This test, devised by the Swiss chemist N. Gerber, is used in Europe and to some extent in this country. Like the Babcock test, it makes use of chemicals and a centrifuge to separate the fat from the sample. In addition to sulfuric acid, isoamyl alcohol is used to hasten the action of the acid and to prevent the formation of charred matter which may occur with acid alone. Special test bottles, called *butyrometers*, are used as well as special pipets and equipment for adding the acid and alcohol to the test bottles (Fig. 20:6). A special centrifuge is used, which, unlike that used for the Babcock test, need not be heated in order to obtain an accurate test. The Gerber test is probably somewhat quicker than the Babcock test and when conducted with care, it is equally accurate. Either test gives results with milk that average within 0.04% of those obtained when the fat is extracted with ether.

The procedure for the Gerber test for fat in milk is as follows:

1. Place 10 milliliters sulfuric acid (Babcock test grade), sp. gr. 1.82, in the butyrometer.
2. Measure, with the special pipet, 11 milliliters of milk into the butyrometer, allowing it to run down the side of the bottle, and form a layer above of the acid.
3. Add 1 milliliter of isoamyl alcohol.
4. Insert the stopper and thoroughly mix until the curd is dissolved.

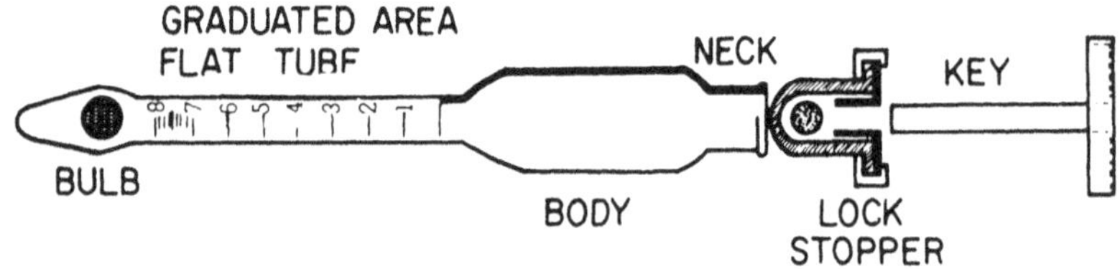

Fig. 20:6. Gerber Bottle: Gerber milk-test bottle, lock stopper and key. (Approximately one-half size)

5. Place the bottles in the centrifuge with the neck toward the center.
6. Centrifuge 4 minutes at 1000 revolutions a minute.
7. Remove the bottles from the centrifuge and place for 5 minutes in a water bath heated to 140°F.
8. Read the fat percentage by subtracting the reading of the lower flat portion of the fat column from the *bottom* of the upper meniscus. By raising or lowering the stopper in the butyrometer, the fat column can be brought to a position to facilitate the reading.

By using special butyrometers and making modifications in the procedure, the Gerber test may be used for estimating fat in cream, skim milk, buttermilk, ice cream, butter, and cheese.

Combined Acid and Detergent Test for Milkfat

The author has used a modified Babcock test for the routine examination of ice cream, ice milk, cottage cheese, chocolate flavored milk, and some varieties of hard cheeses, such as Cheddar. This test makes use of a cationic detergent, glacial acetic acid, and sulfuric acid.

The special reagent used is glacial acetic acid containing 5 grams of a quaternary ammonium compound, an alkyl dimethylbenzylammomium chloride, in one liter of the acid. The quaternary ammonium compound is available under many trade names, such as Roccal, BTC, and Hyamine 3500. If a 10%-solution of the quaternary ammonium compound is used, 50 ml of this is added to one liter of glacial acetic acid. Probably other cationic deterents may be used.

Nine grams of ice cream or ice milk is weighed into an 18-gram Babcock milk test bottle. Ten ml of the acetic acid reagent is added and mixed well with the sample. Next, about 9 ml of regular milk test Babcock sulfuric acid is added and mixed with the sample. The test is completed as for the Babcock milk test. A drop of glymol is added to the fat column before reading the test. The percentage of fat is found by multiplying the fat reading by 2 as a 9-gram sample is used in an 18-gram test bottle.

Cottage cheese is tested by weighing 9 grams of well mixed sample into a 50-ml beaker; 4.5 grams of Cheddar cheese is taken for the test. Ten ml of the glacial acetic reagent is added and the cheese is dissolved by heating on a hot plate. A small glass rod is used to stir the sample. Transfer the dissolved cheese into a Babcock milk test bottle and wash the remaining contents of the beaker into the test bottle with a total of 9 ml of Babcock test sulfuric acid, added in separate small portions. Complete the test in the regular manner, multiplying the reading for fat by 2 to obtain the percentage of fat in Cottage cheese, and by 4 in the case of Cheddar cheese.

The cationic detergent promotes the dispersion of the milk protein in the acid and any sugar or cocoa present forms little or no charred material that rises into the fat column.[29, 30]

Detergents Tests[3]

Many attempts have been made to modify the Babcock test in order to make it quicker and more accurate, or to eliminate the use of sulfuric acid. None of the proposed tests has been universally successful. In 1949, a method was described for the use of detergents in making the fat test. This method did not prove accurate and numerous modifications have since been suggested. Modified detergent tests include the BDI test and the TESA or Banco test.

The author has found that the fat obtained by the BDI procedure is satisfactory to use for the determination of chlorinated hydrocarbon pesticide residues in milk products. The procedure is much quicker and less costly than other extraction methods.[9, 14]

Ether Extraction (Mojonnier Method)

Even though it does yield acceptable results for milk and cream, the Babcock test is not sufficiently accurate for the exact analysis of evaporated milk, ice cream, cheese, and dry milk. An exact determination of the fat is a chemical procedure that demands an ether extraction. The Mojonnier test is a widely used mechanical adaptation of the classical Roese-Gottlieb procedure for ether extraction.[3]

An accurately weighed sample is dissolved in water. Ammonia is added to dissolve the milk proteins, and then alcohol is added to prevent the formation of a thick emulsion, which might occur when the mixture is shaken. The fat is extracted from the solution by shaking with ethyl ether. This ether dissolves a small amount of the aqueous solution also and therefore contains a small amount of non-fat material, such as milk sugar. In order to remove these materials from the ethyl ether an equal amount of petroleum ether is added. The ethers, being lighter than the other liquids present, rise to the top of the mixture, carrying the fat in solution. The ether layer is transferred to an accurately weighed aluminum dish which is placed on an electric hot plate to remove the ethers by evaporation. The vapors are removed by means of a suction fat. The extracted fat remains in the dish which is dried in a vacuum oven and then is cooled in a desiccator. From the weight of the fat in the dish, the percentage of fat in the sample may be calculated.

A modification of the Mojonnier equipment, intended to increase safety as well as better temperature control and more accurate testing has been designed by the Kraft Foods Company of Chicago, Illinois.[24]

Automated Testing

A number of photometric devices have been made to analyze milk. The electronic equipment called the Infra-red Milk Analyzer or IRMA was developed by Dr. John Gould of the National Institute for Dairy Research, Shinfield, England. The equipment is made by Grubb Parsons, Newcastle upon Tyne, England and is also obtainable from the A. Reyrolle Co., Toronto, Canada. It will analyze milk for its fat, protein and lactose content. A beam of infra-red light is passed through a thin layer of milk. The various components absorb different wavelengths of the light and the absorption is measured by a double beam spectrophotometer. Fat absorbs infra-red energy at a wavelength of 5.75 microns, protein at 6.46 microns and lactose at 9.60 microns. The infra-red absorption by water is corrected for by

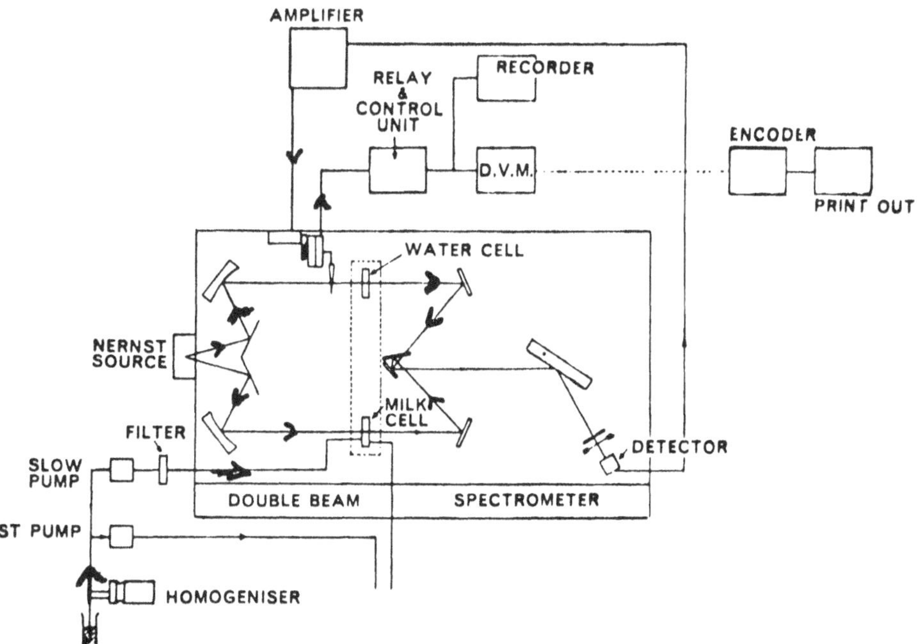

Fig. 20:7. Infra-red Milk Analyser, Mk-2 (Courtesy A. Reyrolle and Co. Ltd.)

having water in the reference cell of the spectrophotometer. Fig. 20:7.

The analytical range for fat is 0 to 9%, protein 0 to 8%, and lactose 0 to 5.5%.

A 15 ml. sample is pumped from an ultrasonic homogenizer to the analysis cell. The measurement is converted to the actual percentage of the component desired and read directly on the meter on the machine or the instrument may be fitted with a chart recorder or a paper-tape punch and numerical read-out. A sample may be analyzed with an accuracy that does not vary more than about 0.05% from chemical tests for fat, protein and lactose. Although no special skill is needed to make a test, a trained technician is needed to calibrate and maintain the equipment. The instrument is costly, but is finding increased use. Several states have approved its use, especially for raw milk.

Milko-Tester

The Milko-Tester, developed by A/S N. Foss of Hillerod, Denmark, was approved in January 1969, by the Association of Analytical Chemists for measuring the fat content of raw milk. Several hundred Milko-Testers are in use by commercial dairies for in-plant testing in the United States (Fig. 20:8).

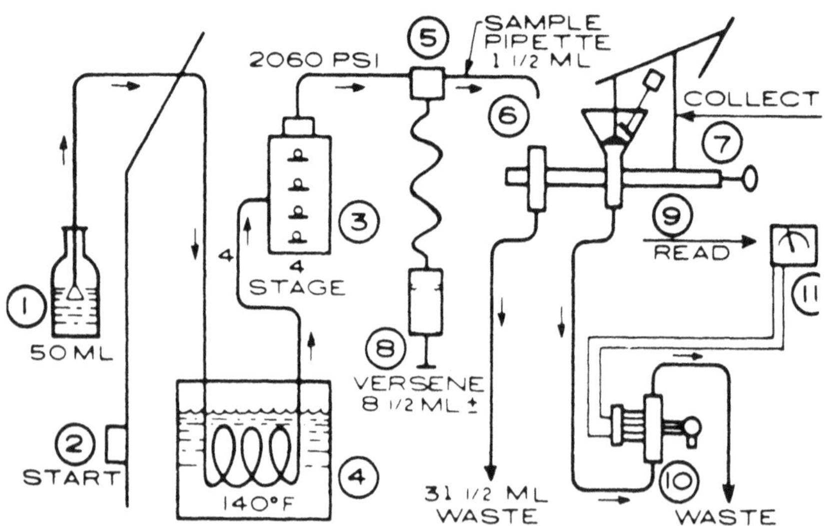

Fig. 20:8. Flow Diagram for Milko-Tester. (Courtesy, Foss America, Fishkill, N.Y.)

The instrument operates by measuring fat on the basis of light scattering by the fat globules in the homogenized sample. The prepared sample is drawn through a filter into the machine where it is warmed to 60°C. In the Milko-Tester MK 111, 1.6 ml. of sample is used. The sample is mixed with a chelating reagent which disperses the milk proteins and prevents precipitation of the calcium salts. It is then homogenized at about 2000 psi. in a 4-stage homogenizing unit. The opacity of the solution is measured by passing a beam of light through it and measuring the absorption on a photometer calibrated to read directly in percentage of fat. After a sample is run, the next sample flushes the previous one out of the machine, making the operation continuous. The test results on the Milko-Tester MK111 are displayed on an illuminated digital read-out or can be used with an automatic print-out unit.

The Milko-Tester may be calibrated with a standardized "Milko-Gel". This is a stable mixture of milk, preservative and a gelatin stabilizer, standardized to simulate milk of either 4% or 7% fat content.

A test may be completed in about thirty seconds. The machine will measure the fat content of milk ranging from 0 to 9.3%. By diluting the sample the range may be extended to 40% fat. The accuracy is comparable to that of the Babcock or Gerber test.[21] Care must be taken to avoid carry-over effects when consecutive samples are measured which have a fat content ranging from high to low.

The tester may be placed "in line" and automatically check the fat content of raw milk every 20 seconds. If the test exceeds a preset level, controls automatically add low-fat milk.

The composition of the chelating agent is as follows: Sodium (tetra) ethylenediamine tetraacetate, 45 grams; polyoxyethylene sorbitan monolaurate (Tween 20) 10 ml.; sodium hydroxide, 7.6 gm and water, 10 liters.

Total Solids and Solids-not-Fat

Many procedures have been proposed for determining the solids content of milk.[10] The approximate content of total solids or solids-not-fat usually is determined from its fat content and specific gravity. The specific gravity is found by means of a special form of hydrometer called the *lactometer*. The Quevenne lactometer, used with milk at 60°F, and the Watson lactometer, for use at 102°F, are most generally employed. The instruments are read in terms of lactometer degrees; a lactometer degree is 1000 times the specific gravity minus 1000. The usual Quevenne type of lactometer has a scale graduated from 15 to 40, corresponding to specific gravities of

1.015 to 1.040. The Watson instrument reads from 26 to 37. The best instruments have a large bulb and a narrow neck and often are provided with a built-in thermometer.

The reading on the Quevenne lactometer is correct only when taken at a temperature of 60°F, and corrections must be applied if the reading is made at some other temperature. On the average, between 50 and 70°F, the reading increases 0.1 lactometer degree for each degree of temperature below 60°F, and decreases a like amount for each degree above 60°F. Temperature correction for the Watson lactometer is more limited. The milk should not vary more than 2°F from 102°, in which case a correction of ±0.2 of a lactometer degree is made for each degree of variation in temperature.

The use of the lactometer at 60°F has been criticized, because the specific gravity of milk is not constant. At the time of milking, the fat globules are liquid and if the milk is not cooled, the globules will remain liquid for 12 to 15 hours. As they cool and gradually solidify, they contract and become smaller in volume. As a result, the volume of milk will also decrease and so its specific gravity will increase, because the same weight of milk will occupy less volume. This is known as the *Recknagle effect.* If this factor is not taken into consideration, a solids-not-fat determination on relatively fresh milk may be 0.1% or more too low. Lactometer readings made at 102°F are intended to avoid this effect. The results obtained in many comparisons between readings, made on the same milk at 60° and 102°, indicate that there is no practical difference in results. In either case, variations from gravimetric tests for solids may be as much as 0.2 too high or too low.[12]

Use of Lactometer

To take a lactometer reading, place the well mixed milk sample in a cylinder or container large enough to float the lactometer and having a diameter greater than that of the bulb of the lactometer. The container should be filled so that it will overflow when the lactometer is placed in it. Let the lactometer float freely and after one-half minute take the reading. Any bubbles on the surface of the milk should be blown off before reading the lactometer. The scale is read at the highest point reached by the milk meniscus around the stem of the lactometer, not at the principal surface of the milk.[12]

By means of a formula based upon the fat content and the lactometer reading, the total solids or the solids-not-fat in a sample of milk may be easily calculated. The Babcock formulas for use with the Quevenne lactometer, read at 60°F, are:

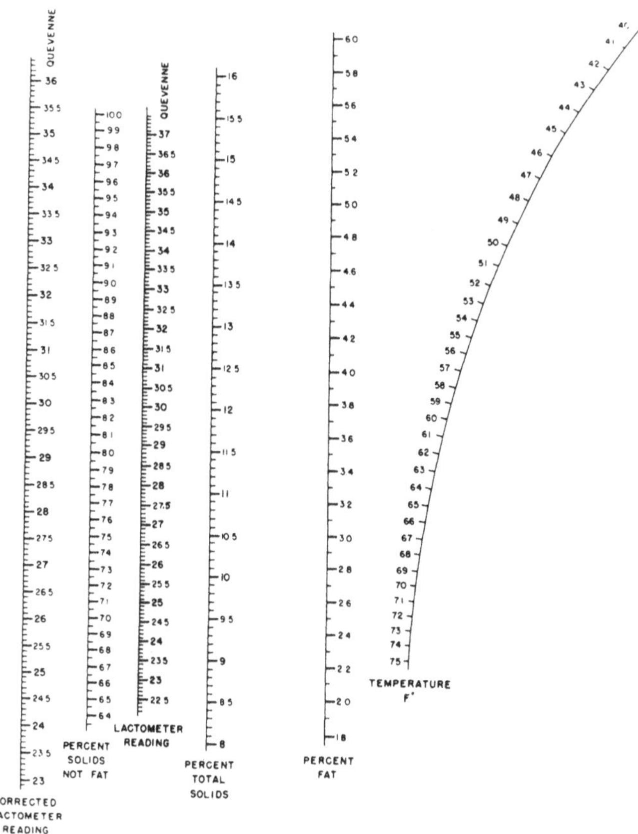

Fig. 20:9. Nomograph for Quevenne Lactometer.

$$\text{Solids-not-fat} = \frac{\text{Fat \%}}{5} + \frac{\text{Corrected lactometer reading}}{4}$$

$$\text{Total Solids} = 1.2 \text{ Fat \%} + \frac{\text{Corrected lactometer reading}}{4}$$

Tables and slide rules for this calculation have also been constructed.

The author's nomograph shown in Figure 20:9 gives temperature corrections, as well as total solids and solids-not-fat for milk of a given lactometer reading and fat content.[13] If the lactometer is read at a temperature other than 60°F, a straight line on the chart connecting the *temperature* with the *observed lactometer reading* will pass through the *corrected lactometer scale* at a point that indicates the corrected reading. If this point is then connected by a straight line with the fat content of the sample, as shown on the *percent fat* scale, the line will pass through a point on the appropriate scales showing the *percent solids-not-fat* and *total solids* in the sample. For routine use of the chart, a transparent plastic straight edge is most convenient. The number opposite the *percent fat* figure shows the average solids-not-fat content of milk of a given fat content. These values generally are within 0.2 of gravimetric results. The formula for use with the 102° lactometer is

$$\%\text{Total solids} = 1.33\ F + \frac{273\ L}{L + 1000} - 0.40$$

where F is the percent fat and L the lactometer reading.

Tables have been published for use with the 102° lactometer. The author's nomograph for use with this lactometer is shown in Figure 20:7.

Approximate specific gravity readings may be quickly made by means of the Golding Plastic Bead Test.[6] Ten plastic beads, which cover a specific gravity range between 25 and 34 lactometer degrees, are used. Milk is added to the container with the beads, and the number of beads left on the bottom of the container is counted. The solids-not-fat of the milk sample is obtained from the formula

$$SNF = 913 - 0.279B + 0.307F,$$

where B is the number of beads on the bottom of the jar.

Published results indicate that this method is less accurate than the lactometric determination.[4]

The lactometer is used as an aid in detecting milk to which water might have been added. Normal milk rarely has a specific gravity, at 60°F, less than 1.030 (lactometer reading, 30) and therefore a lower lactometer reading would justify suspicion that the milk had been adulterated. A more accurate test, such as the freezing point, or cryoscope test should be used to confirm the presence of added water.

Gravimetric Test

Total solids in milk products may be determined by weighing a small sample of the product in an aluminum dish of known weight. The dish then is heated upon a steam bath or hot plate until most of the moisture in the sample has evaporated. The dish then is placed in

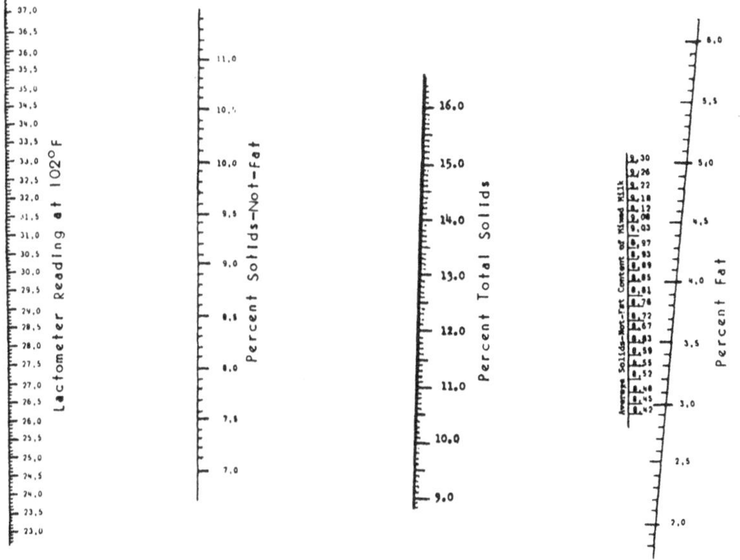

Fig. 20:10. Nomograph for 102° Lactometer

a vacuum oven, at 100-105°C for the final drying. After cooling, it is again weighed and from the loss of weight the amount of water evaporated is calculated. By subtracting the weight of water from the original weight of the sample, the amount of dry matter or total solids is obtained.

Protein Analysis by Dye Binding

The protein content of milk is an important item in cheesemaking as well as being a factor in determining the nutritive value of dairy products. In Holland and other places milk is paid for on the basis of its fat and protein content. The traditional chemical procedures for protein analysis are complicated and time consuming. They are not practical for routine use when hundreds of samples are to be examined.

The use of dye binding has changed this situation and is now widely used.[26] The polar groups in proteins, when on the acid side of the iso-electric point carry a positive charge and can combine with dyes which carry an opposite charge. Such a dye- protein compound is insoluble. In the dye binding procedure an excess of dye is used and the amount of dye not bound to the protein is determined by a measure of the optical density of the remaining solution.

The dyes commonly used are orange G and amido black 10B. The procedure used is as follows:

1 ml. of milk is mixed with 20 ml. of amido black solution. The mixture is centrifuged to precipitate the dye-protein complex. The optical density of the supernatant liquid is measured and the reading converted directly to percentage of protein. All of these steps may be automated. As a control measure, a Kjeldahl nitrogen determination is made frequently on a sample to check the calibration of the equipment. The method has been approved for official use in the United States.[20]

An instrumental procedure, the "Pro-Milk", developed in Denmark by the makers of the Milko-Tester, was found suitable for the routine, protein determination in milk.[27] It uses the amido black dye-binding method, giving results generally slightly higher than the Kjeldahl analysis. The Fiske "Prot-O-Mat" is another dye-binding instrumental method.

It has been noted that milk of low protein content has a lower dye-binding capacity than that of bulk milk of the same protein content. Milk from high-protein producing cows shows a higher protein content than bulk milk of the same protein content. Mixed or bulk milk from the supply area should be used to calibrate the instrument.

HYDROGEN-ION CONCENTRATION: pH

Acidity

Fresh milk or cream does not contain any appreciable amount of lactic acid and therefore an increase in its acidity is a rough measure of its age and bacterial activity. In sour milk, the acidity usually is more than 0.3%, expressed as lactic acid. This acidity is determined by titration; that is, the acid is exactly neutralized by the addition of a measured amount of standard alkali solution, usually 0.1 normal sodium hydroxide solution. Each milliliter of 0.1 normal alkali will neutralize 0.009 gram of lactic acid. When 9 grams of milk, diluted with twice its volume of distilled water, are titrated, each ml of 0.1N alkali used is equivalent to 0.1% of the lactic acid. A 1% solution of phenolphthalein in alcohol is used as the indicator for the titration. Phenolphthalein is colorless in acid solution and pink in the presence of alkali. Two ml of the indicator are added to the sample and the standard alkali solution is added gradually from a burette. As soon as the acid present is neutralized, the first drop of alkali solution in excess turns the indicator pink, thus showing that the end point has been reached. The percentage of acidity may be found by means of the formula:

$$\% \text{ Acidity} = \frac{\text{ml } 0.1 \text{ N alkali} \times 0.009}{\text{Weight of sample}} \times 100$$

If milk is diluted with more than twice its volume of water before titrating the water causes a change in the milk salts with a decrease in their acidity, so the test for acidity will be lower.

When fresh milk is titrated with a standard solution of alkali, its acidity is equivalent to 0.13 to 0.18% lactic acid. As fresh milk contains no lactic acid, this acidity is an apparent acidity and is a measure of the amount of alkali that combines with the protein and mineral salts in the milk. Fresh milk of high normal total solids will show a higher titratable acidity than fresh milk of low total solids. When bacterial fermentation takes place, lactic acid is formed and the acidity of the milk is increased.

The acidity of acids is due to the protons or hydrogen ions. Alkalies or basic substances are characterized by the presence of hydroxyl ions. Both hydrogen ions (expressed by H) and hydroxyl ions (OH) are present in water. In pure water as many H or acid ions are present as OH or alkali ions. Water, therefore, is neither acid nor alkaline, but neutral. Actually it is very difficult to prepare water so pure that it is exactly neutral, because many substances, especially the carbon dioxide of the air, are readily dissolved in it.

For technical reasons, chemists state that a solution in which the acid ions are equal in number to the basic or alkaline ions, has a *p*H of 7 and is neutral in reaction. Different numerical values indicate different degrees of acidity or alkalinity, much as the numbers on the thermometer scale indicate degrees of temperature. A *p*H of less than 7 indicates acidity, a number greater than 7 indicates alkalinity. The *p*H numbers represent logarithmic values and each whole number in the scale means an increase or decrease in the number of hydrogen ions by ten. Thus a *p*H of 2 means ten times the number of hydrogen ions as at *p*H 3, and 100 times as many as at *p*H 4. A similar relationship holds true for the alkaline side; *p*H 8 means ten times as

TABLE 20:2
Approximate pH Value of Some Common Materials

Material	pH	Material	pH
Blood	7.3-7.5	Lemon Juice	2.2-2.4
Bread, White	5.0-6.0	Milk, Cow's	6.5-6.8
Cheese, Cheddar	5.6-6.0	Milk, Evaporated	5.9-6.2
Cheese, Process	5.7-6.2	Milk, Human	6.8-7.2
Cheese, Roquefort	4.8	Sauerkraut	3.5
Cheese, Swiss	5.7-5.9	Water, Distilled	5.8
Cider	2.9-3.3	Water, Purified	6.8-7.0

many hydroxyl ions as at *p*H 7, and *p*H 10 indicates 100 times the hydroxyl ion concentration as at *p*H 8.

The approximate *p*H values of some substances is given in Table 20:2.

In general, two methods are used to measure the *p*H value of a solution. Very accurate measurements can be made with electrical apparatus or by colorimetric determinations with indicator solutions. Various indicators, which change color at different *p*H values, are available. The color obtained is compared with a standard color to determine the *p*H of the solution.[15]

Buffers

Certain substances, known as *buffers* have the property of resisting a change of *p*H in their solutions, that is, they tend to maintain a constant *p*H. Among these materials are soluble borates, acetates, phosphates, and citrates. The phosphates and other mineral salts in milk act as buffers. The *p*H of milk is about 6.6, being only very slightly acid. When bacterial activity results in the formation of lactic acid in milk, the *p*H does not change rapidly because of the buffers acting in the milk. The acid may increase in amount until a break in buffer action occurs, at which point the *p*H of the milk drops to about 4.5 and it curdles. An acid, such as acetic acid or sulfuric acid, may be added to milk until a *p*H of about 4.5 is reached before curdling occurs.

Test for Sediment

The presence of foreign matter in milk is objectionable not only on account of the dirt itself but also because it indicates carelessness during processing. Visible dirt also indicates the probability that other foreign matter has gone into solution in the milk.

When bottled milk is allowed to stand undisturbed for a length of time, insoluble extraneous matter will, in general, sink to the bottom and become visible when the milk is examined through the bottom of the bottle. See Chapter 1.

Special equipment is used by inspection services to examine milk or cream for the presence of sediment. A pint sample is forced by air pressure or vacuum, through a cotton filter disc. The sediment is collected on the disc with an exposed area 0.40 inches in diameter. As a rule, milk conforming to a No. 1, or No. 2 disc is considered satisfactory. Milk containing sediment in excess of a No. 3 disc is condemned for human consumption. A No. 1 disc is free of sediment, a No. 2 disc does not have in excess of 0.5 mg. of sediment. The No. 3 disc represents 2.5 mg. of sediment.

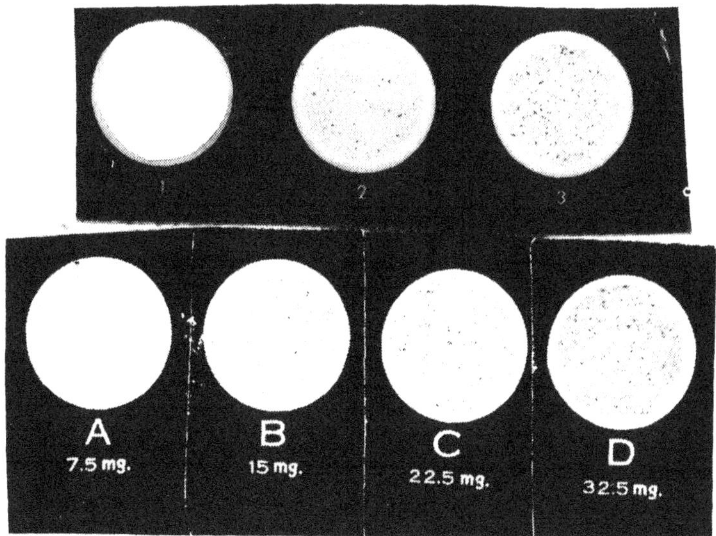

Fig. 20:11. Standards for Milk Sediment: Top—Samples of raw milk, unstirred, from bottom of can 1) Clean; 2) Passable; 3) Dirty Bottom Scorched Particle Standards for Dry Milks (see Chapter 17)

For practical purposes, a one gallon mixed sample from a farm bulk tank is taken as equivalent to one pint of milk taken from the bottom of a five or ten gallon can. (Fig. 20:11)

Test for Phosphatase

Raw milk contains an enzyme, phosphatase, which in the presence of alkali can decompose certain organic compounds of phosphoric acid. A test for phosphatase is very important in milk control work because it enables the plant operator or milk control officer to determine whether a milk product had been properly pasteurized, or whether it contains a detectable amount of raw product. The test is of value in public health work because milk heated suffciently to give a negative phosphatase test also has been heated sufficiently to destroy any pathogenic organisms that may have been present.[24]

The compound commonly used in the test for phosphatase is disodium phenyl phosphate, a combination of phenol and a sodium salt of phosphoric acid. This preparation does not contain any free or uncombined phenol, but when acted upon by phosphatase it is split into its components and phenol is liberated. Extremely small amounts of phenol may be detected by chemical means. The commonly used reagent is CQC solution (a 0.4% alcoholic solution of 2,6, dichloroquinone-chloroimide, which gives a deep blue color with phenol.

When milk is pasteurized, practically all of the phosphatase is destroyed and the milk is unable to act upon the phenol-phosphoric acid compound added to it. As a result, properly pasteurized milk gives little or no color in the test. If the milk is underheated or contains as little as 0.1% raw milk, enough phosphatase is present to liberate phenol and to yield a blue color indicative of improperly pasteurized milk.

The dialysis phosphatase test is a simplified procedure applicable to all milk products.[24] In this test, 10 ml. of buffer solution is placed into a cellulose dialysis tube. The buffer solution contains 11.5 g. anhydrous sodium carbonate, 10.2 g. sodium bicarbonate, and 1.1 g. disodium phenyl phosphate in one liter of distilled water.

5 ml. of a fluid or 1 gram of a solid dairy product is added to the substrate solution in the cellulose bag. The bag is then placed in a test tube containing 10 ml. of a solution made by adding 0.1 g. copper sulfate to 1 liter distilled water. This is incubated at 37°C. for one hour. After incubation, the bag with its contents is removed from the test tube and two drops of CQC solution added. After five minutes, 5 ml. of n-butyl alcohol is added, the tube is inverted five times, and let stand one minute. The color is then measured visually, or with a colorimeter. A phenol content in excess of 1.0 microgram of phenol in 0.5 ml. of milk or 0.25 of solid dairy product is taken as indicative of improper pasteurization.

Compact units, containing the equipment and reagents for making routine phosphatase tests are available commercially.

Milk products, especially those of high fat content, that have been pasteurized by a high temperature process (163°F or higher) may show a positive phosphatase test after being held, even though a test is negative at the time of pasteurization. The pasteurized products are sterile or have a low bacterial count, indicating adequate bacterial destruction. A later positive phosphatase test is attributed to "reactivation" of the enzyme.[24] An increase in the holding time during high-temperature pasteurization reduces the degree of reactivation that may occur.

A method is available to differentiate reactivated phosphatase from that originally present in raw or underpasteurized products.[24] The addition of a small amount of magnesium chloride to a product with reactivated phosphatase increases its activity about ten times, whereas no such increase is shown in the phosphatase present in underpasteurized products.

Reactivated phosphatase is reported to be more sensitive to heat than normal phosphatase, and to be destroyed in five minutes at 55°C (131°F).[18] Work done in Holland indicates that the reactivitated enzyme is not identical to the original phosphatase.[17]

Test for Added Water

The detection of added water is a practical problem to the dairy plant operator. Water may have been added deliberately or through careless production practices. Added water represents an economic loss to the buyer and reduces the quality of the product; it also constitutes adulteration. Care must be taken to drain all milk lines, pumps, vats and other equipment after cleaning. The first bottles or containers of milk during processing should be carefully examined.

Due to the variation in specific gravity of normal milk, as much as 10% of added water may be present in milk without being detected by use of the lactometer. The most accurate means of detecting added water is to measure the freezing point of the milk sample.

Many investigators have shown that milk has a fairly constant freezing point, varying but a few thousandths of a degree among different samples of pure milk.[11] The freezing point is independent of the variation in composition of milk as it comes from the cow. Neither the breed of cow nor her feed have any significant effect upon the freezing point of the milk. Milk from cows that have drunk an excessive amount of water may have an abnormally high freezing point. In the author's experience, cows that had been on pasture all day without access to water except upon return to the barn at night, gave morning's milk with a freezing point of —0.525°C. The evening's milk had a normal freezing point of —0.548.

Pure water, as is well known, freezes at 0°C (32°F) and pure milk freezes around —0.545°C (31°F). If water is added to milk, the freezing point of the mixture rises. The more water is added, the more closely does the freezing point of the milk approach that of pure water. The temperature is measured with a very sensitive thermometer which can be read to 0.001°C. Under carefully controlled conditions, the presence of 0.1% added water may be detected in milk. The apparatus commonly used for this test is known as the Hortvet cryoscope.[3]

Electronic equipment that makes use of calibrated thermistors instead of a thermometer also is used to measure the freezing point of milk.[7, 22]

The effect of pasteurization upon the freezing point of milk is discussed in Chapter 11.

Owing to small variations in the normal freezing point of milk, regulatory officials usually regard a freezing point below —0.530°C as normal. Recently the recommendation has been made that a freezing point below —0.525 should be presumed to indicate that no added water is present in a sample of milk.

MISCELLANEOUS TESTS

Testing of Butter

Butter is tested for its content of fat, moisture, and salt. Ten grams of a well mixed sample is placed in a small, aluminum dish of known weight, which then is treated as follows:

Moisture:

Heat the sample on a hot plate or over a flame until the water is evaporated. Cool, reweigh, and calculate the amount of moisture. As a 10-gram sample is used, the loss in weight times 10 gives the percent of moisture in the sample

As a rule, the water is evaporated when the melted fat has acquired an amber color and the salt and curd have collected on the bottom of the dish. Very few, if any, bubbles should be present on the surface of the melted fat.

Fat:

To the dry residue in the dish, add light gasoline or petroleum ether. Stir thoroughly to dissolve the fat, then allow the salt and curd to settle. Decant the solution, taking care that none of the residue is lost. Repeat the extraction three times in all. Dry the residue, preferably on a hot plate, taking care that none of the curd is lost by spattering. Cool and reweigh. The difference between this weight and the previous one represents the fat in the sample. Multiply by 10 to obtain the percentage.

Salt (Sodium Chloride):

Dissolve the residue in distilled water, making the total volume 250 milliliters. Titrate 25 milliliters with a standard silver nitrate solution, using potassium chromate as indicator.

Curd:

Curd represents the non-fat milk solids present in butter. It is determined by subtracting the percentage of moisture, fat, and salt from 100.

Tests to Distinguish Goat's Milk from Cow's Milk

This is a comparative test, based upon the relative insolubility of the protein in goat's milk as compared to that of cow's milk when dissolved in an alkaline solution.

To 10 ml milk add 0.5 ml of 10% sodium hydroxide solution (10 g of sodium hydroxide in 100 ml of water). Mix well and add 13 ml. of ethyl alcohol. Next add 25 ml. of a mixture of equal parts of ethyl ether and petroleum ether and shake well. Allow the ether layer to separate and after standing about five minutes, compare the aqueous layer with that of a sample of cow's milk treated in the same manner.

Cow's milk gives a clear, transparent aqueous layer, goat's milk an opaque aqueous layer. As little as 10% of goat's milk in cow's milk may be detected by the opacity shown by this test.

The above procedure is not suitable for the detection of cow's milk in goat's milk. A relatively simple method to do this depends upon the use of gel electrophoresis.[2] The method will detect as little as 1% of cow's milk in goat's milk.

Storch Test

This test is used, especially in some European countries, to detect milk that had been heated to 172°F (77.8°C) or higher. To 10 ml. of milk add 2 drops of dilute hydrogen peroxide solution (one part of hydrogen peroxide, 14 parts of water and 0.1% sulfuric acid by volume). Add 2 drops of an aqueous solution of para-phenylenediamine. Mix well. A blue color indicates milk not heated to at least 172°F. No color develops in milk heated to this temperature or higher.

Test for Bakery-Type Nonfat Dry Milk

As explained in Chapter 17, nonfat dry milk to be used in bread baking is given a high heat treatment. The laboratory method to determine the undenatured whey protein nitrogen is a modification of the Harland—Ashford method.[28] A simplified, qualitative test is given in the Handbook for Bakers, No. 107, American Dry Milk Institute. In this test, 9 ml. of the reconstituted milk (20 grams to 200 ml water) is added to 17.6 ml. of test solution (3200 ml. water, 525 gms, salt and 14 ml. concentrated hydrochloric acid). Use pure salt, not table salt.

After mixing, the solution is filtered. Ten ml. of filtrate is placed into a test tube and heated in a boiling water bath for 10 minutes. After standing 30 minutes, the precipitate is observed. A clear solution or slight cloudiness indicates nonfat milk that has been heated sufficiently to give good baking results. A medium or heavy cloudines indicates milk that may give poor baking performance.

A simplified quantitative procedure is given in Bulletin 916 (1971) of the American Dry Milk Institute, Inc.

Sucrose in Dietetic Milk Products

At times it may be advisable to examine a sucrose-free dietetic milk product to make certain that it is as labeled. Some flavors and fruit syrups that contain sucrose should not be used in a sucrose-free mix. During manufacturing operations it is possible that some normal, sucrose-containing mix may have been added to, or become mixed with, the dietetic product. The dietetic product may be incorrectly labeled and may actually be, for example, a normal ice cream or ice milk.

The author's test, based upon the well-known Seliwanoff reaction, is as follows: To 10 ml of mix or melted ice cream add 10 ml or a 20% solution of trichloroacetic acid. If a chocolate or colored product is tested, add about 1 gram of decolorizing charcoal. Filter the mixture through a rapid filter paper. If the filtrate is colored, pour it back through the filter paper. Place 5 ml of the clear filtrate into a test tube and add 1 ml of a mixture made by adding 1 gram of resorcinol to 10 ml of concentrated hydrochloric acid. Place the test tube into a boiling water bath for five mintues. In the presence of sucrose a rose-to-red color appears. Lactose and dextrose do not react. A pink or orange color should be disregarded. As little as 0.2% surcose may be detected.

The red color may be extracted by mixing the contents of the test tube with 5 ml of butyl alcohol. The amount may be quantitatively estimated by measuring the intensity of color against standards made with known amounts of sugar.

DAIRY ARITHMETIC

A number of problems that require the application of arithmetic frequently confront the dairyman and milk processor. Some of these are of sufficiently wide interest to be mentioned here; others, such as the calculations needed in the manufacture of ice cream mix, are too involved to be considered here and information on them should be obtained from special books.[1]

To Calculate the Average Percentage of Fat

The average percentage of fat contained in two or more lots of milk, cream, or other milk products is obtained by multiplying the number of pounds of each product by its corresponding percent of fat in order to obtain the total weight of fat. The weight of fat divided by the total weight of milk or cream, multiplied by 100, gives the average percent of fat in the entire lot.

Example:

Find the average percent of fat contained in 40 gallons of milk testing 3.5% fat, 70 gallons of cream of 30% fat content, and 15 gallons of milk of 3.8% fat content.

Adding the individual fat percentages and dividing by three gives a fallacious result since the weight of the individual lots of milk and cream is not taken into consideration.

Solution:

First determine the weight of each lot. Table 20:4 shows that 1 gallon of milk weighs 8.6 pounds, and 1 gallon of cream containing 30% of fat weighs 8.4 pounds.

40 X 8.6 = 344 pounds of 3.5% milk.
70 X 8.4 = 588 pounds of 30% cream.
15 X 8.6 = 129 pounds of 3.8% milk.
1061 pounds—total weight of milk and cream.

344 X 0.035 = 12.040 pounds fat in the 3.5% milk.
588 X 0.30 = 176.400 pounds fat in the 30% cream.
129 X 0.038 = 4.902 pounds fat in the 3.8% milk.
193.342 pounds fat in the entire lot.

$$\frac{193.342}{1061} \times 100 = 18.22\%$$ fat in the entire lot of milk and cream mix.

To Standardize Milk and Cream

In order that a milk product should have a uniform fat content it generally is necessary to adjust the percentage of fat to a definite amount, this process being known at *standardization*. The calculation involved is to determine the amount of cream that must be added to increase the fat content of the material being standardized or the amount of skim milk to add if the fat content is too high. A simple method, devised by Prof. R. A. Pearson of Cornell University, and known as the *Pearson Square* is used in the computation. The procedure is as follows:

Draw a square or rectangle and place in its center the percentage of milk fat desired. At the left-hand corners place the percentages of milk fat in the materials to be mixed. From the larger number on the left-hand size, subtract the number in the center. Place the remainder on the right-hand side of the square and at the corner diagonally opposite from the number used on the left-hand side. Next subtract the smaller number on the left-hand corner from the number in the center, and place the remainder on the diagonally opposite corner on the right-hand side of the square. The figure on the upper right-hand corner refers to the product represented by the figure on the upper left-hand corner of the square and the figure on the lower right-hand corner refers to the product represented by the figure on the lower left-hand corner. Examples will make the use of the Pearson square clear.

Example:

How much 30% cream must be mixed with 3% milk to prepare milk that tests 4.5%?

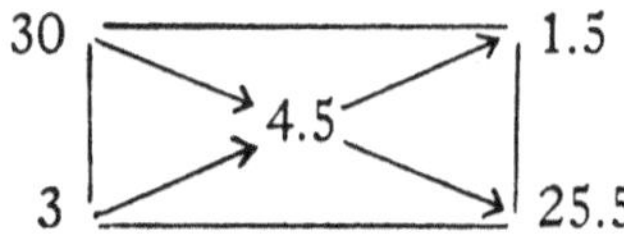

This means that 1.5 pounds of 30% cream must be mixed with 25.5 pounds of 3% milk to yield milk testing 4.5% fat. The figures on the right hand side give the proportions to be used and so any multiple of them can be used; for example, 3 pounds of cream and 51 pounds of milk, to make a total of 54 pounds of 4.5% milk.

Example:

500 pounds of 40% cream is to be reduced to 25% fat content. How much milk of 3% fat content must be added?

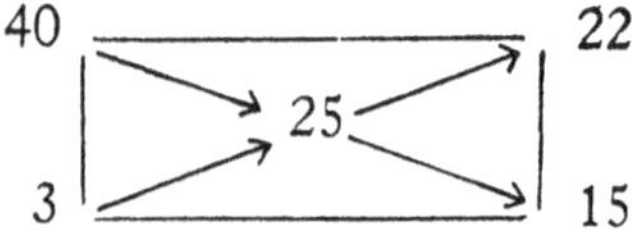

According to the diagram, 22 pounds of 40% cream is needed for every 15 pounds of 3% milk in order that the mixture contains 25% milk fat. This mixture would weigh 37 pounds, but according to the problem, 500 pounds is needed. In order to obtain a 500-pound mixture it is necessary to multiply both the 22 and 15 by the number of times their sum, 37, is contained in 500, that is by 500/37 or 13.513.

$$22 \times 13.513 = 297.286$$
$$15 \times 13.513 = 202.695$$

or

$$22:37 = x:500 \text{ or } x = 297.29$$
$$15:37 = x:500 \text{ or } x = 202.7$$

The result shows that 297.3 pounds of 40% cream must be mixed with 202.7 pounds of 3% milk in order that 500 pounds of 25% cream be obtained.

TABLE 20:3
Approximate Weight of One Gallon of Milk or Cream
At Various Temperatures

Temperature, °F	*Skim Milk lb*	*4% Milk lb*	*20% Cream lb*	*25% Cream lb*	*30% Cream lb*	*35% Cream lb*	*40% Cream lb*
40	8.660	8.610	8.540	8.500	8.470	8.450	8.420
45	8.665	8.605	8.525	8.485	8.450	8.425	8.395
50	8.650	8.600	8.510	8.470	8.430	8.400	8.370
55	8.645	8.595	8.495	8.450	8.410	8.375	8.345
60	8.640	8.590	8.480	8.435	8.390	8.350	8.320
70	8.630	8.580	8.450	8.400	8.350	8.310	8.270
80	8.620	8.565	8.420	8.370	8.315	8.270	8.230
90	8.600	8.550	8.390	8.340	8.280	8.230	8.190
100	8.580	8.520	8.370	8.310	8.250	8.200	8.150
145	8.470	8.420	8.270	8.220	8.150	8.100	8.045

For all practical purposes, milk testing from 3 to 5% milk fat may be assumed to weigh 8.6 pounds per gallon at 60-70°F.

To Calculate Weight per Gallon

At times it is necessary to know the specific gravity or weight per gallon of a product such as milk, ice cream mix, or evaporated milk. A practical and relatively accurate formula for frozen desserts is as follows—[32]

$$\text{Sp.Gr.} = 1 + \left\{ \frac{(4.87 \times \%\text{Sugar}) + (4.41 \times \%\text{Serum Solids}) - (0.88 \times \%\text{Fat}) - 6.26}{1{,}000} \right.$$

For a temperature between 5 and 15.6°C, a correction should be made by subtracting (T°C—5)X0.0003 from the result found above.

The specific gravityX8.34=weight per gallon.

Example:

Find the weight per gallon at 8°C of a mix that contains 10.5% fat, 11% serum solids, 14% sugar, and 0.15% emulsifier and stabilizer. The emulsifier, stabilizer, and any egg yolk may be considered as part of the serum solids.

Solution:

$$\text{Sp. Gr.} = 1 + \left\{ \frac{(4.87 \times 14) + (4.41 \times 11.15) - (0.88 \times 10.5) - 6.26}{1{,}000} \right\} - (8 - 5) \times 0.0003$$

$$= 1 + (68.18 \ 49.17 - 9.24 - 6.26 - 0.0009) = 1.102$$

$$\text{Wt/per gal} = 1.102 \times 8.34 \text{ or } 9.19 \text{ lb at } 8°\text{C}.$$

The approximate weight on one gallon of various milk products is given in Table 20:4.

TABLE 20:4
Approximate Weight of One Gallon of Various Milk Products At 60°F

(Water	8.34)
Evaporated Milk (7.9% Fat)	8.89
Condensed Milk (9.5% Fat; no Sugar)	9.05
Condensed Milk (Sweetened)	11.00
Condensed Skim Milk (Unsweetened)	9.18
Condensed Skim Milk (Sweetened)	11.20
Ice Cream Mix* (8% Fat)	9.05
Ice Cream Mix (10% Fat)	9.15
Ice Cream Mix (12% Fat)	9.07
Ice Cream Mix (14% Fat)	8.97
Skim Milk	8.64
Milk Fat (Butter Oil)	7.5

*The finished ice cream with 100% overrun will weigh one-half that of the ice cream mix.

Temperature Conversion Factors

$$C° = (F° - 32) \times 5/9$$
$$F° = (C° \times 9/5) + 32$$

To Estimate Composition of Milk

Sometimes it is desirable to know the approximate compositon of milk without having recourse to a chemical analysis. Studies have shown that the total solids and the protein content of mixed milk obtained from a number of cows closely parallels its fat content. The following arithmetical forumlas may be used to obtain these values, usually to within 0.2%.

Protein % = 1.597 plus 0.446 times fat %.

Jacobson's formula, which is widely used, states that

Solids-not-fat % = 7.07 plus 0.4 times fat %.

Other relationships of the fat to the solids-not-fat content, based upon milk of different fat contents are as follows:[19]

% Fat	% Solids-not-fat
2.75-3.33	0.3151 times fat % plus 7.3672
3.30-4.25	0.70 times fat % plus 6.10
4.25-5.85	0.3846 times fat % plus 7.44
5.90-6.75	0.2457 times fat % plus 8.2340

TABLE 20:5
Table of Equivalent Temperatures

Cent.	*Fahr.*	*Cent.*	*Fahr.*
—50°	—58°	75°	167°
—45°	—49°	80°	176°
—40°	—40°	85°	185°
—35°	—31°	90°	194°
—30°	—22°	95°	203°
—25°	—13°	100°	212°
—20°	— 4°	105°	221°
—15°	5°	110°	230°
—10°	14°	115°	239°
— 5°	23°	120°	248°
0°	32°	125°	257°
5°	41°	130°	266°
10°	50°	135°	275°
15°	59°	140°	284°
20°	68°	145°	293°
25°	77°	150°	302°
30°	86°	155°	311°
35°	95°	160°	320°
40°	104°	165°	329°
45°	113°	170°	338°
50°	122°	175°	347°
55°	131°	180°	356°
60°	140°	185°	365°
65°	149°	190°	374°
70°	158°	195°	383°

Table of Values for Interpolation in Above

1°C = 1.8°F	4°C = 7.2°F	7°C = 12.6°F
2°C = 3.6°F	5°C = 9.0°F	8°C = 14.4°F
3°C = 5.4°F	6°C = 10.8°F	9°C = 16.2°F
1°F = 0.55°C	4°F = 2.22°C	7°F = 3.88°C
2°F = 1.11°C	5°F = 2.77°C	8°F = 4.44°C
3°F = 1.66°C	6°F = 3.33°C	9°F = 5.00°C

MILK EQUIVALENTS

The amount of milk, containing 3.5% fat, that is required to prepare 1 pound of a milk product is approximately as follows:

TABLE 20:6
Milk Equivalents

Product	*Pounds of Milk Required*
Butter	22
Cheddar Cheese	10
Evaporated Milk	2.1
Sweetened Condensed Milk	2.9
Nonfat Dry Milk	11.2
Ice Cream (per gallon)	13
Dry Milk, whole	7.8
Dry Cream	19
Malted Milk	2.3
Cottage Cheese, non-fat	6.25

Milk equivalent means the amount of milk of standard milk fat content that would be needed to produce the product concerned. The actual amount needed varies in different areas and to some extent, with the season.

REFERENCES

1. Arbuckle. W. S. Ice Cream, Second Edition, Avi Publishing Co., Inc., Westport, Conn., 1972.
2. Aschaffenburg, R. and Dance, Janet E., Detection of cow's milk in goat's milk by gel electrophoresis, *J. Dairy Res.*, 35:383, (1968).
3. Association of Official Agricultural Chemists, Official methods of analysis of the Association of Official Agr. Chemists, 11th ed. Washington, D.C. 1970.
4. Erb., R. E., Manus, L. J. , Ashworth, U.S., Comparative accuracy of the plastic bead method of Golding and the lactometer method of Watson for routine determinatioin of solids-not-fat in milk, *J. Dairy Sci.*, 43:584-597 (1960).
4a. Federal Register, Vol. 34, October 25, 1969.
5. Frost, W. D., Improved technique for the micro or little plate method of counting bacteria in milk, *J. Infect. Dis.*, 28:176 (1921).
6. Golding, N. S., A solids-not-fat test for milk using density beads as hydrometers, *J. Dairy Sci.*, 42:899 (1959).
7. Henningson, R. W., Thermistor Cryoscopic Determination of the Freezing Point Value of Milk Produced in North America, *J. Assoc. Offc. Anal. Chem.*, 52:142 (1969).
8. Hoyt, C. F., Counting bacteria in ice cream, *Monthly Bull.*, Calif. Dept. Agr., 21:406 (1932).
9. Kroger, M. and Patton, S., Comparison of two milk fat extraction methods for pesticide residue analysis, *J. Dairy Sci.*, 50:324 (1967).
10. Lampert, L. M., Testing for solids-not-fat content, *Am. Milk Rev.*, 24:10 (1962); 25:1 (1963).
11. Lampert, L. M., The freezing point of milk, *J. Assoc. Offic. Agr. Chem.*, 22:768-771 (1939).
12. Lampert, L. M., The many ways of measuring solids-not-fat in milk, *Bulletin, Calif. Dept. Agr.* 50:239 (1961).
13. Lampert, L. M., Nomograph for correction of lactometer readings and calcu-

lation of lactometer readings and calculation of milk solids, *Ind. Eng. Chem.* 12:527 (1940).
14. Lampert, L. M., Rapid separation of fat for pesticide residue analysis of milk products, *J. Dairy Sci.*, 47:1013, (1964).
15. Mattock, G., pH Measurement and Titration; The MacMillan Co., New York, 1961.
16. Pasteurized Milk Ordinance, U.S. Public Health Service, Washington, D.C. 1967.
17. Peereboom, J. W. C., Non-identity of raw and reactivated alkaline phosphatase from cream, *Neth. Milk Dairy J.*, 20:113, 1966.
18. Posthumus, G., Phosphatase reactivation, *Nederland, Zuivelbond. Off., Org.*, 44:450 (1952).
19. Richardson, G. A. and Folger, A. H., Compositional quality of milk. The relationship of the solids-not-fat and fat percentages., *J. Dairy Sci.*, 33:135 (1950).
20. Sherbon, J. W., and Luke, H. A., Dye binding and Kjeldahl, methods for protein analysis of nonfat dry milk and ice cream, *J. Assoc. Official Anal. Chem.*, 52:138 (1969).
21. Shipe, W. F., Collaborative study of the Babcock and Foss Milko-Tester, Methods for measuring fat in raw milk, *J. Assoc. Official Anal. Chem.*, 52:131 (1969).
22. Shipe, W. F., Report on cryoscopy of milk. *J. Assoc. Offic. Agri. Chem.* 41:262 (1958).
23. Soike, K. F., Miller, D. D. Elliker, P. R., Effect of pH of solutions on germicidal activity of quaternary ammonium compounds. *J. Dairy Sci.*, 35:764 (1952).
24. *Standard Methods for the Examination of Dairy Products*, 13th Ed., American Public Health Assoc., Washington, D.C. (1972).
25. Statutory Rules and Orders: The Milk Regulation, Ministry of Health, London, 1960.
26. Tarassuk, N. P., Abe, N., and Moats, W. S., The dye binding of milk proteins., Tech. Bull. No. 1369, Agr. Res. Serv., U.S. Dept. Agr., Washington, D.C. 1967.
27. McNeil, J. E., Stine, C. M. and Hedrick, T. I., Pro-Milk dye binding and Kjeldahl methods for determining protein content of cow's milk, *Michigan Quar. Bul.*, 46; No. 2:162 (1966).
28. Kuramoto, Jenness, Coulter and Choi, A modification of the Harland-Ashworth Test, *J. Dairy Sci.*, 42:28, (1959).
29. Putnam, F. and Neurath, H., The precipitation of proteins by synthetic detergents, *J. Am. Chem. Soc.*, 66:692 (1944).
30. Wildasin, H. L. and Anderson, E. O., A modified Babcock Test for homogenized milk using cationic detergents, *Storrs Agr. Expt. Sta., Bull.* 282, Storrs, Conn. (1951).
31. Ward, G. E., and Schultz, L. H., Estimation of Somatic Cells in Milk by Filter Deoxyribonucleic Acid Method with Indole. *J. Dairy Sci.*, 56:1097-1101 (1973).
32. Hahn, A. J. and Tracy, P. H., How to Calculate Gallon Weights of Dairy Products, Food Industries, (June, 1942).

chapter 21

Imitation Milk Products

Milk Substitutes

Milk and dairy product substitutes, such as filled and imitation milk, coffee whiteners, and whipped toppings, have gained wide acceptance by the consumer. These products may be made by the milk processor in his existing equipment and offer him and the distributor an increased profit. The lower cost of these products appeals to the consumer, as well as the fact that they may offer long shelf-life, packaging convenience, ease of use, and possibly some desirable dietetic advantage.

The makers of dairy product substitutes give considerable attention to the improvement of their products. As a result, margarine, whipped toppings and coffee whiteners are being used to a greater extent than the dairy product imitated. Ingredients of a non-dairy vegetable product usually can be chosen on the basis of the function they have in the product, so that the desired characteristics may be obtained. On the other hand, the manufacturer or processor of a dairy product is limited by the composition of the product and legal restrictions.

Except for the fat content, which in itself may not be nutritionally equivalent to milk fat, most imitation dairy products are nutritionally deficient. These deficiencies are shown in the comparisons between the composition of dairy products and the imitation products in Table 21:1 [1]

All substitute milk products, whether called filled or imitation milk, milk analog, or "white drink" should meet the same standards of sanitation and quality as market milk. The American Medical Association has stated that these products "should not be considered nutritionally equivalent to whole milk".

According to a proposal of the United States Department of Agriculture, a food is to be considered to be an imitation if it is a substitute for and resembles another food but is nutritionally inferior to the food imitated. A food that is a substitute for and resembles another food is not deemed to be an imitation if it is not nutritionally inferior to the food for which it substitutes and resembles.

The physical properties of the fats used are an important factor in the production of an acceptable milk product substitute. They should in general, have a melting point near the body temperature or, as described in the section on margarine, a low S.F.I. Such a melting point will provide a tactile sensation in the mouth which is pleasant and not greasy. The composition of some vegetable oils used in imitation dairy products is given in Chapter 2, Table 2:2.

If nonfat milk is used, it must first be heated sufficiently high to inactivate any lipase that may be present, else a rancid flavor may develop before processing is completed. A low heat-treated nonfat dry milk, such as is used for making cottage cheese, may be used. The nonfat milk or the reconstituted product should have 9% solids content. The milk is heated to about 90°F and the melted vegetable fat added. The mixture is continually agitated. When vat pasteurized, a temperature of 155-160°F is used for 30 minutes; with the HTST method a temperature of 172-180°F is used for at least 16 seconds. The product is homogenized immediately after pasteurization at 2000 and 500 psi. It is then quickly cooled and packaged.

The commercial vegetable fat products usually used for the manufacture of filled and imitation milk contain in addition to the fat, the needed emulsifiers, stabilizer, and color. Mono-and diglycerides are generally the emulsifiers used; sodium caseinate, carragenin, potassium and sodium phosphates are among the stabilizers used.

Filled Milk Products[3]

A filled milk is defined as "a combination of any milk, cream, or skimmed milk (whether or not condensed, evaporated, concentrated, powdered, dried, or desiccated) with any fat or oil other than milk fat, so that the resulting product is in imitation or semblance of milk, cream, or skimmed milk". Between 1923 and 1973 the shipment of any filled milk product in interstate commerce was prohibited under the Filled Milk Act. This Act was enacted by the United States Congress, and was not a ruling of the Food and Drug Administration. It limited the sale of filled milk products to the state wherein they were made. The Act has been declared unconstitutional and filled milk products may be lawfully shipped in interstate commerce.

In order that a filled milk product be considered nutritionally equivalent to its counterpart milk product the Food and Drug Administration proposal states that it must contain a specified quantity of minimum milk solids-not-fat and vitamins A, D, and E.

TABLE 21:1
Comparison of Certain Dairy Products and Their Simulated Counterparts[1]
(Constituents per 100 g dry matter)

		Gross Composition				Minerals				Vitamins		
		Carbohydrates	*Fat*	*Protein*	*Ash*	*Calcium*	*Phosphorus*	*Sodium*	*Potassium*	*A*	*Riboflavin*	*Total Solids*
		grams				mg				*IU*	*mcg*	*g*
Milk, Fluid, Whole		38.9	27.8	27.8	5.6	936	738	396	1143	1112	1350	12.6
White Beverages, (Imitation Milks)	A	52.8	31.7	7.4	4.0	25	248					12.1
	B	61.8	27.7	6.8	4.5	136	673	636	3270			11.0
Coffee (Lt.) cream		14.6	72.7	10.5	2.2	353	280	182	331	3020	509	27.5
Coffee Whiteners												
Dry	A	48.5	39.9	4.9	2.8	12	718	293	788	110	0	98.9
	B	49.1	36.5	5.0	2.7	16	625	258	768	110	0	98.9
	C	46.1	37.2	5.0	2.7	46	561	290	606	440	108	98.9
	D	48.7	35.8	4.9	3.0	12	62	146	1040	200	219	98.5
											0	
Liquid	A	50.2	47.9	3.0	1.5	23	30	543	6	2170	0	26.7
	B	42.0	48.0	5.0	2.0	70	155	245	300	250	0	20.0
	C	49.1	40.5	8.6	2.7	72	212	496	121	0		22.2
	D	37.9	52.6	2.6	1.1	52	57					21.1
Whipping (Heavy) cream		7.8	85.3	5.6	1.2	190	149	98	134	3510	268	41.0
Whipped Toppings												
Dry	A	40.8	43.2	4.6	1.0	17	32	100	48	800	0	99.9
	B	40.6	45.4	5.7	0.6	12	46	97	6	1370	0	98.9
Liquid	A	25.7	58.5	0.0	0.5	15	3	198	13	1040	0	39.3
	B	31.6	55.3	2.0	0.2	14	20	20	2	2220	0	49.1
Aerosols												
Dairy Base	A	22.0	58.0	10.7	2.3	310	276	255	310	2540	145	35.5
	B	20.4	63.3	8.5	2.3	274	267	413	209	2650	189	41.2
"Non-Dairy"	A	29.2	58.7	0.0	0.5	10	3	243	11	1130	0	39.1
	B	23.9	67.7	7.6	0.5	23	71	173	6	1370	0	39.3

The FDA recognizes that some vegetable fats are deficient in certain polyunsaturated fatty acids. Provision is made for the polyunsaturated fatty acids of the linoleic series to be present at a level equivalent to 4 percent of the fat in the filled product.[3] To date filled milk products have not been shown to be nutritionally equivalent to or comparable with milk.

Imitation Milk

The processing of an imitation milk follows that used for filled milk, except that no dairy product may enter into its composition. Sodium caseinate, demineralized whey solids and lactose are not considered milk products, but are held to be chemical products derived from milk. Unlike filled milk, an imitation milk may enter interstate commerce, although its manufacture is illegal in some areas. The ingredients of an imitation milk are shown in Table 21:1. An imitation milk of higher protein content has the approximate composition shown in Table 21:2.

TABLE 21:2
Imitation Milk

Ingredients	*Percent*
Water	87.5
Demineralized Whey Solids	6.5
Vegetable Fat	3.25
Sodium Caseinate	1.0
Sucrose	1.0
Emulsifier	0.4
Disodium Phosphate	0.2
Stabilizer	0.15

The water used for an imitation milk is heated to 140-160°F. and the nonfat ingredients are dissolved in it. The vegetable fat is added in a molten condition, at 140°F. or higher, with constant agitation. The mixture is pasteurized as for milk. Two stages of homogenization generally are used at 2500 and 500 psi. The product is cooled to 40°F. or lower and packaged. The composition of one type of imitation milk is shown in Table 21:2; another type of low protein content has the composition shown in Table 21:3.

Coffee Whiteners[5,6]

The use of coffee whiteners has taken over most of the market once held by coffee cream. Cost savings and long shelf life without refrigeration has made these products popular for restaurant and

TABLE 21:3
Percentage Composition of Dairy Product Substitutes[5]

	Filled Milk	*Imitation Milk*	*Coffee Whitener*	*Imitation Sour Cream*	*Whipped Topping W/Milk Solids*	*Whipped Topping W/Sodium Caseinate*
Vegetable Fat*	3.5	3.5	8-11	14-18	24-28	24-28
Corn Syrup Solids	—	1.5-3.0	9-10	—	3-5	—
Sucrose	—	3-5	—	—	7-10	10-15
Stabilizer	—	0.3-0.6	0.3-0.5	0.5-0.75	0.2-0.4	0.2-0.4
Nonfat Milk Solids	10.0	—	—	8-10	5-7	—
Sodium Caseinate or Soy Protein	—	1.0-3.5	1-2	—	—	1-2
Sodium or Potassium Phosphates	—	0.1-0.2	0.25-0.4	—	—	—
Emulsifiers	0.1-0.3	0.1-0.3	0.2-0.4	0.2-5.0	0.5-1.0	0.5-1.0
Water**						

Flavor, Color, Vitamins Added As Required

*—The S.F.I. values are given in Table 21:4
**—Sufficient water to make 100

similar institutional users. The homemaker has found the coffee whitener an acceptable product. In many vending machines the product used is a coffee whitener rather than a milk product.

A coffee whitener, like coffee cream or half and half, may serve functions other than inparting a desirable color to coffee. It adds body to the coffee and reduces the acrid flavor that often is present in coffee.

Coffee whiteners are obtainable in liquid, dry, or frozen form. Their controlled composition favors a uniform whitening in coffee. Unlike coffee cream, "feathering" or oiling-off is rarely encountered.

A typical formula for a coffee whitener is given in Table 21:3.

In a consumer survey, it was found that 43% of the people who tried coffee whiteners reported that they would continue its use.[8]

Whip Topping

For a number of years bakers have sought a product which would be more stable than whipped cream. The development of the whipped topping has met this requirement both for bakers' and home use.

The whipped toppings have a uniformity of composition not present in whipped cream. A higher overrun may be obtained during whipping. They yield a foam which has good stability and which resists serum leakage or "weeping". The basic formula for a topping is similar to that of a coffee whitener except for a higher fat and sugar content. A typical formula for a whipped topping is given in Table 21:3.

A lower homogenizer pressure than for other imitation products is used in making a whip-topping in order to avoid emulsion inversion.

In a consumer survey it was found that 60.4% of the people who tried whip topping reported that they would continue its use.[8]

Imitation Ice Cream—Mellorine

Although imitation ice cream and related frozen products have been known for many years and are much used in England and other European countries, their sale is illegal in most States in this country. The increased use of margarine during and following World War II had a counterpart in the use of frozen desserts made with vegetable fats. The product called Mellorine, first made in Corsicana, Texas, in 1942, was soon made in those neighboring States where its manufacture was permitted. In 1972 nearly one half of the United State production of Mellorine was made in Texas. The name "Mellorine" was patented, but with the provision that it may be used by anyone without restriction or royalty payment.

Mellorine is a frozen, pasteurized product, which, except for its fat is similar in composition to ice cream or milk. The fat used is a mixture of vegetable fats which has a melting point very similar to that of milk fat. Hydrogenated coconut oil is preferred and its bland flavor is more desirable, but economic factors favor the use of domestic oils, such as hydrogenated corn and soybean oils. Safflower oil is used when a product high in unsaturated fatty acids is desired.

It has been noted that the processing of Mellorine differs slightly from that of ice cream. More stabilizer is used and about 300-500 pounds more pressure is used for homogenization. Owing to the bland flavor of the vegetable fats about 10% more flavoring material is necessary. Some states require the addition of not less than 8400 U.S.P. units of vitamin A to the gallon, based on a fat content of 10 percent. A proportional increase is required if the fat content is over 10 percent.

In most areas Mellorine must be labeled Imitation Ice Cream. It usually is made in ice cream plants as the same equipment can serve for both products. In 1972 the United States production was 45 million gallons.

The composition of a typical imitation ice cream is given in Table 21:4.

TABLE 21:4
Imitation Ice Cream (Mellorine)

Vegetable Fat	10-12%
Nonfat milk solids	10-12
Sucrose	10-11
Corn Syrup Solids (42 D.E.)	8.5-8
Emulsifier	0.0-0.3
Stabilizer	0.2-0.3

MARGARINE

Margarine (Oleomargarine) is not a dairy product and is not sold as one. It was invented in 1870 by the French chemist, Hyppolyte Mège Mouriés, following the search by Napoleon III for a butter substitute during the Franco-Prussian War. Mège first called it "buerre economique" (economical butter), but had to change the name because a reference to butter was illegal. Margarine is named after *margaric acid*, at one time supposed to be a constituent of fats but which has since been shown to be a mixture of fatty acids.[7]

From its inception, legal restrictions and special taxes on the manufacture of margarine have been made to protect the dairy industry, not only against possible adulteration of butter but also to deter competition with a dairy product.

Margarine has been regarded as a substitute for butter, especially since in most areas it is obtainable at a lower price than butter. The price usually follows that of butter, going up when the cost of butter is high and dropping when it is cheaper. Even today, most margarine is packaged as is butter, being put up in flat packages where butter is so marketed and in square units in other areas. Legislation has become more liberal and technology has so changed the character of margarine that it may be considered a distinct foodstuff, rather than an imitation of another product. A manufacturer of margarine is not so restricted by law or tradition that he is unable to develop a product that may be continually improved. Unlike the making of butter, which is essentially the same everywhere, the margarine manufacturer is under competitive pressure to improve his product.

For reasons such as the following, margarine manufacturers believe that their product will gain increasing importance.[5] Butter making is a costly operation relative to margarine manufacture and margarine improvement is progressive compared to that of butter. The prejudice against margarine is greatest amongst the older age groups, so that in time, the population change will favor the use of margarine relative to the demand for butter.

Manufacture of Margarine

A number of fats and oils are used in the manufacture of margarine. In 1973, in the United States, the principal ones were soybean oil (1,458 million pounds) corn oil (194 million pounds) and cottonseed oil (66 million pounds). Safflower seed and peanut oil are also used. Some lower-cost margarines use animal fats alone or in combination with vegetable oils. About 130 million pounds were used in such margarines in 1973. The milk used in margarine was equivalent to about 33 million pounds of nonfat dry milk.

In Canada, rapeseed oil and sunflower seed oil are used while in some European countries marine oils, such as whale oil, find use. The margarine fat usually is a blend of fats, a hard stock which may or may not have been hydrogenated and a soft stock of a liquid oil high in linoleic acid content.

An analytical method based upon dilatometry gives the "S.F.I." or solid fat index of the fat. This is a percentage value based upon the gradual expansion of a fat when it is warmed from the solid to the liquid phase. The S.F.I. at 92°F. gives a measure of the acceptability of a fat when it is tasted. The S.F.I. at 50°F. indicates its ease of spreading when taken from the household refrigerator. At

TABLE 21:5
Typical S.F.I. Values[3]

Temp. F°	Milkfat	Coconut Oil	Mellorine	Fat Blends for Imitation Milk	Dry Coffee Whitener
50	34	53.9	24.38	15	61
70	13	26.4	13-23	11	50
80	8	0.05	10-20	10	46
92	0.5	—	4-14	7	28

intermediate temperatures the S.F.I. at 70-80°F. measures the ability of the fat to retain its body and not "oil-off" at room or higher temperature.

Some typical S.F.I. values are given in Table 21:5.

The refining of edible oil is an important industry but does not pertain directly to this discussion. Many margarine factories conduct their own oil refining and hydrogenation processes.

The manufacture of margarine for table and cooking use by a batch process is giving way to a continuous process. The batch process is still being used to make special products for baking and confectionary use. In a batch system, the oil stock is mixed in a vat with the aqueous phase. The aqueous portion usually is pasteurized skim milk or reconstituted skim milk ripened with a lactic acid culture such as is used in the manufacture of butter. Sometimes, only water is used. To this phase, salt and other water soluble constituents, such as sodium benzoate and sorbic acid are added. The oil-soluble constituents, such as the emulsifiers, color, biacetyl, lactones, and other flavor agents, and the vitamins are mixed into the product by vigorous agitations. The emulsion is cooled quickly, sometimes by contact with ice water, and the mixture is then worked in the manner used for the manufacture of butter.

More often, a steel chill-roll is used to cool the margarine mixture. The fat emulsion is fed into a trough from which it contacts the internally refrigerated roller. The chilled margarine is scraped from the roll and is conveyed to an extruder and fed into the hopper of a packaging machine.

The continuous flow process now in common use, makes use of "Votators", externally cooled cylinders within which scraper-blades rotate[2] (Fig. 16:5). This is not unlike the freezing unit used for making ice-cream. The agitation produced in the cooling tubes results in such dispersion of the water phase that the droplets rarely exceed one to five microns in diameter. The margarine emulsion is quickly cooled as it is pumped through the unit and is scraped off the cooling surface, emerging at about 53 to 57°F. This cold emulsion is then further agitated to promote crystallization of the fats and is then molded by automatic machinery and packaged, much as is done for butter.

Salt content of a margarine may vary. In the United States it usually is about 1.5%. In Germany and France, as little as 0.2% salt is added, whereas in England and Scandinavia, 2% or more of salt may be used. Most margarines contain some preservative, especially if the salt content is low. Sodium benzoate or sorbic acid is commonly used. The use of emulsifying agents is common, but modern equipment has made their use less important. The presence of an emulsifier such as lecithin or glyceryl monostearate, has an indirect advantage in that they may assist in the baking of cakes, when the margarine is used as part of the batter. Another advantage of an emulsifier is that it holds the water droplets in suspension and prevents spattering when foods are fried with margarine.

Even when made with milk, margarine does not have the true flavor of butter. Among the ingredients used to impart a butter flavor are diacetyl and delta-lactones. A few parts per million of these is sufficient. To remedy any shortage of nonfat dry milk or skim milk which could cause an increase in the cost of manufacture, the Food and Drug Administration permits the margarine manufacturer to use a "suitable edible protein", such as soy bean protein.

Much variation exists in the color of margarine. In the United States B-carotene is used to give a color similar to that of butter. The β-carotene also adds to the vitamin A content of the product. In other countries, such as France, no color or flavor may be added. In Canada, most provinces except British Columbia, require margarine to have an orange color, whereas in Quebec, it must not be colored.

An antioxidant may be added but with the increasing use of liquid oils, sufficient tocopherol may be normally present to provide antioxidant property, as well as vitamin E activity. Vitamins are added to margarine. In the United States, a margarine must contain not less than 15,000 International Units of Vitamin A to the pound. Vitamin D may also be added and such addition is mandatory in Newfoundland. Margarine made for the use in the Phillipine Islands and some other Southeast Asiatic areas has thiamine added (5 mg to the pound). In Europe, vitamin E often is added.

By controlling the fat blend in the margarine, the manufacturer can regulate its consistency, thus soft margarines high in liquid fat content are marketed in small tubs. These are made of plastic or paper laminated with aluminum foil. These products may be spread easily, even when used directly from the household refrigerator.

Imitation Margarine

In order to produce a low-calorie product, an imitation margarine is made. Margarine is emulsified with water in order to lower its fat content to about 40-50%. Owing to its soft body, imitation margarine is packaged in small tubs.

Composition

The average over-all composition of a margarine is about as follows:

Water	Fat	Salt	Nonfat milk solids and other components
16.40	80.2	1.5	1.9

In the United States, margarine must contain not less than 80% of fat. Most States and the Federal government no longer impose a tax upon the sale of margarine.

Consumption of Margarine

In 1972, the United States produced 2,361 million pounds of margarine. The per capita consumption was 11.3 pounds compared to 5.0 pounds of butter. In Canada in 1967, the per capita consumption was 9 pounds for margarine, and 17 pounds for butter.

Some European countries have a relatively large per capita consumption of margarine, for example 42 pounds in the Netherlands, 39 pounds in Denmark, in contrast to 6.5 pounds in France, and 12 pounds in the United Kingdom.

Imitation Sour Cream

The composition of an imitation sour cream is given in Table 21:3. The mixture after homogenization and cooling is acidified with 2.5% of lactic acid (30%) solution. The mixture is stirred until smooth and then packaged. A lactic acid culture may be used as in the making of genuine sour cream. Food acids other than lactic acid may be used.

Filled Cheese

A filled cheese is one made from skim milk and a fat other than milkfat. A filled cottage cheese or cream cheese has met with consumer acceptance. Other types of filled cheeses have not met the favorable acceptance given to other non-dairy products. One reason is that the flavor is inferior to that of a cheese which contains milkfat. The volatile and other flavor producing compounds derived during the ripening of a cheese depend in large part upon the short-chain fatty acids present in milkfat. The addition of synthetic flavor has not yet provided for an acceptable filled cheese product.

Filled cheese may be shipped in interstate commerce. The Federal government levies a tax of one cent per pound on the product, in addition the manufacturer must pay an annual tax of $400. The wholesaler and retail dealer also must pay a special tax.

REFERENCES

1. Comparison of dairy, filled, and non-dairy products/. nutritionally speaking, Coulter, S. T., and Manning, P. B., *The Milk Dealer*, March, 1968.
2. Devine, J. and Williams, P. N., Eds. The Chemistry and Technology of Edible Oils and Fats, Pergamon Press, New York and London 1961.
3. Food and Food Products. Definitions, Identity, and Label Statements. Federal Register, 38:20702, Aug. 2, 1973.
4. Graham, I. C., Technical application of vegetable oils in the more recent substitute dairy products; in Symposium on Margarine and New Edible Oil Products, Canadian Institute of Food Technology, 1968.
5. Hamilton, H. D., Manufacture and Quality of Dairy Substitutes, ibid., 1968.
6. Knightly, W. H., The role of ingredients in the formulation of coffee whiteners, *Food Technology*, 23, No. 2:37-48 (1969).
7. van Stuyvenberg, J. H., Ed. Margarine, Liverpool Univ. Press, Liverpool, England, 1969.
8. Herrmann, R. O., et al, Consumer Adoption and Rejection of Imitation Food Products, Pennsylvania Agricultural Exp. Station, University Park, Bul. 779(1972).

Index

C

www.ingramcontent.com/pod-product-compliance
Lightning Source LLC
Chambersburg PA
CBHW021025210326
41598CB00016B/909